P9-CEG-729

NEED TO KNOW

ALSO BY NICHOLAS REYNOLDS

Writer, Sailor, Soldier, Spy: Ernest Hemingway's Secret Adventures, 1935–1961

Treason Was No Crime: Ludwig Beck, Chief of the German General Staff

NEED TO KNOW

WORLD WAR II AND THE RISE OF AMERICAN INTELLIGENCE

NICHOLAS REYNOLDS

MARINER BOOKS
New York Boston

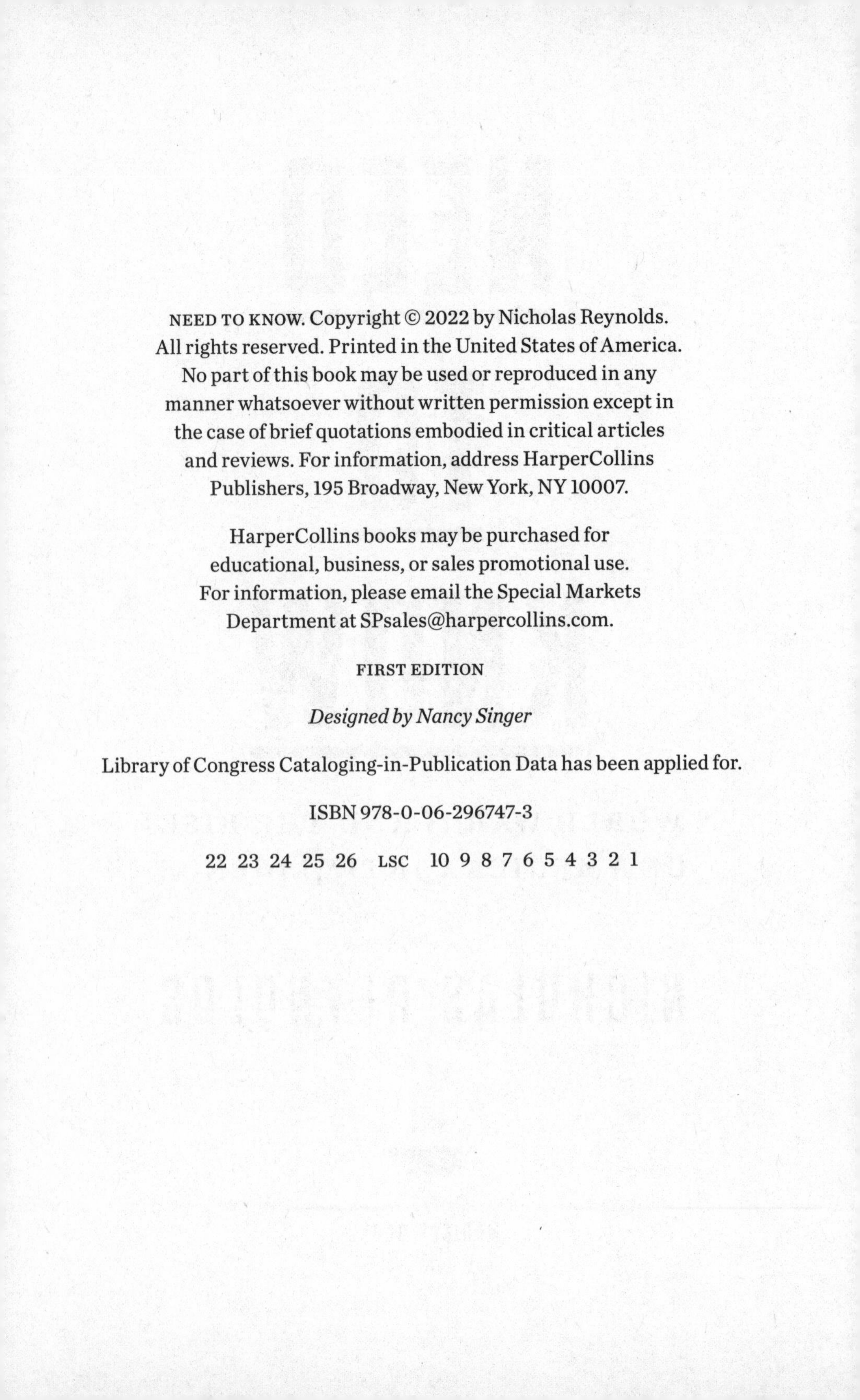

NEED TO KNOW. Copyright © 2022 by Nicholas Reynolds.
All rights reserved. Printed in the United States of America.
No part of this book may be used or reproduced in any manner whatsoever without written permission except in the case of brief quotations embodied in critical articles and reviews. For information, address HarperCollins Publishers, 195 Broadway, New York, NY 10007.

HarperCollins books may be purchased for educational, business, or sales promotional use.
For information, please email the Special Markets Department at SPsales@harpercollins.com.

FIRST EDITION

Designed by Nancy Singer

Library of Congress Cataloging-in-Publication Data has been applied for.

ISBN 978-0-06-296747-3

22 23 24 25 26 LSC 10 9 8 7 6 5 4 3 2 1

This book is for Rebecca Reynolds

Real intelligence work . . . will never cease. It's absolutely essential that we have it. . . . It brings to a strong government what it needs to know.

—John le Carré, 1997

CONTENTS

PRINCIPAL CHARACTERS

Vincent Astor—Scion of a wealthy family and FDR intimate who ran the Room, a private intelligence agency based in New York City

Adolf A. Berle Jr.—Assistant secretary of state who served as the point man at the Department of State for intelligence matters during World War II; New Dealer with strong ties to FDR

David K. E. Bruce—One of the founders of the Secret Intelligence Branch of OSS, then chief of the OSS base in London in 1943 and 1944, a senior diplomat after 1945

John Franklin Carter—Sometime diplomat, journalist, spy novelist, and FDR confidant who ran a private spy bureau for the president during World War II

William J. Casey—New York lawyer who joined OSS and learned how to run operations against the German homeland in 1944–45; future director of CIA

Carter W. Clarke—Career army officer who supported the work of Alfred T. McCormack in signals intelligence

A. G. Denniston—The first wartime head of the Government Code and Cypher School at Bletchley Park, Britain's codebreaking establishment

William J. Donovan—World War I hero, Wall Street lawyer, and Republican internationalist who founded COI and then OSS in World War II

Allen W. Dulles—New York lawyer and former diplomat who ran the OSS base in Bern, Switzerland, during World War II; future director of CIA

William A. Eddy—Arabist and educator who served in the Marine Corps in World War I and directed intelligence operations for OSS in North Africa in World War II

Carl F. Eifler—US Army reservist who created OSS Detachment 101 in the China-Burma-India Theater, pioneer of OSS special operations

James Russell Forgan—Chief of the OSS in Europe following David K. E. Bruce

William F. Friedman—The grand old man of army cryptology who laid the groundwork for its success in World War II

Hans Bernd Gisevius—German lawyer and Abwehr officer stationed in Switzerland who, as part of the German Resistance to Hitler, met frequently with Allen Dulles

John H. Godfrey—Royal Navy admiral and director of Naval intelligence early in World War II; with his aide Ian Fleming, instrumental in promoting Donovan's fortunes

Colin M. Gubbins—British Army officer who guided the growth and operations of the Special Operations Executive in London, the rough equivalent of OSS Special Operations Branch

Thomas Holcomb—Commandant of the US Marine Corps during World War II, overseeing its growth from a small landing force to a fourth armed service

J. Edgar Hoover—Longtime FBI director, mainstay of many a Washington intrigue, who directed counterespionage operations at home and in Latin America during World War II

Cordell Hull—American secretary of state for most of World War II

Joseph P. Kennedy—American ambassador to London, remembered for his defeatism and for being the father of future president John F. Kennedy

Ernest J. King—Strong-willed US Navy admiral who served as both chief of Naval operations and commander in chief of the fleet during World War II

Kenneth A. Knowles—US Navy officer who, with British help, stood up a successful operations center in Washington to combat U-boats in the Atlantic

Frank Knox—Newspaper publisher, Republican politician, and supporter of William J. Donovan while secretary of the navy in World War II

Fritz Kolbe—One of the great spies of World War II, this midlevel bureaucrat in the German Foreign Ministry passed original documents to Allen Dulles in Switzerland

Edwin T. Layton—US Navy officer who served as the fleet intelligence officer at Pearl Harbor for most of World War II

William J. Leahy—US Navy admiral who, during World War II, served as ambassador to Vichy France and then as chief of staff to Presidents Roosevelt and Truman, making him the most senior and influential officer in the US military

Duncan C. Lee—Rhodes scholar, Wall Street lawyer, and Soviet spy who served as an aide to Donovan at OSS; despite overwhelming evidence against him, he never confessed to betraying his country

John Magruder—US Army officer who was one of Donovan's deputies at OSS and head of the successor organization, the Army's Strategic Services Unit

George C. Marshall—Chief of staff of the US Army during World War II, a pivotal figure in strategic decision-making and a patron of signals intelligence

Joseph O. Mauborgne—Multitalented US Army officer who was a pioneer of cryptology during World War I; future chief signal officer of the army; supportive of Friedman and his work

Alfred T. McCormack—New York lawyer who shaped the system for processing and delivering army signals intelligence in World War II

Stewart Menzies—During World War II, head of Britain's MI6, the agency responsible both for human intelligence and signals intelligence

Chester W. Nimitz—US Navy admiral in command of the Pacific Fleet for most of World War II

John and Joseph Redman—Brothers who were senior officers in the Office of Naval Communications, remembered for centralizing signals intelligence as well as downplaying the role of Joseph Rochefort

Joseph J. Rochefort—US Navy officer responsible for "radio intelligence" at Pearl Harbor in 1941 and 1942, remembered for predicting Japanese movements during the Battle of Midway

James G. Rogers—Distinguished lawyer, educator, and mountain climber who served as a strategic planner for OSS

Frank Rowlett—Cryptologist who was one of Friedman's early hires in army signals intelligence; remembered for his groundbreaking work on Japan's Purple code

Laurance Safford—US Navy officer who, in the 1920s and 1930s, laid the foundations for navy cryptology; the rough equivalent of the army's William F. Friedman

William Stephenson—British spy impresario based in New York during World War II, responsible for operations to increase American support for Britain, which included promoting Donovan as US intelligence chief

Henry L. Stimson—Prominent lawyer and Republican statesman who served both as secretary of state and twice as secretary of war, progressed from believing that "gentlemen do not read each other's mail" to supporting signals intelligence in World War II

George V. Strong—US Army general who promoted signals intelligence and collaboration with Britain but clashed with Donovan and OSS

Telford Taylor—Brilliant legal scholar who served in army intelligence under Carter W. Clarke and liaised with Bletchley Park; joined and eventually headed the prosecution at the trials of German war criminals after the war

Joseph N. Wenger—US Navy officer and cryptologist responsible for strategic planning for the navy's Op-20-G

Rodger Winn—Brilliant British lawyer and wartime Royal Navy officer who created a very successful operations center in London for the war at sea against U-boats

Sir William Wiseman—British spy impresario based in New York in World War I who set precedents for William Stephenson by winning President Wilson's confidence

Karl Wolff—SS general who met with Allen Dulles and arranged the "Secret Surrender" in Italy in 1945

Herbert O. Yardley—One of the pioneers of American codemaking and codebreaking in World War I, head of the Black Chamber in New York during the 1920s, the US government's off-the-books codebreaking enterprise

INTRODUCTION

World War II has always been in my blood. It was the defining event of my parents' lives, and from early childhood I started to absorb their stories of the great cataclysm. My father was in London after D-day, enduring the V-1 "buzz bombs" and the V-2 "doodlebugs," German missiles that would strike the city hard and at random, each with almost a ton of high explosives. A junior member of the American Foreign Service, he was preparing to deploy to Germany on a joint US State Department–British Foreign Office team to capture the Third Reich's Foreign Office documents before they were destroyed. My mother was one thousand miles away in Hungary, enduring the siege of Budapest and the Soviet occupation that followed. What they wanted for their children after the war was a more peaceful existence, one without incoming German bombs or trigger-happy Red Army soldiers standing on every street corner. But what I wanted from childhood on was to understand what it was like to be in the thick of world events.

Few parts of the war have been more captivating to me than the European Resistance to Adolf Hitler. Dumb luck got me started. In Berlin, through my father, I happened to meet Fabian von Schlabrendorff, who contrived to smuggle a bomb (that did not go off) onto Hitler's plane in 1943. One contact and fascinating story led to another. While still in my twenties, I met and interviewed German resisters, as well as British intelligence officers who had spied on Nazi

Germany, and tried to understand the Resistance. The end result was my first book, *Treason Was No Crime*, a biography of Ludwig Beck, the former chief of the German General Staff who fought Hitler and his policies from 1938 to 1944.

As time went on, I was able to read more about wartime intelligence, by which I mean (a) the collection and processing of secret information about the enemy, or (b) special operations against the enemy.[1] In the more than seventy-five years since it ended, historians and participants have written more about this slice of World War II than most people can read in a lifetime. A handful of memoirists started the process in earnest in the 1970s; it would only accelerate as time went on. The US government released hitherto secret files, and the British government followed suit. Vast collections have been digitized and made available online. There are now so many books that a reader can easily specialize, choosing to read about only one particular type of wartime intelligence, and still amass a stack of books that will overwhelm even the sturdiest of nightstands. This is true whether the choice is to read about brilliant cryptanalysts in dingy offices on the Mall in Washington, navy codebreakers in an airless basement at Pearl Harbor, or Wall Street lawyers descending on the nation's capital to take control of American intelligence—to say nothing of amateur spies spread across the globe and young officers leading guerrillas behind the lines in Europe and Asia.

My own reading influenced my career choices and vice versa, creating my own seemingly inescapable vortex of history and intelligence. I served in the Marine Corps (like the World War II generation, I believed—and still believe—in national service); I worked for the CIA at home and abroad. This included everything from running individual cases to collecting sensitive information and evaluating major programs as a member of the Office of Inspector General. I taught history and case studies in intelligence at the Naval War College, I researched and wrote history for the CIA Museum, styled "the best museum you never saw." I wondered about origins and dynamics, especially considering how things got to be the way they are. I found many of the answers in the history of World War II.

American spies, codebreakers, and guerrillas created organizations and set precedents during the war. What happened between 1940 and 1945 changed the landscape forever. Before the war, American intelligence was a cottage industry, with a handful of craftsmen working on their own, mostly isolated from one another.[2] They fabricated only a few products, some of them enormously valuable but not always valued. Neither the craftsmen nor their customers knew what they did not know; almost no one grasped what was missing. Change came only after the surprise attack on Pearl Harbor on December 7, 1941, jolted the nation awake.

But hasn't the story been told? What could be missing from the American intelligence bookshelf?

The answer is straightforward. Even among all the books, it is rare to find a "crossover"—a book that tackles more than one kind of intelligence or agency, asking how each relates to the other and what, together or separately, each contributed to victory. Unanswered questions abound. It is not unusual to read that codebreakers shortened the war by giving American—and Allied—leaders a figurative seat at the enemy's table. If that is true, what does it mean for all of the other Americans who worked in intelligence? What was their defining value? Winston Churchill said that his codebreakers were like the geese that laid golden eggs and didn't cackle. Does that mean that other spies laid only ordinary brown eggs and contributed less to victory? (Many of them eventually did cackle, judging by the volume of memoirs on both sides of the Atlantic.)

This crossover book, then, is a look at the main threads of American intelligence in World War II and how they were developed, particularly how they related to each other, and where they were positioned at the end of the war. It is mostly about the kinds of intelligence that focused on the foreign field. Whenever possible, I try to look through the practitioners' own eyes to piece together a useful overview for specialists and generalists alike, for students of history as well as intelligence professionals. Rather than offer an exhaustive history, I have chosen topics that are representative and illustrate major trends.

Organizing the wonderful plethora of material at my disposal from libraries and archives was a challenge.

First I decided that context was essential. Intelligence history does not stand alone; it is a part of a much larger whole. Modern American intelligence was more of a reaction to events overseas than a homegrown phenomenon. The war in Europe during and after the spring of 1940 is enormously important to this story. But so too is what had happened before World War II. While American intelligence has been around since the Revolution, the seeds of modern American intelligence were planted in World War I and started to sprout in the interwar period.

Next I looked for unifying threads that run through the history of our *secrets war* (to borrow a phrase from a National Archives symposium). One of the most important threads is the organization of intelligence, and how institutions developed. Especially in Washington, who were the main decision-makers and how did they organize (or fail to organize) the work? What did individual organizations produce? How much did each contribute to victory?

Winston Churchill and Franklin D. Roosevelt both play leading roles in the story that follows. Compared to Churchill, Roosevelt was an amateur when it came to intelligence, but it fell to him to make the ground rules in Washington—which he did by deftly balancing competing interests rather than creating strong institutions. The president shared the stage with a nearly unbelievable cast of characters: old-school British spies, US Army and Navy codebreakers, navy officers who had studied the Japanese Navy and language for decades, not to mention the astonishing number of lawyers who put on the uniform for the duration and gave everything they had—keen insight, high energy, fresh perspective—to defeat the enemy. Just offstage was the NKVD, the Soviet intelligence service, countered by the FBI, the first working for Joseph Stalin and Soviet interests, the second working for J. Edgar Hoover and the United States.

Among the many lawyers was a Republican internationalist and World War I hero named William J. Donovan. Roosevelt allowed him to set up two offices to coordinate and run intelligence, COI and OSS, even though no one, except perhaps the British, was entirely sure what either should do. Donovan turned out to be as interested in literally going to the front as in leading change. Though primarily committed to special operations, the paramilitary side of the business, he did not neglect his Research and Analysis Branch and its distinguished academics who turned out world-class studies. Finally, he created the Secret Intelligence Branch that became the cradle of modern American espionage. At the end of the war, he pressed the campaign to create a permanent civilian central intelligence agency out of this incredible cast of characters.

My book is more about strategy than tactics, the view from thirty thousand feet rather than from the forward edge of a foxhole. The fall of France in 1940 looms large, as does the Battle of Britain. So, too, do Pearl Harbor, Midway, and Operations Torch and Overlord, the invasions of North Africa and France, respectively. Intelligence success more than balanced out intelligence failure, almost as if a narrative arc started on June 14, 1940, when the Germans marched into Paris, and spanned the years to June 6, 1944, when the Allies came ashore at D-day. The amazing growth of American intelligence occurred between those dates and allowed it to contribute to victory in Europe as well as the Pacific. When the guns fell silent in August 1945, the armed forces started to demobilize, but for the first time in American history, many of the wartime intelligence outfits stayed in business or took new shapes after a brief hiatus. The machinery forged to fight the greatest war mankind had ever known was about to become the foundation for a new American enterprise on a hitherto unimagined scale.

Nicholas Reynolds

NEED TO KNOW

1

FRIENDS IN DESPERATE NEED

Awakening to the Threat

On June 14, 1940, with no one to bar their way, German troops marched into Paris. It was a clear and sunny summer day, but the streets were largely deserted, millions of Parisians having fled south along with the French Army and government, which was en route to signing a humiliating armistice on Hitler's terms.

By afternoon, German soldiers had climbed the tiny spiral staircase to the roof of the Arc de Triomphe and unfurled an enormous black-and-red flag with a swastika in the middle. One hundred and fifty feet below at street level, Wehrmacht bands set a jaunty tone for the seemingly never-ending procession of troops who marched smartly down the broad, tree-lined avenues to the circle around the most famous war memorial in the world. They goose-stepped past the tomb of the unknown soldier under the arch, then continued on their way to the west and the south, where the fighting was not yet over.[1] Around 3:30 p.m., the Eighty-Seventh Infantry Division, which would stay in Paris, paraded nearby on the Place de la Concorde, coming within a few feet of the American embassy—as if they wanted to send a message—before presenting honors to their general, a man with the memorable name of Bogislav von Studnitz. The US diplomats drew their curtains and refused to publicly observe the celebration, a small but fitting gesture given the country they represented.[2]

It seemed that America wanted to shut her eyes and ears to this war. Most of her citizens hoped that Britain and France would be able to deal with Hitler on their own. This was now well-nigh impossible. The swastika that flew over the City of Light, long a symbol of liberty, proclaimed Nazi domination over Europe. The Royal Navy still ruled the waves, but without France and her once-powerful army, Britain had little hope of prevailing on the continent. That left America. But, in the summer of 1940, she lacked the means to intervene even if she wanted to. With most of her navy in the Pacific and fewer than two hundred thousand soldiers on active duty anywhere, she had precious little force that she could bring to bear in the European theater. Worse, the country had only a tiny, diffuse intelligence apparatus. It was unable to tell her government much about potential enemies and what they could do, let alone warn of any impending attacks, a disastrous shortcoming. Not only did the fall of Paris signal that America could no longer sidestep events in Europe, but also that she needed to develop new capabilities and institutions, especially in the field of intelligence.

A Difficult Year

For Britain and France, the past year had brought one disaster after another. On August 23, 1939, Germany and the Soviet Union signed a nonaggression pact, safeguarding Germany from her main rival in the east, which would free her to do as she pleased in the west a few months later. What she pleased to do first was invade Poland on September 1, 1939. Having guaranteed Poland's borders, France and Britain declared war a day and a half later, but could do little to help the Poles. By early October, the Germans controlled Poland while the perfidious Soviets seized the eastern provinces of the country.

The night of October 14, 1939, was "one of the most miserable" of the war for officers at the Royal Navy's operational intelligence center in the basement of the Admiralty in London.[3] Earlier in the day, a daring young German captain named Günther Prien maneuvered his U-boat through a set of narrows, past wrecks sunk to keep submarines like his out, into the anchorage at Scapa Flow at the tip of Scotland, the northern home of the British fleet, and sank the elegant old battleship *Royal Oak* while she swung at anchor in the cold, still water. The daring attack demonstrated that Britain's defenses were in such disarray that even its most valuable ships were unsafe in home waters. The ensuing German propaganda victory was almost as embarrassing for the Admiralty as the actual loss of the *Royal Oak*.

Less than a month later, in November, two officers from MI6, Britain's civilian spy service, traveled to a meeting in the quiet town of Venlo, Holland, on the Dutch-German border.[4] With London's blessing, Richard Henry Stevens, the station chief at The Hague, and Sigismund Payne Best, who was both the principal of an import-export company and an undercover MI6 "resident officer," went to meet German officers who claimed that they were plotting to overthrow Hitler. Typically sporting an iron monocle and spats to protect his shoes, Best looked the part of a character in one of the send-ups of the British upper class by Evelyn Waugh or P. G. Wodehouse. The meeting turned out to be a trap. Perhaps worst of all, Best arrived with a list of agents' names and addresses in his pocket. The Germans got the list when they forced him and Stevens across the border at gunpoint. The Nazis turned the affair into another propaganda coup, reporting, accurately enough, that they had obtained a great deal of information from Stevens and Best; German radio even broadcast the names of senior MI6 officers and where they worked, anathema to the men of a service that cloaked itself in mystery.[5] It was a compromise that would make Dutch spies suspicious of English operatives for years, and bring MI6's activities on the continent to a virtual standstill.[6]

The "Venlo Affair" served as a painful reminder of the service's limitations. Created in 1909 to spy on Britain's rivals—especially the upstart German empire with its dreams of hegemony—MI6's principal purpose was to steal secrets from foreign countries. This it would do primarily by recruiting and handling spies overseas, the almost always illegal and often messy business that the military preferred to leave to someone else. MI6 expanded in World War I but contracted between the wars, maintaining small stations embedded in British consulates, mostly in Europe. Each was manned by a small cadre of case officers, typically former army or navy officers who had little formal training and learned to spy on the job. In many places, there were only one or two professionals supported by a secretary or two. In the 1930s, MI6's European stable of assets—recruited foreigners who reported secret information—produced some workmanlike reports on the growth of the German Navy and Air Force. One agent focused on shipbuilding, a vital subject for Britain. But there was very little information on political or military strategy. MI6 was trying to remedy this shortcoming by sending Stevens and Best to Venlo. Since the agency's abilities fell far short of the challenge, the results were disastrous, like "a supposedly healthy body experiencing a heart attack."[7] The MI6 mystique had outstripped reality. At the turn of the century, British novelists had started conjuring up a highly capable cadre of gentleman spies; it was now painfully clear that they existed mostly in fiction.[8]

Almost half a year later, in April 1940, the Germans attacked in the west, starting with Denmark and Norway. The British fought back in Norway, but, caught by surprise, would lose men and ships, including the aircraft carrier HMS *Glorious*, when they evacuated the country. In one American observer's eyes, the defeat in Norway was a staggering blow to the British.[9] When the Wehrmacht turned to Belgium, Holland, and France, it again caught the Allies flat-footed—neither British nor French intelligence provided clear warning—and defeated the numerically superior French and British Armies in a textbook demonstration of what became known as *Blitzkrieg*, or "lightning (fast) war."[10]

Some ten divisions strong, the British Expeditionary Force (BEF), the cream of the British Army, ceased to exist as a fighting force. In some places, whole units surrendered intact while, at the Dunkirk beachhead, close to two hundred thousand British soldiers were saved by the Royal Navy with the help of a ragtag fleet of civilian boats that crossed the Channel in late May and early June.[11] Either way, surrendered or evacuated, BEF units lost almost all of their materiel, including two trainloads of tanks in pristine condition that no one had had time to unload. The campaign ended when the French Army stopped fighting under the terms of the armistice dictated by Hitler and signed on June 22. The next day, the dictator savored his victory on an early morning tour through the empty streets of Paris that lasted from 6 to 9 a.m. He stopped for a photo at the Eiffel Tower and, conqueror to conqueror, gazed down at Napoleon's Tomb.

The "miracle of Dunkirk" was more a reprieve than a victory. "Wars," the new prime minister Winston S. Churchill admonished Parliament on June 4, "are not won by evacuations."[12] They could do only so much to brighten future prospects. Ten days later, when the Germans entered Paris, Alexander Cadogan, the dour aristocrat who ran the British Foreign Office, noted in his diary that "all [is] lost," everything is "as black as black" can be.[13] He had never imagined that "one could endure such a nightmare."

The Italians, who had declared war on Britain on June 10 despite a last-minute plea from Churchill, now reinforced the triumphant Germans. The Soviets, still happily digesting the eastern provinces of Poland under their pact with the Nazis, had no incentive to turn against the Third Reich for the time being. The ultranationalist Spanish dictator Francisco Franco was far more sympathetic to Hitler than to Churchill, and could easily seize the tiny British territory of Gibraltar that guarded the entrance to the Mediterranean. Half a world away in Asia, the "menace of Japan [was] measureless upon the horizon," as her empire threatened prized parts of the British Empire, especially the crown jewels Singapore and Hong Kong.[14] None of the Dominions across the seas—Canada, New Zealand, South Africa, or Australia—could send decisive aid. The United States was neutral by act of Congress and constrained by isolationists.

To make matters worse for the British, the American ambassador to the Court of St. James's, Joseph P. Kennedy, was now an outspoken defeatist. Ivan Maisky, the Soviet ambassador in London and an astute judge of capitalists, found the entrepreneur-turned-diplomat to be "tall, strong, with red hair, energetic gestures, a loud voice, and booming infectious laughter—a real embodiment of the type of healthy and vigorous businessman that is so abundant in the USA, a man without psychological complications and lofty dreams."[15]

In 1938, Kennedy supported then prime minister Neville Chamberlain's policy of appeasement that compelled Czechoslovakia to cede the Sudetenland (an area with a large German-speaking population) to Germany. In return Britain and France received a commitment for "peace in our time" under the Munich Agreement. Now, almost two years later, Chamberlain no longer believed in negotiating with Hitler, but Kennedy did, fearing that Britain would lose everything if she continued to fight the powerful Germans. He let all and sundry know that he favored "an early peace, at almost any price," to avert the prolonged war that would lead to "general ruin."[16]

Kennedy's statements, so inconsistent with Churchill's determination to fight on, disturbed His Majesty's government. The Foreign Office had already opened a secret file on Kennedy, keeping tabs on him as if he were an enemy operative.[17] With classic understatement, the Foreign Secretary himself, Edward Wood, the First Earl of Halifax, would confide in his diary that Kennedy was "not . . . a very good fellow."[18] Sir Robert Vansittart, a senior diplomat known for his hardline views on Germany, was more direct, minuting that "Mr. Kennedy is a very foul specimen of double-crosser and defeatist."[19]

Even though they were as incompatible personally as they were politically, the prime minister was too good a politician to confront Kennedy directly, and they would maintain civil, if not cordial, relations so long as the American ambassador remained in London. In the meantime, Churchill would work around Kennedy and defend his country as energetically and imaginatively as he could. One of the stronger weapons in his arsenal was his oratory. The ambassador's son, the future president John F. Kennedy, later observed that Churchill "mobilized the English language and sent it into battle," delivering a series of speeches in the spring and summer of 1940 that are still stirring to read more than seventy-five years later.[20]

On June 18, Churchill rose in the House of Commons to declare his "inflexible resolve to continue the war."[21] Now that the "Battle of France" was over, the "Battle of Britain" was about to begin. The stakes could not be higher. If Britain failed, "the whole world . . . including all that we have known and cared for, will sink into the abyss of a new Dark Age." Nazi Germany was not just another country with legitimate aspirations, but one with a monstrous agenda of dictatorship, racism, and conquest. The specter of a Nazi Europe was close to reality. But if Britain could stand up to Hitler, "all Europe" might one day be free again, enabling its countries to advance into "broad, sunlit uplands":

> Let us therefore brace ourselves to our duties, and so bear ourselves that, if the British Empire and its Commonwealth last for a thousand years, men will still say, "This was their finest hour."

Winston Churchill and British Intelligence

Churchill had emerged as the man of the hour. In British politics, he had been both an insider and a renegade. A member of Parliament since 1900, he had held senior office before, during, and after the First World War, advocating imaginative but sometimes disastrous initiatives, including the ill-fated Gallipoli campaign in Turkey. He had also changed parties twice. Now a member of the Conservative Party, he was scorned by more conventional Tories for his eccentricity, including his heavy drinking, which started many days at breakfast with an "eye opener"; his penchant for unconventional dress, like the dressing gowns that he wore in public and the one-of-a-kind zip-up "siren suits" that he would wear during the war; and his odd, lisping manner of speech—to say nothing of his scathing criticism of his fellow conservative Neville Chamberlain for trying to appease Hitler. At sixty-five, he was round and short, a mere five foot six, and hardly looked the part of a charismatic leader. But the man who seemed almost to enjoy the life-or-death struggle that his country faced, who could call on enormous reserves of energy coupled with a fertile imagination, was uniquely suited for this crisis. It was almost as if he had been preparing all his life for this moment.

One of the duties Churchill assigned to himself was to cultivate the United States, with its untapped manpower, natural resources, and prodigious industry. Britain could come up with a good-enough plan to avoid defeat, but without America she could not win the war. The problem with Churchill's developing plan was that something like three-fourths of voting Americans still could not countenance joining in another European war.[22] More than fifty thousand Americans had died in World War I, "the war to end all wars" that had clearly not ended much. Why intervene again?

Churchill knew that Franklin D. Roosevelt did not side with his isolationist countrymen. The American president was willing to edge toward an Anglo-American alliance against Nazi Germany, moving as fast as he thought politics at home would permit. As early as September 1939, Roosevelt took the remarkable step of opening a private channel of communication to Churchill, then first lord of the Admiralty. What made this action notable was that Roosevelt was the commander in chief and head of state of the United States, while Churchill at that time was the rough equivalent of the secretary of the navy, several rungs lower on the ladder of power. If the president had a message for the first lord, the proper channel was through the American ambassador in London, who might share its substance with the foreign minister or even the prime minister, or through the British ambassador in Washington. But, almost as if he believed in the future of an undervalued stock, Roosevelt wanted to establish a direct personal connection with the man who might ascend to the most senior level of his government. In his first letter, he told Churchill that he would "at all times welcome it if you will keep me in touch personally with anything you want me to know about."[23] He could always send "sealed letters" through diplomatic channels—that is, letters that no one else, especially Kennedy, would read. Churchill responded "with alacrity," "delighted" with Roosevelt's proposal and what it promised.

Though they shared an affinity for secrets, the prime minister had a much firmer grasp on the profession—and potential—of intelligence than Roosevelt. Indeed, for a political leader in 1940, Churchill had matchless experience and ability in this field—one of many life skills that he possessed. He had almost literally been present at the creation of Britain's domestic and foreign civilian intelligence services, MI5 and MI6, before the First World War. As first lord of the Admiralty in World War I, Churchill had presided over the Naval Intelligence Division and its director, then-captain Reginald Hall. Though nicknamed "Blinker" on account of a nervous tick that caused him to blink his eyes like a ship's signal lamp, Hall was calculating and effective, turning his directorate into the preeminent British intelligence agency of World War I. Under Hall, NID was involved in a range of activities from espionage to codebreaking to political action far beyond the remit of the average naval intelligence office.

Immediately after the Great War, Churchill held the position of secretary of state for air and war, becoming the political head of the British Army and the Royal Air Force. He did not hold any other cabinet posts until 1939, but friends kept Churchill informed. After he returned to the Admiralty, Churchill supervised another brilliant director of Naval Intelligence, then-captain John H. Godfrey, who took his cues from Hall, and even lived in Hall's London apartment at 36 Curzon Street, a ten-minute walk from the Admiralty in downtown London. From the day he was appointed, Godfrey was to play a key role in leading not only naval but also other kinds of intelligence. Like most British politicians and officials, he accepted, almost as a matter of course, the need for all of this machinery to run the empire, and for its constituent parts to work together.

When he became prime minister, Churchill took immediate steps to assert control over the intelligence system,[24] shifting it into wartime tempo. He installed a personal private secretary for intelligence, Maj. Desmond Morton, an old friend who had served as the third-highest-paid member of MI6 for some fifteen years. Part of the power base of the British establishment—having gone to the right schools, served in the right regiments, and joined the right clubs—Morton could communicate effortlessly with the men who administered British intelligence, especially the head of MI6, Sir Stewart Menzies. Morton would be the expert referent, finding and relaying vital information to and from 10 Downing Street.[25] In Churchill's own words, "He became, and continued during the war to be, one of my most intimate advisors."[26] Within a week of becoming prime minister, Churchill made intelligence a part of day-to-day decision-making by invigorating the coordinating body known as the Joint Intelligence Sub-Committee (usually abbreviated as JIC) and directing that it forward urgent reports to him, at any time of the day or night. Within four weeks, he fired the director of MI5, Sir Vernon Kell, for not moving fast enough to meet wartime challenges. Kell had been in office so long—since 1909—that Britain's internal service seemed like a family business impervious to change.

Franklin Roosevelt and American Intelligence

While Churchill seemed to shift effortlessly from one profession to another—writer to politician to soldier to strategist—his American counterpart was more singularly focused. Roosevelt was a supremely gifted politician, a student of power since his cousin Theodore had been president at the turn of the century. Like few others in American history, he understood the presidency and how to exercise power. But he was not a great intelligence manager or leader. His prior experience stemmed from his World War I service as assistant secretary of the navy. He had enthusiastically supported investigations of German sabotage in the United States and, in the summer of 1918, made his way across the Atlantic to experience the war firsthand. His trip included a visit with "Blinker" Hall, who was happy to brief his visitor on his accomplishments, some of them exaggerated to impress, which led Roosevelt to conclude that the Royal Navy's "intelligence department is far more developed than ours."[27]

In 1921, Roosevelt was stricken with polio and spent the first part of the decade recovering in body and spirit. The tall—six foot two—handsome man whose gait had been an energetic stride would, for the rest of his life, wear cumbersome steel leg braces and use a wheelchair. Through an act of will, he learned how to swing his legs in a kind of walk and almost always insisted on standing when delivering his speeches. His triumph over his infirmity was such that voters saw him not as disabled, but rather as confident and capable. They elected him governor of New York in 1928 and then again in 1930. In 1932 and 1936, his energy and optimism during the Great Depression made him the people's choice for president.

The Depression made America look inward rather than outward. One individual exception was Vincent Astor, Roosevelt's fabulously wealthy neighbor in upstate New York. Happy to share his heated indoor swimming pool with Roosevelt, Astor had become a close friend during FDR's recovery from polio. In the late 1920s, Astor convened an informal group called "the Room," and later, "the Club," that was part social club and part information exchange, not unlike a secret society at an Ivy League university. Acquiring an apartment on the east side of Manhattan to use as its clubhouse, the group would meet once a month for dinner to discuss financial, political, and foreign affairs. Including Wall Street bankers and lawyers, as well as explorers and writers, its members had impeccable East Coast credentials, as well as great wealth and influence. The information they exchanged might have come from privileged sources or from their travels, but it was not just aimed at ferreting out investment opportunities or protecting assets. Astor had a social conscience as well as curiosity about the world, and the men at his table discussed many of the great questions of the day.[28]

Roosevelt was not a member of the Room, although his cousins Kermit and Theodore Roosevelt Jr., President Theodore Roosevelt's sons, were. Along with other mutual friends and acquaintances, FDR frequently embarked on Astor's yacht *Nourmahal,* the flagship of the New York Yacht Club, described by the president's son James as a "sleek white yacht, an ocean liner in miniature" with its handsome German-built furnishings and walnut, mahogany, and teak accents.[29] Roosevelt enjoyed the time on the water, sometimes for days at a stretch—fishing in Long Island Sound or cruising up to Newport to watch the yacht races, all the while absorbing information. In 1938, as one amateur to another, he went so far as to task Astor with sailing *Nourmahal* through Japanese-held islands in the Pacific on a spy mission to see what he could find.[30]

In Washington a man named John Franklin Carter operated like Astor but on a smaller scale. A Yale graduate and former diplomat, he gave up the Foreign Service to devote himself to the Bureau of Current Political Intelligence (CPI). The CPI did not actually exist; it was a figment of Carter's imagination, a feature of the moderately successful spy novels that he was writing. The novels had irked his superiors at State enough to force him to choose between spy fiction and diplomacy. It was an easy choice. During the 1930s Carter wrote more novels, traveled widely, mostly in Europe, and launched "We, the People," a newspaper column that supported the

New Deal. Working out of the National Press Building a few blocks from the White House, he cultivated a relationship with Roosevelt, who was happy to meet with him from time to time. In private the two men shared political gossip and talked about foreign affairs; Roosevelt got in the habit of asking for information and favors from Carter, which included shading or even killing an unfavorable column or two.[31]

Astor and Carter are significant for their roles in shaping Roosevelt's approach to intelligence. When he wanted to learn about a subject, FDR's instinct was to rely on members of his social class who were amateurs like him. He seldom felt a need for more. Even if he had wanted to call on professionals, he would not have found anything in Washington that rivaled the British intelligence machine. There was no American equivalent of MI6. The army and the navy each had small intelligence bureaus, the Military Intelligence Division (MID) and the Office of Naval Intelligence (ONI), both dating back to the late nineteenth century. They existed mainly to collect information about foreign armies and navies, but with rare exceptions only operated "above board," through military and naval attachés, meaning that there was little or no spying.[32] Neither had much bureaucratic clout.[33] At Justice, the FBI had long been fighting crime and, more recently, keeping watch on various threats to American security, including the occasional foreign spy, but this called for a different set of skills. Coordination among ONI, MID, and FBI was haphazard at best.[34] Their capabilities tended to atrophy in peacetime. America had a long history of spying in wartime—George Washington was one of the first to appreciate its value—but the nation was not entirely comfortable with the idea. She entertained thoughts that she was not like a European power, whose aristocratic leaders were constantly manipulating and intriguing against enemies for gain, but something different, perhaps better: a democratic republic that answered to the will of the people.[35]

Frank Knox, William J. Donovan, and the Fifth Column

During the 1930s, the democracies watched the rise of right-wing nationalism with concern. Their primary focus was on Nazi Germany, but imperial Japan, fascist Italy, and even Spain were also in the picture. In 1936, reactionary Spanish Army officers rebelled against the democratic republic and gave rise to a haunting idea. When he moved against Madrid in October 1936, the nationalist general Emilio Mola declared in a radio broadcast that while four columns of troops were advancing on the city, a fifth column of spies and saboteurs was already inside its walls. Supposedly they were undermining the enemy from within, betraying military secrets, perhaps preparing to rise up to support the other four columns. The Nationalists did not actually have a well-organized fifth column in Madrid, but the phrase—and the irrational fear it represented—caught on.[36]

How strongly governments believed in the fifth column showed in their policy toward aliens on their soil. Although France had attracted political refugees through September 1939, enemy aliens *and* many neutrals were rounded up and placed in camps once hostilities broke out. It often made no difference that the enemy alien happened to be an active anti-Nazi on the run from the Gestapo. Britain was not far behind the French in its policies toward foreign citizens, interning them en masse.

By the summer of 1940, the concept seemed to explain how Hitler had been able to rack up so many victories in such a short period of time. It was never clear exactly who the members of the fifth column were or what they did, only that they seemed to be everywhere. Had they passed secret maps to the enemy? Or undermined the morale of Allied troops? As the German divisions came on, and French armies seemed to melt away in front of them, many French citizens were convinced that they had somehow been betrayed. Clare Boothe of *Life* magazine heard the word "*trahi*" ("betrayed") again and again in France in May and June: "At first it was no more than a whisper, like the little winds . . . before the hurricane. . . . And then the whisper became a great wail that swept through" the country.[37] On the eve of the German occupation of Paris, Edgar Ansel Mowrer of the *Chicago Daily News* heard the word as well, and would conclude that Hitler could not have overrun France without the aid of a fifth column.[38] Even the usually well-informed Churchill fretted about "Fifth-Columnists," directing that much thought be given to countering enemy tricks, like impersonating British soldiers.[39] By summer, spy mania had taken hold in the home islands and upset social norms. Many British citizens and police constables heard rumors that German soldiers had infiltrated the Netherlands in various guises, including paratroopers supposedly disguised as nuns. The result was an awkward moment or two for English nuns who were stopped and searched to see what they might be hiding under their habits.[40]

One of the prominent Americans who worried about the fifth column was Frank Knox, a Republican in the mold of Theodore Roosevelt who shared his belief in the virtues of a strenuous life.[41] Born in 1874 in Boston to a family of modest means (his father was a grocer), Knox started to sell newspapers to help support the family at the age of eleven and continued to work at a breakneck pace for the rest of his life. During the Spanish-American war, he enlisted in the army, and with Roosevelt's Rough Riders fought his way to the top of San Juan Hill in Cuba in July 1898. Though a private soldier, Knox left an impression on his commander. In 1911, Theodore Roosevelt wrote a simple note to recommend him to another Republican stalwart, Henry L. Stimson, as "just our type."[42] By that he presumably meant an idealist who was willing to fight for his beliefs; the Rough Riders were not just adventurers, they believed that they were part of the American mission to free other nations from oppression.

During World War I, Knox interrupted his career as a newspaperman, the work he loved most, to rejoin the army at age forty-three and serve as an artillery officer. After the war, he would make his mark as the publisher of the *Chicago Daily News*, a newspaper that was critical of the New Deal and big government. Still, he was a liberal Republican, an internationalist who believed in a strong defense. Once described as "a large, florid man with a friendly eye and a tongue which . . . [was] too large for his mouth," he was plain-spoken, sometimes direct to a fault and prone to overreacting.[43] In 1936, he ran for vice president on the Landon-Knox ticket. Not a winning combination, the ticket carried only Maine and Vermont, losing the electoral vote by a margin of 523 to eight. (This led Roosevelt campaign manager Jim Farley to quip, "As Maine goes, so goes Vermont," a play on the epigram that the easternmost state was a bellwether for the rest of the country.) After the war in Europe broke out in 1939, Roosevelt started courting Knox, asking him to consider joining a cabinet of national unity, one that comprised both Republicans and Democrats.

When Roosevelt summoned Knox to a private meeting at the White House in December 1939, he offered him a "non-political" cabinet post, one like Navy or War that was not part of the New Deal. Even so, Knox declined, telling the president he did not want to be a latter-day Benedict Arnold.[44] But he left open the possibility that he might serve in case of a genuine emergency. That condition was met when the Germans occupied Paris on June 14. The next day, observing that "poor France" was "done in," Knox worried that the French fleet could fall into German hands, and hoped that "the whole [American] people" would now be ready to do everything they could to help the Allies.[45] He concluded that it was "foolish to talk about keeping out of war as if we had a choice in the matter."

In early July, when Knox made his way to Washington to prepare for the confirmation hearings that Roosevelt's offer had set in train, he found his friend William J. Donovan waiting.[46] Donovan and Knox had much in common, and appear to have genuinely liked and respected each other. Both started out life in the northeast, and had spent decades putting distance between themselves and their modest origins. In Donovan's case, his family was emerging from a humble Irish Catholic neighborhood in Buffalo, New York, where his father was working his way into management at a local rail yard. Knox and Donovan had both served in the army during the Great War and were politically compatible. Like Knox, Donovan was a Republican and an internationalist. Each believed America's future was bound up with that of Western Europe.

But there were differences. Knox was nearly a decade older than Donovan, who was born on New Year's Day in 1883. Photographs of Knox typically show him in a dark suit and a formal pose, like the senior executive that he was. He had devoted himself wholeheartedly to one profession and to one woman, his college sweetheart, Annie Reid. Donovan's essence is harder to distill. For much of his life, he seemed to be striving for something he never could quite grasp. He was not desperately poor when young, but felt the sting of prejudice against Catholics, especially those of Irish origin. His hard work in preparatory school had earned him a place at Columbia University in New York City, where he obtained both undergraduate and graduate degrees, becoming a lawyer after two years of law school and returning to practice in Buffalo. His 1914 marriage to Ruth Rumsey, the daughter of a wealthy Protestant businessman, went against social norms. Whether it was this marriage or marriage in general that did not suit Donovan is open to debate. In either case, Ruth and Bill Donovan started to grow apart within a few years.

The problem was that Donovan was the opposite of a homebody. He would spend much of his life searching for adventure, as a British friend observed, "with . . . the daring of an arrested adolescent."[47] At five feet, eight inches tall, trim in youth and early middle age, his face a handsome rectangle, he had a full head of light brown hair, occasionally a dashing mustache, and piercing blue eyes that stand out even on a black-and-white photo. Most photographs show him stylishly and carefully dressed, with jacket and tie, "impeccably tailored and fastidiously neat," in the words of one associate.[48] With his good looks and quiet, easy charm, he had no difficulty attracting women, often for brief trysts, and occasionally for longer relationships with some of the great beauties of the day.[49]

Military service—which in his case was voluntary—suited Donovan's needs perfectly. Like many restless men, he found what he was looking for in combat, serving in the trenches in France in 1918. Though he was one of the older (and, at first, less experienced) infantry commanders, he learned quickly and led energetically. The army decorated him repeatedly for his frontline leadership. He emerged from the war with one of the army's most senior medals, the Distinguished Service Cross, upgraded in 1923 to the nation's highest award for valor, the Medal of Honor. Upon return to the States, he led his troops up Fifth Avenue in the 1919 victory parade, flags flying and bands playing in the background. Seeing him for the first time, a young New York socialite would never forget how glamorous he looked in his uniform with all of his medals. Adding to the glamor was his reputation: "something like [that of] a pirate."[50]

After being demobilized, Donovan returned to the law, maintaining his private practice while simultaneously serving as a US attorney.[51] No one has ever claimed that Donovan was passionate about the law, and he kept looking for fulfillment outside a conventional law practice. He happily accepted political appointments, but found they did not quite suit; he seemed to always have his eye on some other job, typically at the next higher level. In the Calvin Coolidge administration, he moved to Washington to serve as assistant attorney general. But after Coolidge's successor, Herbert Hoover, failed to elevate him to attorney general, Donovan left Washington to found the firm of Donovan, Leisure on Wall Street. It flourished during the Depression by handling the legal fallout from the stock market crash. In 1932 he ran for governor of New York on the Republican ticket, but he turned out to be a lackluster and disorganized candidate, better suited to the courtroom than the stump. One reporter said that he was more like a debater than a politician. He lost by an embarrassing 18 percentage points.[52]

No matter what he accomplished at Justice, or how much money he made on Wall Street, he could not recapture the intensity found in combat. He continued to embark on long trips, seeking out war zones whenever possible. In 1935, he talked Mussolini into allowing him to observe the war that the Italian dictator had started in Abyssinia. In early 1939, he witnessed the final battles of the Spanish Civil War, and then attended maneuvers of the German Army. The onset of World War II was like a call to arms for him. In 1940, he wrote a newspaper article arguing that older men—those in their fifties—were suited, indeed even needed, for modern warfare.[53] (Years later, in 1943, standing on an invasion beach in Sicily, Donovan would agree with Gen. George S. Patton that what mattered most in life was sex and combat, in that order.[54])

Donovan's interest in foreign affairs gave him a broader, more cosmopolitan outlook than many of his peers in America.[55] He believed in American democracy and wanted to protect the middle ground between Nazism and communism. He studied power and wanted to use it to make a difference for good. In his pursuits he was energetic to a fault. "If there is a loose football on the field," a senior army officer warned, "Wild Bill will pick it up and run with it."[56] By 1940, he had rivals and even a few enemies. They tended to be politicians on the other side of the aisle or bureaucrats who had crossed swords (or exchanged memos) with him, but they were outnumbered by those who appreciated his ability to kindle enthusiasm among his friends and followers.[57]

Frank Knox wanted Donovan's friendship and support when the newspaperman came to Washington in 1940. With his Washington experience, Donovan could put his legal acumen to work to help prepare Knox for his confirmation hearings. He even insisted that Knox move in with him until he could find his own place to live. The property that Donovan had acquired in 1924 was certainly inviting and spacious enough: the elegant two-and-a-half-story house at 1647 Thirtieth Street NW in fashionable Georgetown, surrounded by an acre and a half of grounds, looked like a small French chateau. Donovan lived there mostly by himself when he was in Washington.[58] The three servants (who kept house even when he was in New York) would see to his every need. Apparently worried about wearing out his welcome, the Midwesterner Knox hesitated until Donovan agreed to let him pay his way. When, on July 11, 1940, Knox was sworn in as secretary of the navy at the White House, he celebrated with friends at Donovan's table that day at lunch, and sat down with him again the next day for dinner. Even after Knox moved out, Donovan would insist that the secretary keep some of his clothes at the house just in case.[59]

Knox was willing to promote Donovan's fortunes as well as his own. At least twice, once in December 1939 and again in June 1940, he had recommended his friend to Roosevelt as a Republican willing to serve in a national unity government.[60] Until Knox came to Washington, FDR's reaction was lukewarm, noncommittal at best. He claimed in a letter to Knox on December 29, 1939, that "Bill Donovan is an old friend of mine," but was reluctant to offer him a cabinet post, especially secretary of war, as Knox had suggested.[61] Roosevelt was exaggerating; they were not personal or professional friends but acquaintances. It was true that he had known Donovan for decades. They had attended Columbia Law at the same time and had mutual acquaintances like the members of Theodore Roosevelt's family, but they moved in different social circles.[62] (Besides which, the TRs and the FDRs were often barely on speaking terms.) They had both run for office in New York State—but hardly on the same platform. In 1932, Eleanor Roosevelt had even spoken out against then-candidate Donovan, assailing his half-truths and linking him to heartless Republican policies that put thousands out of work in the name of fiscal responsibility.[63]

Knox kept looking for opportunities for Donovan. On the evening of July 9, 1940, he told the British ambassador to the United States, Philip Kerr, the 11th Marquess of Lothian, how "anxious [he was] to make [a] survey of the Fifth Column methods, as they have been disclosed in . . . the Low Countries [and] France . . . in order to warn the American public."[64] He himself could not go, but wanted to send two well-qualified surrogates: Donovan and Edgar Mowrer, who had been the *Daily News* bureau chief in Paris until the Germans marched down the Champs-Élysées. Two days later, Cordell Hull, the American secretary of state, cabled Ambassador Kennedy in London that Donovan was on his way "for a brief survey" and wanted reservations at Claridge's, one of the most luxurious hotels in London. Donovan would assess "certain aspects of the British defense situation" in addition to the fifth column.[65] Hull added that the president had approved the trip.[66] Knox wrote letters of introduction, announcing that Donovan was traveling "on an official mission for me."[67]

Donovan in London

At midnight on July 13, Donovan left Washington for New York on one of the night trains favored by officials who split their time between the political and financial capitals. Donovan waited until the last minute to telephone his wife, Ruth, and even then only to tell her that he was going abroad on a secret mission.[68] In the early afternoon of the fourteenth, he made his way to the Maritime Terminal at LaGuardia, where Pan Am was preparing a Boeing 314 flying boat—the largest airliner of the day—for the transatlantic crossing, first to the Azores and then on to Lisbon, where he would transfer to a British airliner. That day, only seven passengers were on board, ensuring a high standard of service and as much room to walk around as in the lobby of a small hotel. Donovan refused to discuss the nature of his business—but someone at the *Times* learned from "customs men" that he was carrying "a special passport from Washington."[69] The accompanying photo did not flatter Donovan, showing the fifty-seven-year-old in a dress shirt that was too small, constricting his double chin. The full head of brown hair was now gone, replaced by thinning gray hair. His expression was harried.

England was staring into the abyss when Donovan and Mowrer landed. On everyone's mind was the parlous state of the British Army and the threat of German invasion. In July there were sporadic raids by the Luftwaffe, usually at night. This was not yet the Blitz that would devastate London in September, but the raids were dangerous and worrisome enough, especially when the Germans dropped bombs on civilians. In London, you saw sandbags, billows of barbed wire, and guards with bayonets affixed to old American rifles in front of government buildings—some of them the aging bureaucrats who occupied desks during the day and donned the uniform of the Home Guard at night.[70] Even the king had set up a small rifle range behind Buckingham Palace and started to carry a pistol.[71] If you looked up into the brilliant blue sky near Dover, a short drive southeast of London, you saw the contrails of German and British fighters trying to destroy one another. At ground level, you saw old cars and buses lined up in open fields to discourage German gliders from landing. Edward R. Murrow, whose CBS radio broadcasts soon would bring the Blitz into American living rooms and make him famous before he was thirty-three, told his audience back home in his distinctive gravelly voice that there were no longer any road signs near Dover, and that he had to find his way back to London by driving toward the densest concentration of barrage balloons—tethered obstacles for bombers that looked like blimps and were clustered over the heart of the British capital.[72]

Mowrer had been able to secure a room at the Dorchester, newer and sturdier than Claridge's, but still one of the city's luxury hotels, home to many members of the British elite while they were in London. It was also convenient to the American embassy on Grosvenor Square. This made it easy for the two men to meet and work together. "From the first," Mowrer later wrote, "I liked 'Wild Bill': plenty of will behind a smiling but realistic shrewdness."[73] He had to ask Donovan what their mission was. Donovan explained that "at Knox's request he and I were to collect and publish information on the [Germans'] 'Fifth Column,'" and explore what the British were doing to defend themselves from it. Donovan added they had another important mission: to find out whether the British Isles could hold out.

Knox had made no secret of the fact that Donovan and Mowrer were traveling for him. But in the space of a few days, the two visitors' status rose in British eyes, thanks to a misrepresentation. On July 15, the MI6 office in New York sent a cable to London stating that Donovan was traveling for the president as well as for Knox and FBI director J. Edgar Hoover, a new twist that would have been news to Hoover and Donovan.[74] Motivated by the cable from New York, MI6 director Menzies took a leading role in arranging for him to meet "all leading Government officials and ministers."[75] In a note to Churchill, the senior diplomat Vansittart stretched the truth a little further, claiming that Donovan, "a friend of Roosevelt," had been sent to London "by consent of the two political parties in the United States."[76] The result was that, by the end of July, Churchill and his ministers would focus more on Donovan. That the president cared enough to send a personal representative like Donovan was the message a government desperate for American support wanted to hear, especially one tired of Ambassador Kennedy's pessimism. It opened many doors and turned Donovan into a celebrity.

On July 25, Churchill set the tone by inviting Mowrer and Donovan to a late lunch at 10 Downing Street, the sprawling but elegant 250-year-old building that was both home and office on a street named for a notorious seventeenth-century spy, Sir George Downing. Mowrer remembered how, seated at lunch, Churchill spoke "incessantly and gaily," wanting to know about Roosevelt's next moves and drawing on Mowrer's knowledge of the continent.[77] He could tell that the old warrior clearly relished the challenge, even finding Britain's isolation "rather exhilarating." If worse came to worst—if the Germans invaded—he would defend Britain to the last, then continue the fight from Canada.

Together and separately, Mowrer and Donovan proceeded to meet with various members of the cabinet and senior officers from all three services, and socialized with the social and political elite, including (briefly) the king and queen.[78] Mowrer found time to consult with the internal security service, MI5, which seemed taken aback by the Americans' sudden interest in the fifth column.[79] They toured air bases, watched army exercises, and boarded navy ships. The British held back the country's most sensitive secrets, but otherwise senior officers like Gen. Alan Brooke, in charge of home defense, were amazingly open, realistic but upbeat.[80] Donovan even found time for two of London's great beauties, the movie star Lady Diana Cooper and Baba Metcalfe. He sent a bouquet of yellow roses to Cooper, who wrote a thank-you note to "my dear wild colonel," telling him about the lovely young ladies she had arranged for him to meet who, sadly, had been disappointed when he did not appear for dinner.[81]

Even those who did not particularly care for Americans proved willing to meet with Donovan and Mowrer. On August 1, Donovan met twice with the foreign secretary, Lord Halifax, a strange looking man, tall—six foot five—gaunt, and one-handed. One detractor characterized him as "a boring giraffe."[82] Yet he was one of the pillars of the British establishment. For six years Halifax had been viceroy of India, the de facto emperor of the subcontinent. In 1938, he had been in favor of appeasement. In 1940, he had been many conservatives' preferred candidate for prime minister over the sometime rebel Churchill, who had decided it would be wise to keep him in the cabinet.

Reacting as if he had discovered a new species, Halifax found Donovan to be "an intelligent American . . . [who] had a good deal to say."[83] Donovan averred that his countrymen were more interested in the Atlantic than the Pacific, and "more frightened of Hitler than [the British] were," to the point of readying their defenses for any German parachutists who might drop into "the Middle West."

Reflecting a common prejudice, the somewhat perplexed foreign secretary confided in his diary that he thought Americans were "really . . . quite incredible people."

The two principal British directors of intelligence, Stewart Menzies of MI6 and John Godfrey of Naval Intelligence, understood far better than Halifax the need to put their prejudices aside. Both met with Donovan at length and focused more on what Britain needed from America than on intelligence. Though he used blue stationery and green ink to sign the initial *C*, reserved for the chief by tradition, Menzies was colorless compared to Churchill or even Halifax, with little to distinguish him physically. To look at him, he might have been a banker instead of a spy. He was not brilliant. But he had the right credentials for His Majesty's Secret Service, having attended Eton and distinguished himself on the battlefield in World War I before migrating to intelligence work. He had a sixth sense for bureaucratic survival and maintained the social ties that mattered—which included entertaining at his country home outside London and riding to the hounds. He had impeccable manners. A junior officer remembered him as "an essentially nice man" with a quick, shy smile.[84]

Menzies welcomed Donovan to his downtown London headquarters. Though only a few minutes' walk from the Houses of Parliament and Westminster Abbey, this was a place where few foreigners were ever admitted. Built in 1924, the eight-story Broadway Building was already well used by 1940 in a way that visiting Americans would come to know well during the war. The floor was worn linoleum, the walls an off-white color that one employee described as "a mucky gray/white/cream," the rooms lit mostly with bare bulbs.[85] Only senior officers were granted desk lamps against the prevailing murk. Donovan had to walk through the warren of cubicles defined by wooden partitions to the elevator that seemed ancient to British employees and must have seemed preindustrial to an American. The deferential elevator man would have conveyed him to the fourth floor, where Menzies did his best to win him over.[86]

Godfrey, recently promoted to rear admiral, met with Donovan at least three times: at his office in the Admiralty in the heart of Whitehall, part of the ornate complex of government offices built to run an empire; for dinner at the venerable, wood-paneled Royal Thames Yacht Club in Knightsbridge; and at Braddocks, his comfortable home in Sevenoaks, a town of Tudor cottages and stately old homes some twenty miles from downtown London. He was a brilliant leader who could be ruthlessly demanding even though he looked more like an Oxford don than a naval officer; one portrait shows him in full uniform, with cap, sitting in a chair, legs crossed, not quite erect, with a bemused look on his face. His unofficial nickname was "Uncle John"—because he was so unlike an uncle.

Godfrey leveraged the navy-to-navy origins of Donovan's visit. He answered Donovan's many questions about the fifth column and the war. But above all, Godfrey seized the opportunity to reinforce British requests for overage American destroyers, desperately needed to fight the Germans at sea.[87] Donovan eased the way by confessing that he had been pleasantly surprised by the warmth of his reception; Godfrey found it easy to persuade Donovan that, despite the challenges she faced, Britain would never give in, and deserved American support. "Undeceived by appearances," Godfrey would write, "he quickly became aware of the spiritual qualities of the British race—the imponderables that make for victory."[88] Despite the long string of recent disasters, and the shortages of almost every sort, moral fiber would see the country through. Donovan promised to impress that message upon Knox and Roosevelt.

Donovan spent his last afternoon and evening in England with Godfrey at Braddocks, talking until 2 a.m. He left with the admiral's shopping list and, after a few hours of sleep, met Brig. Gen. Raymond E. Lee, the American military attaché, for breakfast at Claridge's.[89] Donovan told Lee he had concluded that the British had a sixty-forty chance of holding out against the Germans. It was a sobering calculus of the odds for the survival of a country that, a few months earlier, had believed itself able to project power around the globe with ease.

In the afternoon, Lee escorted Donovan to the south coast of England, where the colonel boarded *Clare*, a British Overseas Airways Corporation flying boat camouflaged with green and blue patches. The landlubber Lee watched, transfixed, as the ungainly craft taxied, slowly gathered speed, and took off; "it was," he wrote Donovan, "a fine sight to see the plane leave the water."[90] As long as they could, British warplanes flew alongside to protect the slow-moving airliner from any marauding German fighters.

The pace was agonizingly slow—the transatlantic crossing from Ireland to Newfoundland alone was sixteen hours—and the restless Donovan must have found the long trip hard to bear. But thanks to Churchill's parliamentary private secretary, the ebullient Irish-born Brendan Bracken, he had a few books "to mitigate the tedium."[91] In the handwritten note that accompanied the thoughtful gift, Bracken captured the importance of Donovan's visit to the British and the desperate tenor of the times. "Bless you," Bracken wrote, for "all you are trying to do for this little island of ours." A few months or weeks hence, Claridge's might be a ruin, and Bracken himself might be dead. When he returned, Donovan would either find "the battered relics of a race which never surrendered," or one that, against the odds, had "triumphed over the most bestial tyranny the world has ever known."

2

THE BRITISH COME CALLING

The News from London

Night was beginning to fall when the first British airliner to fly on a scheduled run since October 1939 landed in New York on Sunday, August 4, 1940. At 7:30 p.m., *Clare* disgorged her only passenger. Colonel Donovan was quickly surrounded by waiting reporters. Once again, he refused to comment, telling his questioners, a little gruffly, that he had gone abroad "for the Secretary of the Navy" and, if they wanted to know more, they would have to ask him.[1]

By noon on Monday, he was in Washington behind closed doors with Secretary Knox at Main Navy, the plain white concrete, three-story "temporary" building on the National Mall between the Washington and Lincoln Memorials that was left over from the First World War. The continuing aura of secrecy piqued the interest of at least one reporter who wrote that "the mission on which Colonel Donovan journeyed overseas is one of the present-day mysteries of the Navy."[2] The secretary did nothing to clarify the matter; he cared more about Britain than making a reporter or two happy. Acting with urgency—Britain's fate was at stake and he was eager to make his mark—he convoked a meeting over dinner the same day on the

Sequoia, the hundred-foot-long, sometime presidential yacht docked at the Washington Navy Yard that the secretary was able to use as his interim residence.

Among the handpicked guests who gathered around the long table in the yacht's mahogany-trimmed dining room were Adm. Harold R. Stark, the well-mannered, good-natured chief of naval operations who never shed the nonsensical nickname "Betty" that he had acquired at the Naval Academy (and even used to sign memos to the president);[3] then brigadier general Sherman Miles, the director of military intelligence, son of the old Indian fighter Nelson A. Miles and a distant relative of William Tecumseh Sherman; and the pugnacious, brilliant undersecretary of the navy, James Forrestal, Roosevelt's former neighbor from upstate New York and, more recently, his special assistant. Lingering for hours, the guests listened with rapt attention as Donovan and Mowrer recounted their adventures. The most vital point that the travelers made was that they believed the British would be able to defeat a German landing. Though they lacked equipment, the defenders' morale was high. Within two months, they would have made good the material losses sustained during the Battle of France. When he described the dinner to his wife, Knox added that he would do his part, promoting the "destroyer matter," hopefully fast enough for the ships to arrive in time to "fend off the German assault."[4]

Over the next few days, Donovan continued to spread the word through official Washington, meeting with congressmen and cabinet members, stressing three points: the importance of sending surplus American destroyers to Britain, introducing conscription in the United States, and British prospects for survival. One of the key meetings was with Secretary of War Stimson, the other Republican internationalist Roosevelt had brought into the cabinet in June to help prepare for war. Stimson invited Donovan to tell his story over a small dinner at his home on August 6.[5]

With similar careers and interests, Stimson and Donovan were good, almost close, friends. Stimson had the right pedigree and Ivy

League credentials. In 1940, he came into office with broad experience. He first served as secretary of war from 1911 to 1913, then as an artillery officer in World War I, followed by an appointment as governor general of the Philippines from 1927 to 1929, and by service as secretary of state from 1929 to 1933—to say nothing of his time as a US attorney and Wall Street lawyer. He has been characterized as the quintessential gray man, a conservative who did not appreciate the value of intelligence.[6] He might have looked the part, his face a narrow oval, its only adornment a carefully trimmed mustache. The old-fashioned, rimless pince-nez eyeglasses perched on his pointed nose spoke to function over form. But Stimson was much more than a gray man; he was one of the wise leaders who guided American policy at a time of unparalleled crisis. He had a strong sense of duty, tempered by common sense and good judgment, as well as amazing stamina. He was seventy-two when he answered Roosevelt's call, and he would serve tirelessly for more than five years in one of the most demanding jobs in Washington with little more than an occasional massage or horseback ride for relaxation. Stimson could take that horseback ride on one of his estates: Highhold on Long Island or Woodley in Washington. Set on seventeen acres in DC, which made it one of the largest private properties within city limits, Woodley was less than three miles from the White House. The white mansion on the knoll overlooking Rock Creek Park was nothing if not stately, and had welcomed many prominent Americans over the years, including four presidents and now Donovan.

After dining at Woodley, Donovan met with the president himself. Knox arranged for the colonel to accompany him and Roosevelt for part of a tour of New England that started at 10:30 p.m. on August 9 at the small-town, two-track train station in Hyde Park, New York, along the shores of the Hudson and a little more than two miles from Roosevelt's family home in Dutchess County.[7] There they boarded the special train that, by 1940, included a purpose-built Pullman car for this president: the "Ferdinand Magellan" featured its own kitchen, dining, and living spaces, as well as four private bedrooms with their own washrooms, perfect for accommodating important visitors.[8] The next morning found the party at the Portsmouth Navy Yard in New Hampshire. Two hours later, Roosevelt, Knox, and Donovan were on the 165-foot-long USS *Potomac*, the larger presidential yacht that Roosevelt favored, bound for Boston, where Donovan disembarked in the afternoon.

During their time together, Roosevelt (characteristically) delivered one of his lengthy monologues. From the White House logs it is clear that other guests were competing for the president's attention at the same time, including a Senator Walsh, an Admiral Wainwright, and even a newborn grandson, presented in Portsmouth by Roosevelt's son John. Still, Donovan had been able to deliver his report to the chief executive, summarizing his impressions and urging the president to support the British cause. He stressed their need for American destroyers.[9] The president welcomed Donovan's message and asked him to publish his findings about the fifth column.[10]

The Republican lawyer with no official position was becoming ever more prominent. But it was not just an exercise in self-promotion. He felt that he was doing his patriotic duty by supporting his friend Knox and lobbying for Britain at a crucial time. He did not want to be rewarded with high office for his efforts. Instead, true to form, he hoped for another chance to fight on the front lines in the coming war. In September, he would explain to his English friend Sir Robert Vansittart that he had refused the Republican nomination for a Senate seat from New York. What he wanted to do was to "go with [the] troops." With luck, he would spend the winter in Alabama training a division.[11]

Donovan's hope sprang from a conversation with Stimson. At their dinner on August 6, Stimson had floated the idea: would Donovan consider taking command at one of the training camps that were being reactivated in the South? He would not say no. The camp would house the New York National Guard unit, the Twenty-Seventh Infantry Division, that was now home to Donovan's World War I regiment, the Fighting Sixty-Ninth. Stimson was delighted with Donovan's reply. He had not changed: he was still "the same old Bill Donovan that we have all known" and liked, a man "determined to get into the war some way or other."[12]

For months Donovan would hear nothing from Stimson about returning to active duty, and so remained a free agent. In mid-August, as part of a program to sensitize the American public to impending threats, Knox started circulating articles about the fifth column with a Donovan and Mowrer byline.[13] Donovan covered much the same ground in a radio address.[14] A few weeks later, on September 2, Donovan was happy to learn that Roosevelt formally approved the Destroyers for Bases Agreement between Great Britain and the United States.[15] Within days, the World War I–era warships started to sail for British ports, and both Ambassador Lothian and Admiral Godfrey expressed their appreciation for Donovan's work.[16] For the rest of the month, Donovan split his time between his law practice and travel with Knox, giving more of his time to Knox than to the law. He spent one day in Wyoming on a legal matter, then joined Knox on the Pacific frontier. The two men spent two weeks observing fleet maneuvers and visiting more naval bases. A flight to Pearl Harbor from an aircraft carrier offshore was one of the highlights of the trip.[17]

The New British Spy in New York

Stimson and his colleagues in Washington proved slower to find Donovan a job than the British. The MI6 representative in New York, William Stephenson, took the lead. Stephenson had exaggerated Donovan's status before the trip in July, and now he was at it again. On September 4, he cabled Menzies, telling the director to "give [himself] fifty pats on [the] back," for his support to the Donovan-Mowrer mission that, in turn, led to the president's decision to release the fifty destroyers. "Without Colonel [*sic*]," Stephenson wrote, "it could not possibly have happened at this time."[18]

During the summer of 1940, Stephenson was emerging as an important actor in a set of transactions that would lead to the birth of modern American intelligence.[19] Born in the Canadian province of Manitoba in 1896, he joined the Corps of Royal Canadian Engineers at the outbreak of World War I, serving at the front in France until he was badly gassed in 1915. Only partially healed, but impatient to get back into the war, the short—five foot seven—slight young man volunteered for the Royal Flying Corps, amazing his doubters with skill and aggressiveness by shooting down some twenty German aircraft over France before being shot down and captured himself. He spent the 1920s in Britain, becoming a British citizen but taking an American wife. Always a self-made and largely self-educated man, he blossomed as an inventor and entrepreneur, first in the field of radio transmission and then by acquiring companies as diverse as Sound City Films and Pressed Steel. The economic intelligence that he acquired as a by-product of his work eventually led to a meeting with Adm. Sir Hugh Sinclair, the director of MI6 during the 1930s.

In the spring of 1940, Menzies—Sinclair's chosen successor—started to explore Stephenson's potential. An odd string of coincidences gave Stephenson a rare advantage despite his limited experience in the field of intelligence. At the end of World War I he was not only a fighter ace but also a highly rated amateur boxer despite (or perhaps because of) his small size. He had fought in the same ring in Amiens, France, as the American champion Gene Tunney, then serving as a US Marine. Stephenson and Tunney stayed in touch. Tunney, it turned out, was connected with FBI director J. Edgar Hoover; he would be happy to broker an introduction.[20] "C" wanted Stephenson to try out for the role of intermediary between the FBI and MI6.

In April 1940 Stephenson met with Hoover to discuss "arrangements for co-operation."[21] He could hardly contain his enthusiasm when Hoover reacted positively, and even invited him to "procure official position" in the United States to strengthen their budding relationship.[22] With Stephenson's input in hand, Menzies came to believe that Hoover was close to Roosevelt, seeing him "daily." He concluded that "this may prove of great value to the Foreign Office in the future."[23]

Stephenson and Menzies did not understand the inner workings of the US government. By 1940, Hoover had a good working relationship with Roosevelt, but the two men were not close. FDR reserved his private time for political and social intimates, not middle-class bureaucrats. The president and the director exchanged notes, usually through aides, and occasionally met face-to-face. Roosevelt called on Hoover for official favors; in return, Hoover used the credits that he built up by doing those favors.[24] Nevertheless, for the British, the prospect of access to the chief executive was too good to pass up, and Menzies decided to put Stephenson on his rolls, appointing him principal passport control officer in New York, a light cover that MI6 routinely used.[25]

Stephenson took up his new post on June 21, 1940, exactly a week after the Germans marched into Paris. Thanks to Menzies, the forty-four-year-old businessman was now able to list his occupation as "civil servant" and claim diplomatic status as he disembarked from steamship MV *Britannic* in New York Harbor on his way to the Waldorf Astoria Hotel. On the landing card the INS clerk wrote "ind"—presumably for "indefinite"—for length of intended stay.[26] Stephenson planned to stay as long as it took to accomplish his mission.

A Lesson from the Kaiser War

On the day Stephenson arrived in New York, Britain was confronting a familiar crisis. As in 1916, she once again desperately needed help in a war against Germany. Like most British officials, Stephenson remembered that year of what Admiral Godfrey liked to call "the Kaiser War."[27] Hard-pressed by the German Army on land in France and at sea by the German Navy's U-boats, Britain was finding it harder and harder to replace the tens of thousands of men killed in action, to say nothing of her dwindling ability to pay for the munitions she had to purchase from America and transport across the Atlantic. The British wanted neutral America to enter the war on their side. Their spies and diplomats went on high alert for opportunities to leverage. By early 1916, the forerunner of MI6 had established a small office in New York City charged mostly with counterespionage.[28] Posing as members of the Transport Department of the Ministry of Munitions, the MI officers did not openly or officially declare their actual purpose.

The undercover officer who became the senior British intelligence officer in North America was an aristocrat named Sir William Wiseman, scion of an ancient family whose pedigree dated back to 1628. Wiseman was a man of many talents who had already lived an eventful life on both sides of the Atlantic. He had attended Cambridge University, where he distinguished himself as a bantamweight boxer against Oxford University; reported for the London *Daily Express*; and, deciding that his future lay in banking, invested in Canadian real estate and Mexican meatpacking. When war broke out in 1914, he had hurried home to join the army. In 1915, he had been gassed and temporarily blinded in France. Though not cleared to return to the front lines, he was fit for other war work. The original "C," a Royal Navy officer named Mansfield Cumming who had served with Wiseman's father, was happy to put him on the rolls and send him overseas. Wiseman's lack of formal training and experience mattered little. When he arrived in New York, Wiseman was a young-looking thirty-one-year-old, with a round face, small mustache, and close-set eyes. Though impeccably dressed and groomed, he looked more like a recent graduate than a secret representative of the Crown. His unprepossessing appearance, along with his manners and pedigree, made it easy for him to make American friends.

In December 1916, Sir Cecil Spring-Rice, the old-school British ambassador in Washington, asked Wiseman to deliver a note to Col. Edward M. House, President Woodrow Wilson's closest advisor. The transplanted Texan was not a military man—the title was honorific, a reward for political services—but a wealthy businessman who liked to operate behind the scenes. Slender, almost elfin, he gave the impression of being small but was of average height at about five feet, nine inches tall. He had moved to New York because he wanted to escape the unhealthy climate in Houston. Wiseman found the colonel from Texas in his comfortable apartment on the ninth floor of 145 East Thirty-Fifth Street in Manhattan. A five-minute errand turned into a long conversation. House found it easy to talk to Wiseman, who seemed more in tune with the administration than Spring-Rice.

Over the next few weeks, Wiseman turned from messenger to confidant and then into neighbor, eventually taking an apartment on the second floor of House's building. The ever-solicitous Wiseman asked the colonel if he minded. On the contrary, House liked having Wiseman close.[29] Wiseman cabled London and pointed out, with unusual delicacy for a spy, that, when in House's apartment, he would inevitably see US government documents that were not intended for foreign eyes. What should he do? MI6 cabled back, with equal delicacy, that Wiseman should use his discretion—implying that some documents were so important that he should not avert his eyes.[30]

House kept Wilson apprised of his "conferences" with Wiseman, promoting him as "the most sensible Englishman that had been connected with the Embassy here since the war began."[31] When House introduced Wiseman to Wilson over tea at the White House, the president was also smitten with the deferential young English aristocrat. Wiseman was now on his way to becoming one of the handful of men and women with unfettered access to both the president and his closest advisor.[32]

In February 1917, America broke diplomatic relations with Germany. By March, Wiseman was helping House draft an overview of American policy that he, Wiseman, shared with London. Later in the month House repaid the favor by briefing Wiseman on the speech that Wilson would give to ask Congress for a declaration of war against Germany. Wiseman now knew what the president intended, in detail, before most members of his own cabinet; House and Wilson had handed Wiseman the kind of exclusive, life-or-death, eve-of-war scoop that most spies only dream about acquiring, and that even a very lucky spy may acquire only once over the course of decades of hard work.[33]

In the summer of 1918, Wiseman went on a weeklong vacation with Wilson and House to the seaside village of Magnolia, Massachusetts, with its luxurious "cottages" that were actually more like the Gilded Age mansions in Newport, Rhode Island. The president and Mrs. Wilson stayed in the Marble Palace, described by Wiseman as "a beautiful colonial house over-looking the sea."[34] The Wilsons as a rule were guests for lunch at the nearby bungalow that House had rented for himself and Wiseman. At dinner, the Wilsons hosted House and Wiseman. If Mrs. Wilson was along, the president might suggest that that the menfolk retire to talk business. Alone they would discuss such topics as the League of Nations, economic policy, and German peace efforts—all of which Wiseman dutifully reported to London.

Wiseman's relationship with Wilson lasted until September 29, 1919, the day after the president and first lady returned to Washington from a trip cut short by his failing health. He had roused himself to deliver brilliant speeches, but could not escape the effects of a series of increasingly serious strokes that were turning him into an invalid. Wiseman asked for an appointment with the president to discuss important business, but when he came to the White House at 11 a.m., he found a disapproving Edith Bolling Wilson blocking his way.[35]

Wiseman had better manners than many spies. He genuinely liked House and took care to protect him. But his attitude toward Wilson was realistic and calculating. The president had, he thought, "something of the chill of the cloister in his manner."[36] He registered that Wilson could be "arbitrary" and "aloof," and was, at times, "a most difficult person to deal with," with a limited ability to compromise.[37] Instinctively Wiseman fit his approach to his target. He waited. He listened. He empathized. In a letter to London, he described how he would let the president engage first, and only then offer his point of view.[38] Like a good intelligence officer, he developed insights into a foreign leader that his government would put to good use.[39]

After the war, Wiseman chose to live in New York as an expatriate. Undeterred by Mrs. Wilson's chilly attitude, on good terms with Colonel House and many other prominent Americans, Wiseman made his fortune in the United States without giving up his British passport. In 1921, he joined the investment bankers Kuhn, Loeb in New York and prospered. By 1929 he had become a partner in the firm and would become prosperous enough to move uptown to a seventeen-room Fifth Avenue penthouse overlooking Central Park.[40] More than once, he appeared in the society pages of New York newspapers.

A few weeks after the second German war started in 1939, Wiseman offered his services to the British government, conferring first with Ambassador Lothian in Washington and then making his way to London to meet with the unflappable Cadogan, the permanent undersecretary at the Foreign Office who seldom smiled and almost always dressed in black. Sitting in Cadogan's office in Whitehall, Wiseman laid out the basic problem: how to work up enthusiasm for the British cause in the United States. The president was ready to support the British, but the American people were not. The only remedy, he argued, was "as in the case of the last war . . . through our Secret Service."[41] The service would be able to arrange for the publication of articles that did not appear to come from His Majesty's government; the appearance of impartiality would make them more compelling. He added that, while the ambassador should be given an overview of this work, "the less Lord Lothian knew directly of what was being done[,] the better." This would make it easier for the ambassador to deny any knowledge of such activities.

Eight months of inaction later, Wiseman was back in London. Now, in early June 1940, with Britain isolated and the Germans twenty miles away across the Channel, he pressed his case: "Our experience in the last war and in this war shows the urgent need for the best possible intelligence service in the United States."[42] He proposed opening an office that would be "entirely unofficial and without any [apparent] responsibility on the part of His Majesty's Government." He would reach out, quietly, to the FBI to guard his flanks. Wiseman implied—amazingly—that he could obtain the president's approval and then "make the necessary arrangements." All he needed was 100,000 pounds sterling in starter money. This was the voice of an old spy eager to get back in the game.

Cadogan invited "C" to his "room" in the ornate, Victorian interior of the Foreign Office in Whitehall to meet with two senior members of the foreign policy establishment, Sir Duff Cooper and Sir Robert Vansittart.[43] They did not disagree with the basic idea—Britain urgently and secretly needed to spread its message throughout the United States. But they decided to proceed without Wiseman. The reasons were that he "was universally known to have been connected with British propaganda and Secret Service work during the last war," that he had "publicly boasted of his former activities," and that he was associated with "the Jewish firm of Kuhn, Loeb." (The ostensible reason that this mattered was that German propaganda could exploit the connection; the other, unspoken explanation is old-fashioned social anti-Semitism.) The foreign secretary, Halifax, would see Wiseman the next day and let him down gently, asking him to support the government's actual choice for the job.[44]

By then Stephenson was already headed to New York on the MV *Britannic* to start work. Cadogan announced Stephenson's appointment to Lothian on the tenth of June, the day the steamship eased up the Hudson to her berth on the west side of Manhattan.[45] From the start, Stephenson knew that Wiseman was the role model to emulate.

During the First World War, Sir William had shown what a British spy could accomplish in the United States. Now, in 1940, faced with similar threats and opportunities, MI6 was sending Stephenson to New York to perform almost exactly the same mission: gather information, protect British property, and, above all, "organize American public opinion in favour of aid to Britain."[46]

And just as Wiseman had cultivated Colonel House, Stephenson would cultivate Colonel Donovan. Though the two men had not met before 1940, Stephenson had been following the reports of his activities from newspapers, radio broadcasts, and of course the secret cables from London that crossed his desk.[47] He knew that Donovan had made an excellent impression in the British capital, and was already doing some of Britain's bidding off his own bat.

3

GENTLEMAN HEADHUNTERS MAKE A PLACEMENT

Stephenson at Work

William Stephenson was not afraid to take risks. Wasting no time, he started laying the foundations for a British intelligence empire in North America, to be known as British Security Co-ordination (BSC).[1] After taking a quick look at the cramped quarters that his predecessor had occupied in the twenty-two-story Cunard Building on Exchange Place near Wall Street, he decided that Midtown Manhattan would suit his purposes far better, and rented space in the newer and more splendid Rockefeller Center, taking the thirty-fifth and thirty-sixth floors of the art-deco RCA Building at 630 Fifth Avenue.

Stephenson's challenge was to change the mood of the country while hiding the British hand. Like Wiseman, he knew that the best propaganda for Britain would appear to be independent. He proceeded to direct BSC officers to set up satellite offices throughout North—and eventually South—America. They would recruit agents to penetrate enemy businesses and embassies; take over a respected Boston radio station known for its impartiality; and flood American journalists with pro-British material.[2] Isolationists would find themselves embarrassed, even compromised, by material that surfaced mysteriously, while interventionists were courted and flattered.

Donovan was high on the second list. Even after the move to Midtown, the distance from Stephenson's new offices at Rockefeller Center to Donovan's law offices at 2 Wall Street was not great physically or otherwise; on the same island, they were also in the same political, social, and professional world.[3] Donovan and Stephenson likely met in Manhattan for the first time in August 1940. Their backgrounds and interests so compatible, they quickly became both professional and personal friends.[4] Donovan would become known as "Big Bill," presumably because he was taller and heavier, while Stephenson would be "Little Bill."

In the short-term, what mattered most was keeping the German enemy from the English door. What could Donovan and Stephenson, together, do for Britain? The island kingdom's dire straits dominated their discussions in the summer and the fall. It was more about politics than intelligence. An important part of the equation was the 1940 presidential election. By July, Roosevelt had reluctantly decided to run for a third term. This was unprecedented in American politics. He had not been inclined to stay in office. On the contrary, the president claimed that he was dreaming of retiring to the country quiet of Hyde Park after two terms. It was the Nazi victory on the continent that changed his mind. He felt a need to ensure continuity in the fight against Nazism, and maneuvered the Democrats into renominating him. In the end, he won by a comfortable margin four months later in November.

The White House would remain in friendly hands, but that was still only part of the solution. Many Americans remained staunchly isolationist. Even those who voted for Roosevelt did not necessarily want to participate in the war in Europe beyond sending munitions across the sea. Under these circumstances, Stephenson reasoned, why not organize another of Donovan's trips, an extension of the first that had been so successful? If the British continued to impress and court the traveling colonel, he would more than repay their favors with his support.

After the election, Stephenson cabled MI6 in London that Donovan could travel to Britain "to repeat his good work of [the] last occasion and also to combat forces of appeasement here which are gaining ground again."[5] Stephenson had Ambassador Kennedy in mind on November 27 when he reported that the envoy "was doing a great deal of harm . . . to our cause . . . undoing Donovan's good work."[6]

Donovan's second trip was organized in much the same way as the first: under the auspices of Knox's Department of the Navy with casual concurrence from the president and grudging acceptance by the Department of State that Donovan would travel to London and go on to tour the Mediterranean.[7] But Secretary Hull officially distanced his department from the initiative, cabling embassies abroad that Donovan was traveling for Knox; department spokesmen would insist they knew nothing about the purpose of the trip.[8] They did not say so openly, but it was clear that they saw no need for another amateur diplomat to visit Europe.

The new trip also had the same last-minute touches as the first. The British embassy in Washington floated the possibility of the trip on November 27, at Knox's explicit request.[9] But this time the British clearly took the leading role. Stephenson would accompany Donovan to London; MI6 would jump-start his talks there and pay many of the expenses.[10] He would stay mostly in British residences, even when American quarters were available, and keep in close touch with London throughout.[11]

Once again, BSC oversold the visitor from America, claiming in the December 6 cable that he had been promoted to major general (a promotion that had been discussed but would not materialize until 1944) and that he exerted a "controlling influence over Knox [and] a strong influence on Stimson." BSC concluded that the lawyer who was still a private citizen was "presently the *strongest* friend . . . we have [in the United States]."[12]

Back to London

As noon approached on December 6, 1940, a Boeing 314 flying boat bobbed at its pier on the Pan American base in Baltimore harbor, fully loaded and awaiting its last passenger. A passenger manifested as "Donald Williams" had not arrived yet. Just before noon, a cab screeched to a halt, disgorging a man who looked very much like Donovan and carried bags that bore the initials "WJD." Inquisitive reporters quickly established who he was by talking to his friends in Washington and learning one of the other two passengers was a "Mr. O'Connell," who was almost certainly Stephenson.[13]

The clipper departed for Bermuda more or less on time with its three passengers and thousands of pounds of mail, landing around 2 p.m. local time only to find that bad weather farther east would impede onward travel. The upshot was that the travelers would be marooned for eight days on the tropical island in the middle of the Atlantic. Stephenson filled the time by introducing Donovan to an unusual British intelligence operation. Many transatlantic flights had to pass through the Crown colony to refuel. British officials took the opportunity to unload and examine massive amounts of mail to look for secret messages from the Western Hemisphere to Europe. Donovan is said to have been particularly taken with the work of a "beautiful, sharp-witted" "censorette" named Nadya Gardner.[14]

It was December 16 before Stephenson and Donovan finally arrived in London and started their rounds in the great city, now in its winter livery. As Ed Murrow reported in his broadcasts from London that month, at night all was "dark and silent . . . the big red busses that roam[ed] the streets . . . nearly empty."[15] German bombers still followed the Thames to the capital, straining emergency services and citizens alike. One hundred thousand Londoners slept in Underground stations every night. There was damage—in some places considerable—but the city was far from a ruin.[16] The British people were stressed but not broken. Even though they could not explain how, they seemed confident that they would one day prevail over their enemy.

For his part, Stephenson kept on overselling Donovan. The World War I paradigm never far from his mind, Stephenson even went so far as to claim that Donovan "has more influence with the President than Colonel House had with Mr. Wilson."[17] That was why the prime minister should again meet with the colonel. Both Churchill and Halifax did just that. During an afternoon chat at his office on December 17, the sepulchral foreign minister found Donovan "very much alive and out to help"; the next day, Churchill sat down for lunch with the colonel at Downing Street.[18] Said by an aide to "know full well the value of Donovan to us," he was starting to like the American who was close to him in age, as well as a veteran of the Great War that had formed them both.[19] The only thing he did not like about Donovan was that sometimes the colonel seemed too glib and breezy, which was not in keeping with the life-or-death threat Britain faced.[20]

Churchill endorsed the developing plan for Donovan's travels throughout the Mediterranean, charging British officers and diplomats with affording "every facility" to the colonel.[21] The trip would be largely at British expense in British military and naval aircraft. Donovan would have a personal escort, Lt. Col. Vivian Dykes, a forty-two-year-old Royal engineer known as "damn-me-Dykes" for "the transports of violent and profane language that he used."[22] But he had more to offer than colorful language; he was a well-rounded officer who had turned down a place at Oxford to attend the Royal Military Academy at Woolwich. He had served on the Committee of Imperial Defense and as a senior planner for the chief of the Imperial General Staff. Brave, dependable, and cheerful, he was said to be the man you wanted at your side in a tight place.

On to the Mediterranean

Donovan and Dykes left London for Plymouth the day after Christmas to begin their odyssey. In a little more than two months, they would cover some 12,000 miles by air, 2,000 by train, and 1,000 by car.[23] Donovan would meet with commanders, ambassadors, and politicians, tour the battlefields, and board the warships. Along the way, he would continue to learn the ins and outs of wartime intelligence.

Donovan could not help but be impressed by what he saw and heard early on in British strongholds like Gibraltar, at the entrance to the Mediterranean, and Malta, the island colony between Spain and Italy. Dykes recorded the details of the onward flight in the two-engine Wellington medium bomber: the cold in the cramped, unpressurized interior, with nowhere for passengers to sit comfortably; the climb through the clouds to find a little warmth from the sun; flying to the surf line where ocean met desert and showed that they had reached the African continent; and finally following that line to the great green fan of the Nile that pointed upriver to a city that had endured for millennia and was now a British military hub.

In Cairo, the two men received general and specific briefings—including a description of the Long Range Desert Group, commandos who roamed behind the lines, gathering intelligence and raiding enemy outposts. Donovan even had a chance to venture out into the Western Desert himself, walking the ground where the British Army was (*finally*) enjoying more success, here against the Italian Army. Back in Egypt proper, he spent an hour with the somewhat eccentric Dudley Clarke, a "sharp little man with bright, quick eyes" who was beginning his run as one of the masters of strategic deception in World War II.[24] After meeting Clarke, Donovan received a briefing on how the British intercepted enemy radio messages and analyzed them—a form of communications intelligence. At Alexandria he took in the most important Royal Navy base in the eastern Mediterranean and went on board the flagship HMS *Warspite,* built in 1913 but still able to carry the fight to the enemy.

From Egypt the pair ventured east to Baghdad and Jerusalem, and then north to the Balkans, where Donovan lost his passport, probably to thieves working for German intelligence. Donovan's minders could not stop him from spending a week on the front lines in Albania, where Greek forces were fighting the Italians and the old warrior once again could hear, smell, and see ground combat. At first, Lincoln MacVeagh, the American ambassador in Athens, was ambivalent about this "unofficial official . . . trotting around loose."[25] Donovan explained that he "was representing the Secretary of the Navy, unofficially . . . observing and talking with all and sundry, with an idea of finding out what is being prepared . . . and incidentally to carry assurances [to foreign leaders] that the United States means business." In the end, MacVeagh softened: "The colonel is a man of well-ponderated judgment, of considerable intelligence, wide interests, and pleasant and rather restful manners. . . . A fine type of American, who has made himself an important person by his intelligence and charm."

Donovan's artful posturing made Churchill happy, and on January 30, 1941, he sent a personal telegram to Donovan through British channels, offering "many congratulations upon all you are doing."[26] Donovan "deeply appreciated [the] . . . gracious message," and, remembering that he did not work for Churchill, added that he just wanted "to make it clear that the President meant what he said."[27] A few days later, Churchill mused in a note to the Foreign Office that a visit by Donovan to neutral Spain "would be opportune."[28]

For a time, Donovan and Dykes had a travel companion. He was the debonair, young Ian Fleming, elegant in the dark blue uniform of the Royal Navy Volunteer Reserve despite the wavy braid on his sleeve, a reminder to all that he was not a regular officer but merely a reservist. Fleming was serving as the principal aide to the director of Naval Intelligence, Admiral Godfrey, Donovan's new friend from his trip to Britain in the summer. After false starts as a foreign correspondent and indifferent stockbroker, Fleming found himself as an intelligence officer. He was happy to show Donovan around Gibraltar and even explain a far-fetched scheme for secreting British observers deep in the Rock should Spain or Germany occupy the small but strategic British territory at the tip of a Spanish peninsula jutting into the sea. Fleming believed that he had made a friend in Donovan and would stay in touch with breezy notes.[29]

Donovan arrived back in London on March 3, 1941, for another intensive round of office calls, many with various military planners, all of whom listened spellbound as he talked about his Balkan and Peninsular tour.[30] Donovan seized an opportunity to learn more about the work of one of Churchill's newest initiatives, the Special Operations Executive (SOE), which he set up in 1940 to send commandos "to set Europe ablaze" through acts of sabotage. It would be distinct from MI6. An idea born out of desperation when Britain's military options were limited, it was still an unrealized vision, but one that appealed to Donovan.

Churchill himself met with the colonel again at 10 Downing Street and, in a singular mark of favor, took him to call on the king on March 4.[31] Churchill followed up by cabling his thanks to Roosevelt "for [the] magnificent work done by Donovan in his prolonged tour of [the] Balkans and Middle East. He has carried with him throughout an animating, heart-warming flame."[32]

Donovan returned to the United States on March 18, and once again spread his message through official Washington and further afield.[33] On March 19, the president and his closest advisor, Harry Hopkins, gave Knox and Donovan fifty minutes at the White House to report on "Europe's trend"—what the latter had seen and heard overseas that could be useful to the United States.[34] On the same day, Stimson welcomed Donovan home. He would write that they spent more than an hour together standing over a map in his study, "talking . . . in the way in which two old friends who are both interested in military affairs can."[35]

Looking for Work

For Donovan, intelligence was still just one part of a larger picture.[36] British intelligence had made an impression, but it was, above all, the conditions on the ground from London to Belgrade that had seized and held his attention. He had seen, and listened, and spread his own wishful interpretation of American foreign policy, one that meshed with British foreign policy. He had solidified official friendships that would have far-reaching implications for American intelligence later on. But for now, what mattered most was that London was still burning.

Back on American soil, Donovan found himself once again in the pool of candidates for wartime jobs. In November 1940, the British government had authorized Ambassador Lothian to hint, diplomatically, that he would make a wonderful replacement for Joseph Kennedy, whose mission had mercifully ended in October.[37] But Roosevelt had chosen another Republican, the gentle giant John G. Winant from New Hampshire. He looked a little bit like Abraham Lincoln but did not dress even as well as the often-disheveled president: his pants seemed permanently wrinkled, his shirts frayed at the cuffs and collars. Londoners found him a welcome antidote to the well-dressed but abrasive and defeatist Kennedy.

In December, before departing for London, Donovan had followed up with Stimson on the job that he actually wanted: returning to active duty in the army.[38] Stimson regretted that he could not interfere with the established chain of command, and was relieved when Donovan took the rejection in good spirit. Roosevelt briefly mulled over using Donovan as some kind of emissary to the Irish government to tamp down simmering tensions between London and Dublin, but in the end sent no one to deal with that nettlesome problem.[39] In the new year, the administration considered Donovan for three other jobs that did not materialize.[40] By the end of May 1941, the loyal Knox was frustrated. He had kept the spotlight on his friend, taking him to the White House, reminding insiders that he was still available, even introducing him to radio audiences who tuned in to hear about his travels—all to no avail.[41]

The British Solution

Stephenson, who by now had been head of British Security Coordination in New York for almost a year, stepped into the vacuum, upping his campaign to persuade Donovan to become an intelligence impresario. Though he already had an impressive set of contacts—from J. Edgar Hoover at the FBI to the president's own amateur spy Vincent Astor to speechwriter Robert E. Sherwood—Stephenson decided to focus not on them but on Donovan. He was gearing up for the "three months of battle and jockeying for position in Washington."[42]

More signs of British influence now appeared in Donovan's work. On April 26, 1941, Donovan wrote to Knox with a description of "the instrumentality through which the British Government gathers its information in foreign countries."[43] This amounted to an argument for the establishment of a centralized American clone. On May 5, 1941, Stephenson cabled London that he had been "attempting [to] manoeuvre Bill into accepting a job of co-ordinating all 48 land [US] intelligence."[44] The colonel was still Britain's greatest friend, and had the makings of an ideal partner. If BSC had an American counterpart more or less in its image, the two organizations could go on the offensive together, pooling resources and information. Still, the colonel himself was not sure. Intelligence work was interesting and important, but it was not his forte; he wanted to go to war, not become a spy chief.[45]

Stephenson was reflecting the general trend in British intelligence. Over the past decade it had been centripetal, moving toward ever more joint operations at the center. MI6 was already responsible for two kinds of intelligence: signals and human—running the Government Code and Cipher School (GC&CS) at Bletchley Park on the one hand and espionage on the other. Military Intelligence and Naval Intelligence had long been independent from one another but, in 1936, they had reduced duplication by establishing the Joint Intelligence Sub-Committee that would later be known as the JIC. Its primary mission was "to evaluate, sift and disseminate intelligence" to make it possible for consumers to receive consolidated reports on any given topic, not competing facts or analyses from, say, the navy and MI6.[46] Why not add American input to the mix?

British and American officials quickly agreed that American neutrality did not preclude the exchange of intelligence and moved forward in that direction. But crafting the mechanism was not simple. It was a nuisance for Britain to cultivate separate relationships with the US Army, US Navy, the FBI, and the Department of State for access to each one's secrets about Germany or Japan. Was there any way to encourage the Americans to create something like their own JIC? Donovan had come over from the United States, received briefings on defense that included intelligence, and, in very general terms, voiced his support for everything the British were doing. Decreeing that "anything we can do in the way of 'preparation' [now] . . . will be of considerable [mutual] benefit if and when" the Americans enter the war, the JIC decided to continue the process by sending two of their best officers to New York and Washington.[47]

Admiral Godfrey and Commander Fleming

In May 1941, Admiral Godfrey and his aide Ian Fleming flew across the Atlantic to make the JIC's case to the Roosevelt administration. Traveling in civilian clothes, they flew one of the circuitous routes that Donovan had followed—London, Lisbon, the Azores, Bermuda, and finally, after delays and rest stops, New York. They arrived on May 25, on the same clipper as the fashionista Elsa Schiaparelli, who attracted a swarm of press photographers. The official visitors on a secret mission tried to sidestep the swarm but with limited success; the photographers caught the pained expression on Godfrey's face in the background when they focused on the famous designer coming up the ramp.[48]

At the top of the ramp, Godfrey likely turned in the direction of Donovan's luxurious apartment at One Beekman Place with a magnificent view of the East River, where the colonel repaid some of the hospitality the admiral had shown him in London.[49] In the days that followed, with help from his British friends, Donovan worked on the draft of a memorandum on intelligence. By the end of May, Donovan had it circulating among officials in Washington.[50] In a cable to London, Godfrey praised Donovan's "energy and drive" that augured well for "full cooperation in [the] realms of intelligence" between Britain and the United States—but only if the reluctant colonel could be persuaded to lobby for himself as well as a new organization.[51] Godfrey went on to suggest that Churchill send a "personal message of exhortation" to Donovan, but JIC chairman William Cavendish-Bentinck rejected the proposal. If it leaked, "it would expose Donovan to the imputation of being a British agent instead of the splendid free-lance that he is."[52] Neither Cavendish nor anyone else on the British side had any objection to having British agents in Washington or manipulating the Americans; it was just that, in Donovan's case, it could be counterproductive.[53]

The results of other early consultations in Washington and New York over the following two weeks were not promising. The welcome was genuine but led nowhere. J. Edgar Hoover granted time, spoke in generalities, and treated his visitors to a demonstration at the FBI's indoor firing range at the Department of Justice on Pennsylvania Avenue, one not particularly likely to impress seasoned naval officers.[54] The reception by the army and the navy was equally cordial, as it was at the Department of State, but the various bureaus were obviously not getting along. Godfrey would remember that they showed "the utmost goodwill towards me and Ian Fleming but very little towards each other."[55]

Looming in the background of Godfrey and Fleming's visit was the specter of the new German superbattleship *Bismarck*. Weighing in at some fifty thousand tons, she was one of the world's most modern and powerful ships, sinister and sleek at the same time. News that she had slipped into the Atlantic on May 21, 1941, along with the heavy cruiser *Prinz Eugen* was immediate cause for alarm in London. Some eleven British-American convoys were either about to sail or already plodding across the Atlantic, and all needed protection from the predators. Churchill cabled Roosevelt to ask that the US Navy "mark them down" if they eluded the Royal Navy and ventured west: "Give us the news and we will finish the job."[56]

America paid little attention to the story until May 25, the day Godfrey arrived in New York and the *Bismarck* sank the *Hood*, a British battle cruiser twenty years older and not as powerful, but still proud and redoubtable. Within five minutes of the first exchange of fire in the Denmark Strait—the cold, foggy body of water in the middle of the Atlantic between Iceland and Greenland—the *Hood*'s aft magazine erupted, breaking her back. The ship's two halves pointed to the sky and slipped beneath the waves. Only three members of the crew of 1,418 survived.

The news of another British naval disaster on a par with the sinking of the *Royal Oak* in October 1939 was met with "stunned silence" in Washington; both the British Embassy on Massachusetts Avenue and the Navy Department downtown were too downhearted to speak to reporters.[57] An American observer noted that the battle had taken place in a zone patrolled by American ships. What if she continued to head west? President Roosevelt himself thought it "not unlikely" that she would continue to the Caribbean.[58] The "Nazi battleship of great striking power," now loose in the Western Hemisphere, worked her way into the draft of the Fireside Chat scheduled for May 27. *Bismarck* would provide additional justification for an important policy shift: the declaration of an unlimited national emergency. The president specifically included a call for action "to assure our internal security against foreign . . . subversion."[59] Hundreds of telegrams, 95 percent of them positive, poured into the White House within hours. Roosevelt was buoyed.[60]

By the time the president was speaking in Washington, the news that the Royal Navy had finally sunk the *Bismarck* was working its way across the Atlantic. The following day Donovan joined the chorus of praise for Roosevelt's hitting "right on the button."[61] At roughly the same time, Godfrey vented his frustration to Stephenson; the future would be grim if the British had to deal with a plethora of "separate [American intelligence] authorities."[62] He was clearly not getting through to the American establishment. Stephenson decided it was time to escalate, to bring his predecessor Wiseman's powers to bear. More than anyone, Sir William knew that the way to break the impasse was to engage the president himself. He could get a message to Mrs. Roosevelt through the publisher of the *New York Times*, Arthur H. Sulzberger.[63]

Thanks to Wiseman, Godfrey soon had an invitation to join Roosevelt for dinner at the White House. Dressed in black tie, he arrived at the White House at 7:30 p.m. on June 10, 1941, and was shown into the dining room with a small assortment of other guests, including the popular songwriter Irving Berlin.[64] After dinner, the president escorted Godfrey into the Oval Office for a private talk. The host placed himself behind the small wooden desk left over from the Hoover administration that he had cluttered with bric-a-brac—one tattered elephant and many donkeys; a lighter, an ashtray, and an engraved case filled with Camel cigarettes; a telephone without a dial. Roosevelt again dominated the conversation, reminiscing about his 1917 meeting with Godfrey's predecessor, Blinker Hall. The admiral finally found an opening during the hourlong meeting, and told the president, more than once, that the United States should have "one intelligence security boss, not three or four." In the cable that Godfrey dispatched to London, he detailed his meeting with "Flywheel"—the not entirely flattering codename the British had assigned to the president—who had listened and "cross-examined [him] closely about [the] co-ordination of intelligence."[65] On the drive back to the home of the naval attaché in Washington, Godfrey wondered whether the elusive Roosevelt had really taken his message on board.

Godfrey did not mention Donovan's name; that would have been too direct. Telling Roosevelt how to organize part of his government already verged on impertinence, and so the British let their American friends take the next step. By June 10, Donovan had finalized his memorandum on intelligence and sent it to White House.[66] He started with theory: "Strategy . . . without information" was "helpless;" information was "useless unless . . . intelligently directed to the strategic purpose." His loosely worded argument led to plea for a body that would collect, consolidate, and analyze information about the enemy for the president, and then use that information to wage psychological warfare, the other vital "element in modern warfare." This was something that Donovan had learned on his fact-finding missions. The Nazis had succeeded in part because they knew how to use propaganda, to give the impression through the media—especially radio—that they were unstoppable.[67] Performing those functions would be a "Coordinator of Strategic Information" who would work for the president. Not unlike the chairman of the British JIC, this coordinator would be assisted by an advisory panel from the FBI, the army, and the navy.

Donovan did not explicitly promote himself, but his signature on the memorandum made him a candidate for the job. One day later, Grace Tully, Roosevelt's secretary, sent word that the president wanted to see Donovan and Benjamin V. Cohen together; Cohen was the brilliant lawyer known as the "architect of the New Deal."[68] Roosevelt had already created more than one hundred federal agencies, many if not most by executive order; Cohen knew the right words to use.

On June 18, 1941, Roosevelt found thirty minutes for Cohen, Knox, and Donovan in the Oval Office. For once, the president seems to have spent more time listening than holding forth; FDR had, Donovan commented after the meeting, accepted his ideas "in totem [*sic*]."[69] That acceptance amounted to a job offer but was maddeningly casual. The president signaled his approval by adding a brief handwritten note to Donovan's June 10 memorandum: "Please set this up confidentially with Ben Cohen-military-not OEM."[70] When Donovan's time was up, Roosevelt went straight into his next meeting with Darryl F. Zanuck, the filmmaker with a social conscience who had produced blockbusters like *The Grapes of Wrath* that depicted life in the Depression head-on.

A few days later, Donovan claimed that he had agreed to start the "defense information bureau" only after receiving the president's promise that he could then rejoin the army and "handle troops."[71] He stressed that this was not the job he wanted; he still wanted to be in uniform, in the field. In the meantime, "his agency . . . [would] coordinate and undertake research on the basis of information already on file with the various government departments." He would report only and directly to the president. "Any department of the Government . . . [would] furnish necessary information to his agency."[72]

A few hours after meeting the president, Donovan caught a plane back to New York from Washington's two-day-old National Airport at Gravelly Point on the far bank of the Potomac; once in New York, he hurried to report to Stephenson. He in turn cabled London within hours, "relieved" that the struggle was over and, in another display of wishful thinking, exulted that "'our man' is in a position of such importance to our efforts."[73] Donovan would become "Coordinator of all forms [of] intelligence." Charles G. Des Graz, an official at the British Consulate General in New York, chimed in a few days later, telling the Foreign Office that "Donovan has been appointed [to] co-ordinate all activities, reporting directly to [the] President," adding that he was "reminding [Donovan] of our requirements on which he has already made representations."[74]

The British again sounded as if the colonel were a British agent of some sort. Building on Wiseman's precedent, they might be forgiven for wanting to have another agent of influence close to the president. But that description did not fit Donovan. He was never a British agent but always his own man. Nevertheless, he was happy when British and American interests coincided, and had followed the British lead time and again in 1940 and 1941. They played the dominant role in promoting the concept of COI, finding the right director, and persuading the president to agree to his appointment. Without the concerted efforts of Churchill, Stephenson, and Godfrey, there might not have been a Coordinator of Information, and Donovan might have found other war work more to his liking.

4

J. EDGAR HOOVER

The G-Man Who Wanted to Spy

Now more than ever, Donovan knew that he would have to reckon with the director of the FBI, J. Edgar Hoover, who had more than a passing interest in the world of intelligence. In the months before he became Coordinator of Information, Donovan had kept in loose touch with Hoover, sending him "Dear Edgar" notes about getting together with Frank Knox to discuss matters of mutual interest. When Donovan proposed "luncheon," Hoover countered with a "Dear Bill" invitation to dinner at his home on Thirtieth Place NW, about five miles north of Donovan's mansion in Georgetown.[1] In late November 1940, before setting off for Britain and the Middle East, Donovan telephoned to let Hoover know about his travel plans, and Hoover offered to share his own contacts overseas.[2] In mid-June 1941, Donovan told Hoover's assistant, Edward A. Tamm, that whatever else happened, he did not intend to interfere with the Bureau's work. The future federal judge, who could be very direct, felt free to remind Donovan that Hoover did not think a coordinator was either necessary or desirable.[3] On the twenty-seventh, after Roosevelt had agreed to create the COI, Donovan called again, wanting to get a message to Hoover but settling for a lengthy phone call with Tamm. He found Donovan ambivalent about becoming the COI and "extremely anxious to allay any fears you [Hoover] might have"—the colonel did not want the director to think that he wanted to intrude on his domain.[4]

Often described as enemies, the word "rivals" more closely captures Donovan and Hoover's relationship before Pearl Harbor. For more than a decade they had maneuvered warily around each other, sometimes making common cause, sometimes opposing or undermining one another while generally maintaining a veneer of civility. It was not always clear which man held the stronger hand. Hoover had a stable power base, rooted in the capital, but Donovan had a broad network of influential friends and acquaintances from the District of Columbia to New York. Washington lore had it that each man kept a dossier of incriminating information on the other.

The differences between Donovan and Hoover are legion. Donovan was a man of the world who had distinguished himself first as a frontline war hero and then as a Wall Street lawyer and foreign affairs aficionado. The innovative freelancer seized opportunities that presented themselves and made his own luck. Something of a bon vivant, Donovan enjoyed the comforts of life and was a near-compulsive philanderer. Hoover shared a birthday with Donovan—both men were born on January 1—but was twelve years younger. A lifelong bachelor, he might never have had a true romance. Born on Capitol Hill in the District of Columbia, he never ventured far from home, living with his mother in the house where he grew up until she died in 1938. Hoover dressed and lived well, but not like the New Yorker. If Donovan's suits were impeccably tailored and obviously expensive, Hoover's were elegant and stylish but only by Washington standards, with an occasional flashy tie or two-tone wingtip shoes. After obtaining two law degrees from a hometown university, George Washington, he went into government service rather than private practice, and showed every intention of entrenching himself at the Department of Justice. He learned to speak and write in a stilted shorthand that fit his drive for control.

One of Hoover's first jobs at Justice was to serve as Attorney General A. Mitchell Palmer's special assistant during the Red Scare of 1919–1920, a wild overreaction to a mix of bona fide and imagined threats. Palmer was, ironically, a Quaker and a graduate of Swarthmore College, where nonviolent teachings suffused the curriculum. But after a bomb—almost certainly planted by anarchists—went off in front of his house, he felt compelled to act forcefully. Federal agents supported by local police went into action, especially against foreign-born radicals. The so-called Palmer Raids were a jumble of policy brutality, mass arrests, and illegal detentions that Palmer's smart, energetic young assistant helped orchestrate.[5] The abuses were so egregious that they stirred a backlash from other cabinet members as well as the prominent citizens—including Helen Keller, Jane Addams, and Felix Frankfurter who were forming the American Civil Liberties Union.[6]

Hoover quickly learned to gloss over his role in the Raids while burnishing his image as a crime fighter and a stickler for legality. He spent the 1920s skillfully working his way up through the bureaucracy.[7] At the age of twenty-nine, he assumed responsibility for the department's small, general-purpose Bureau of Investigation (BOI), whose employees numbered in the tens. First as the BOI's deputy chief and then as its chief, he enjoyed managing criminal investigations as well as keeping comprehensive, well-ordered files. When Donovan came to Justice in 1924 as assistant attorney general of the Criminal Division, he became Hoover's nominal supervisor. Although he himself was developing into a very controlling, even authoritarian, manager, Hoover is said to have resented input from the political appointee who did not act like a bureaucrat.[8]

In 1934, the BOI cemented its reputation for fighting violent crime by tracking down and killing the famous gangster John Dillinger. The BOI—and Hoover himself—were on their way to becoming celebrities in their own right. In 1930s slang, they were "G-men," or government men, powerful and righteous. In 1935 the BOI became known as the FBI (Federal Bureau of Investigation), still under Hoover, who continued to professionalize its ranks. Agents now numbered in the hundreds. By mid-decade, Hoover's sometimes fawning attempts to ingratiate himself with Roosevelt and his official family were bearing fruit.[9]

In August 1936, the president asked Hoover to come to the White House for a private discussion of, in the words of Hoover's notes, "the question of . . . subversive activities in the United States, particularly Fascism and Communism."[10] Roosevelt wanted domestic political intelligence: "a broad picture of . . . activities as may affect the economic and political life of the country as a whole." After delivering a short lecture about the dangers of communism—how it could infiltrate labor unions and even spread to the civil service—Hoover pointed out that no one in the government was producing "general intelligence information upon this subject." When the president cast about for a possible solution, Hoover, not surprisingly, suggested empowering the FBI. Roosevelt agreed, but he wanted to make sure Hoover coordinated his plans with Navy, War, and especially State. On the grounds that "both of these movements [fascism and communism] were international in scope," the president convened a follow-up meeting the next day with Hoover and Secretary of State Cordell Hull.[11] Hoover felt that, under the terms of an act of Congress, State could grant permission to conduct the investigations. This Hull willingly granted.[12] When Hull asked the president for a written directive, he demurred, "since he desired the matter to be handled quite confidentially."

The president's clandestine but technically legal initiative opened the door to domestic spying by the FBI; Hoover quickly moved beyond investigating fascists and communists to investigating all organizations advocating "the overthrow or replacement" of the government by illegal means.[13] This was different from catching criminals, let alone figuring out how to catch foreign spies—the more obvious additional mission for the Bureau.

In 1938, a spy case in New York led to serious embarrassment, the one thing that Hoover hated more than communism. Initially apprehended by New York City police in February, Günther Rumrich turned out to be part of a Nazi spy ring that targeted American industry. Leon G. Turrou, the senior FBI agent assigned to the investigation, persuaded most of the suspects to confess. But then he inadvertently allowed a number of them to escape justice.[14] He also leaked information to the press and signed a contract to syndicate his story. In June, Hoover publicly traded accusations with the prosecutor over who was to blame for the suspects' flight, and fired Turrou, who would have the gall to claim that the director was simply jealous that he was diverting the limelight away from the FBI chief.[15] The matter even came to the attention of the president, who criticized Turrou's self-dealing during a press conference and opined that the affair was a wake-up call to spend more money on "running down spies in this country."[16]

A few months later in October, as Europe edged closer to war, the issue of protecting America from foreign "military and naval spy activity," surfaced again during a presidential press conference.[17] Roosevelt was characteristically cagey about the details: he would not say that espionage was on the increase, but admitted it was "a great deal larger today than it was ten years ago." The problem was that "we have too many organizations that are not specifically tied together," such as: "G-2, the Office of Naval Intelligence . . . the FBI and several organizations in the Treasury Department," as well as the State Department. The US government did not have the proper machinery "for coordinating . . . their work."

Roosevelt tasked Attorney General Homer S. Cummings with convening a committee to inquire into "the so-called espionage situation" and to report on the need for possible "additional appropriation for domestic intelligence."[18] In a glass-half-full assessment, Cummings informed the president that the army, navy, and FBI had a "well-defined" system that met the government's needs but would benefit from additional funding.[19] He appended a plan from his nominal subordinate Hoover, calling for a greatly expanded FBI that would usurp the functions of other agencies in order to guarantee domestic security. Roosevelt endorsed Cummings's recommendations, doling out additional funds, but never as much as Cummings or Hoover had requested. The FBI received half of the $300,000 that it supposedly needed and started hiring more agents and employees. Hoover also added small codebreaking and translation offices to the Bureau.[20]

In June 1939, Roosevelt took the next step. A presidential directive gave exclusive control of "all espionage, counterespionage, and sabotage matters" jointly to the FBI, the MID, and the ONI, and charged the three agencies with establishing a committee "to coordinate their activities."[21] Focusing on threats from abroad, they in turn made formal arrangements to divide up the work. The navy would handle threats to its ships and men, the army would do much the same for itself, and the FBI would take everything else. They formed a Joint Committee on Intelligence Services—sometimes called the Interdepartmental Intelligence Committee—to share information.

As Hitler's armies swept through Europe in the spring of 1940, the Committee felt a new sense of urgency. The largely unsubstantiated fears of a Nazi fifth column in the Western Hemisphere were cropping up in the American press and on the agendas of government conferences. Undersecretary of State Sumner Welles worried that the Germans could attempt to establish a foothold in South America and threaten US interests from the south.[22] Intelligence committee members agreed that the United States needed an American secret service, one that could operate overseas. They even came up with a name, the Secret Intelligence Service (SIS), and opined that it should operate out of New York City, posing as a commercial undertaking without any discoverable connection with the US government. It was essential to establish the SIS in peacetime so that it would be available in war.[23]

In late May, Assistant Secretary of State Adolf A. Berle Jr. sat in on a committee meeting in the FBI director's one-thousand-square-foot ceremonial office and conference room in the Department of Justice at Ninth and Pennsylvania. With twenty-four-foot-high ceilings, hanging Italian-style brass lamps, and ornate floor-to-ceiling windows, the office could have belonged to the Doge of Venice in 1400.[24] At first, Berle was not sure that he fit into the picture—he did not see "what the State Department has got to do with it"—but he quickly became inextricably involved.[25] By default, the brilliant lawyer—a graduate of Harvard, professor at Columbia, World War I intelligence officer, and charter member of Roosevelt's "brain trust"—was his department's expert on intelligence. Given their backgrounds, Berle and Hoover were not natural allies, but they had overlapping interests and found they worked well together.[26] On June 3 and 11, Berle was back at Justice, listening to the Committee's discussions about the need to defend the country against foreign spies and the ability to field its own spies, something that, Berle noted, "every great foreign office in the world has, but we have never touched."[27]

Berle agreed to raise the matter with the president. Over the telephone in late June, with General Miles of G-2 (army intelligence) sitting nearby, Berle described the work of the committee, and asked for guidance "as to the formation of a unit for foreign intelligence work."[28] He gave the president a choice, asking him to assign the task to the FBI, G-2, or the ONI, making one of the three agencies responsible. This was not what Roosevelt chose to do. In a memorandum dated June 24, 1940, addressed to Hoover and the heads of army and navy intelligence, Berle informed them that "the President said he wished the field to be divided."[29] He had decided that the FBI "should be responsible for foreign intelligence work in the Western Hemisphere, on the request of the State Department," and left any spying in the rest of the world to the army and navy. To make a major decision so casually, without involving Congress, and to transmit it through a third party over the telephone was a classic example of Roosevelt's haphazard approach to intelligence.[30]

Since Roosevelt had left room for interpretation, the Bureau felt free to enlarge upon what it called "the Presidential Directive," seeing it as license to collect "all types of information including economic, industrial, financial, and political" while paying particular attention to threats to American interests from Germany or its allies.[31] General Miles, who considered Hoover "a good policeman [who] does not wholly take in the entire situation," opined that the Bureau was overreaching;[32] it was crafting a job description for itself that was "encyclopedic in scope."[33]

Miles was right; it was a breathtaking leap. The Bureau was already the nation's premier law enforcement agency. Thanks in large part to Roosevelt, it had reentered the realm of domestic intelligence that it had left after the Red Scare. (One of Berle's friends, the ACLU lawyer Morris Ernst, called Hoover "the chief of secret police" but, in a backhanded compliment, said that he had run his force "with a minimum of collision with civil liberties," which was "about all you can expect."[34]) In 1938 and 1939, Roosevelt and the military had agreed to let the FBI take the lead in counterintelligence. Now, by secret presidential directive, it would also collect and disseminate other forms of intelligence. In British terms, it would be Scotland Yard, MI5, and MI6 rolled into one, plus a little more: it would catch criminals and spies, spy on radicals at home, and report on developments abroad—at least in Latin America.

The FBI's own history of the program quotes the only slightly exaggerated claim that "within 48 hours . . . the Bureau had a number of efficient men actually en route to Latin American countries."[35] Certainly Hoover moved fast, modifying the title Secret Intelligence Service to "Special Intelligence Service" (still abbreviated SIS) and making it an FBI undertaking. He placed a special agent named Percy E. Foxworth in charge, a sign of the importance he attached to the SIS. The Mississippian, who some said bore a slight resemblance to the young Hoover, had served as the director's administrative assistant from 1935 to 1939, when he was sent to New York to direct the flagship field office. There, he played a minor role in the Rumrich case before establishing a network of contacts among business and social elites. He would find suitable special agents—preferably younger men who could deploy on their own to Latin America, where they would pose as businessmen employed by the powerful companies run by his contacts, including *Newsweek*, United Fruit, and AT&T.[36] Only in the uncommon exception would the SIS agents have any official contact with local American officials, who would not otherwise know their identities. They would report by mail to post office boxes in New York, where a shell known as the "Importers and Exporters Service Company" in the RCA Building at 30 Rockefeller Center would serve as their home base. SIS would disseminate reports to the appropriate agencies in Washington.

There was no blueprint. The Bureau quickly discovered how difficult it was to set up and run this undercover enterprise, let alone meet the ambitious goals Hoover had embraced. Relatively few of the more than one thousand special agents in the FBI were both qualified and willing.[37] Preparing them was a challenge. The Bureau was hard put to come up with background information about Latin America, even the kind found in a good guidebook for tourists. No one knew much about the threats its agents would investigate. While there was "much publicity [about possible Nazi activities] . . . practically all of which was couched in alarmist phraseology," there was no "specific or accurate information."[38] Nor did the agents receive much training from the bona fide companies that would allow them to pose as employees; the companies were reluctant to invest in a cadre in which they had no real stake. Employees at the Service Company in Rockefeller Center found it tricky to field questions from legitimate businessmen, especially persistent salesmen who literally came knocking. Nevertheless, the first agents deployed on July 3, and roughly a dozen more followed in September 1940, taking up residence in countries from Mexico in the north, to Argentina in the south, and Cuba in the Caribbean.

In the fall of 1940, the diligent Foxworth spent two months on the road inspecting his program.[39] What he found was discouraging. The SIS men were dedicated and hardworking. But they had joined the FBI to be lawmen, not spies, and they were floundering, finding it difficult to pose as businessmen, make contacts, and even to send reports to New York. The chosen method was international mail. In their letters they used a combination of formal codes and "double talk," informal codes that supposedly made no sense to anyone but other FBI agents. Letters were often delayed en route because wartime service was inherently slow, and because they sometimes made postal inspectors suspicious, especially the alert British censors who intercepted mail that happened to pass through British colonies like Trinidad and Bermuda. The Departments of State, Navy, and War dismissed SIS reports as little more than rumor and hearsay. Official readers overseas who covered the same topics were angry when the reports crossed their desks: someone was trespassing on their turf.

Foxworth had to admit that SIS was not on track to accomplish its missions, and, in a brutally frank assessment, shared his findings with Hoover in early December 1940.[40] He recommended a series of changes: creating a system to analyze reports before disseminating them, allowing married men to send for their families, and, most important, working with the Department of State in the field. The subtext of Foxworth's report was that, without the changes, the program ran the risk of embarrassing the Bureau, still one of Hoover's greatest fears.

Hoover did not enjoy reading Foxworth's report. He did not contest the facts but saw pitfalls in the recommendations.[41] One worry was that sharing information about SIS more widely, even with other American officials, would lead to public exposure. Only reluctantly did he approve the recommendation to accept support from Foreign Service posts. The program still did not feel quite right, and Hoover considered a drastic change. In March 1941, he recommended to the new attorney general, Robert H. Jackson, that either G-2 or ONI assume the mission.[42] In early April he followed up, writing Jackson that the Bureau was just "marking time" in Latin America, apparently hoping for word from G-2 or ONI before spending any more time and energy on something so troublesome.[43] That word would never come; the armed services remained as indifferent to running spies as they had always been.

At roughly the same time, Berle met with the FBI to discuss putting some special agents on the diplomatic list, or even giving them office space in embassies. He predicted it would be an uphill fight; resistant to change, the Foreign Service would not welcome a new kind of attaché.[44] However, he could exploit an existing opportunity. Spruille Braden, the US ambassador to Colombia, was open to new ways of doing business. At 260 pounds, the Yale graduate, former championship boxer, and graceful ballroom dancer was not a conventional diplomat. He had requested that the FBI dispatch a special agent to Bogotá to take on two roles: handling investigations of subversive activities and generally supporting field operations.[45] A suitable agent would soon arrive at post and become the precursor of the legal attaché. It was a development that, within less than two years, would allow the FBI to establish representative offices in embassies throughout the region and even in other parts of the world.

A hybrid model started to evolve: a combination of SIS men on the outside, living on the economy, developing confidential sources, supported by special agents inside American embassies who could liaise with local authorities, especially police forces. It was a work in progress, with major limitations, but SIS could claim in 1941 to be the first modern American foreign intelligence service. It was a claim that planted the seed for conflict between Donovan and Hoover over the control and nature of American intelligence.

5

THE OIL SLICK PRINCIPLE

Designing the Organization

Roosevelt's decision to create the office of the Coordinator of Information in June 1941 was less like arriving at a destination than taking the first steps in a long journey. At first it was anticlimactic. It was true that the president had finally committed himself. But little was settled beyond the basic intent. The administration had yet to address the details that would determine the actual makeup and duties of the new agency.

The British knew what had to happen next. To get the kind of intelligence organization that they wanted, Admiral Godfrey left his man Ian Fleming in Washington. Dividing his time between Donovan's staid Georgetown mansion on Thirtieth Street NW and British residences where an energetic young bachelor could spread his wings, Fleming wrote two memos for Donovan that were filled with recommendations.[1] The first addressed cooperation between British and American intelligence, advising that the new service designate officers at American embassies overseas to serve apprenticeships under seasoned SIS officers stationed nearby. The other, dated June 27, 1941, was a six-month plan to get the COI on a war footing by Christmas.[2]

Fleming started with the obvious. Donovan would need to enlist "the full help of [the] State Department and FBI by cajolery or other means" and to "dragoon the War and Navy Departments"—that is, to obtain the "full personal cooperation" of General Miles of Military Intelligence and a Captain Kirk of Naval Intelligence. Fleming suggested asking Hoover to offer up office space for the COI. He stressed the need for hiring the very best men, including two very powerful (and very unavailable) candidates for senior positions, Assistant Secretary of War John McCloy and publishing mogul Henry Luce, founder of *Time* and *Life* magazines. Separately, Fleming went so far as to recommend that Donovan appoint a British officer, Capt. Edward G. Hastings, RN, as his director of communications.[3]

Two weeks later, Fleming cabled London that, in order for the new organization to produce "serviceable intelligence[,] . . . guidance . . . will have to be vigorously exercised by Mr. Stephenson with whom Donovan intends to work in close collaboration."[4] To that end, BSC would open a branch office in Washington, run by Stephenson's deputy, the Australian-born MI6 officer Col. Charles H. "Dick" Ellis.[5]

Fleming's recommendations were a remarkable mix of wishful thinking, imperial arrogance, and common sense. They did not take into consideration the messy reality of Washington in 1941. Happy enough with the status quo, Roosevelt would fit COI into the existing framework rather than take this opportunity to reorder it. For that he relied on two insiders, his administrative expert Benjamin Cohen and Bernard Gladieux of the Bureau of the Budget, the part of the White House staff charged with overseeing the executive branch.[6] Cohen and Gladieux were experienced in the ways of the capital and knew how to coordinate with the officials that mattered, especially Stimson at War and Knox at Navy.[7]

Stimson was inclined to support his friend Donovan's initiative and decided to broach the issue with Gen. George C. Marshall, the army chief of staff who had already proven indispensable to making defense and even foreign policy. (He would be at the head of the army from 1939 to 1945.) On this day, Stimson found the general usually known for his self-control not just upset but "extremely angry."[8] If the COI were established by military order, the president would have two military advisors: one from the army and one from the COI. They would not always be in sync. Nor did Military Intelligence like the idea of a "coordinator" who could claim the authority to regulate its work.[9] The coordinator's full title should not include any word like "defense" or "strategy." That was what *they* did. Under pressure, Stimson shifted his position, and called Knox to persuade him to agree to a toned-down draft. Knox sometimes let his temper control his tongue, and now it was the navy secretary who was outraged—"rampant" was the word Stimson used in his diary—that the army wanted to hobble the horse that he had backed for so long before it had even left the barn. But he, too, eventually calmed down and agreed to compromise.[10]

Stimson met with Donovan on July 3, and tacked first in one direction, then in another. The secretary explained why it was important to civilianize the COI order, and Donovan came around. The other tack was to say that if Donovan gave up the COI idea, Stimson would get him back into uniform as a major general. Donovan said he was tempted—he had a theory of guerrilla warfare "which he thought was characteristic of the American frontiersman" that he wanted to develop—but for now at least he was committed to the COI.[11] The meeting ended amicably between the two old Republican warhorses working in the national interest for a Democrat in the White House.

On the same day, July 3, Cohen and Gladieux met with Donovan again to try to clarify the basic concept. Donovan stressed the need to "correlate" information from various sources before riffing on offensive operations—making radio broadcasts and waging "psychological warfare... on all fronts," America's answer to the enemy's fifth column.[12] The two bureaucrats persisted despite the added wrinkles, and eventually produced a draft that would be general (and vague) enough to make everyone happy. Before long, Cohen's secretary telephoned Donovan's office to dictate a draft for concurrence.[13] The new draft was decidedly more civilian. The old draft was for a "military order"—now the word "military" had been crossed out; "national defense" became "national security"; the word "strategic" was gone. But Donovan's three conditions found their way into the document: reporting to the president, using his emergency funds, and directing other departments to cooperate.

The edits stayed in the final version. When Roosevelt signed the three-hundred-word document on July 11, 1941, it did not rise to the level of a military order or a numbered executive order, which would have given it the force of law. Instead, it had no title, simply starting with a bolded descriptor: "Designating a Coordinator of Information."[14] The coordinator's primary duty would be to "collect and analyze all information . . . which may bear upon national security" for the president and his government—that is, research and analysis, not unlike the work of the Library of Congress. The coordinator would also "carry out such supplementary activities as may facilitate the securing of information . . . not now available to the Government," which was a discreet nod in the direction of spying and other secret operations.[15]

On the same day, the White House issued a brief statement announcing Donovan's appointment that seemed to take away some of the powers that the president had just granted: "His work is not intended to supersede or to duplicate, or to involve any direction of or interference with the activities of the General Staff, the regular intelligence services, the Federal Bureau of Investigation or of other existing departments and agencies."[16] Unimpressed by the announcement, the *New York Times* ran the article on page five, explaining that the president had commented to advisors that "the scattered reports that came to his desk often were hopelessly confusing."[17] It sounded as if he were appointing another staff member to manage the paper flow at the White House, not an intelligence czar, let alone creating an intelligence agency.

Starting Up

In the few weeks between June 18, 1941, the day Roosevelt met with Donovan and agreed to create COI, and July 11, when he signed the Presidential Order for COI, the world changed dramatically. Hitler invaded the Soviet Union on June 22 with more than three million troops, and was now locked in a death struggle with his recent ally Stalin, with whom he had come to an understanding two years earlier. Their nonaggression pact had freed the German dictator to conquer country after country in Western Europe. That task accomplished, Hitler now felt free to turn back east.[18] His armies were soon winning massive battles of encirclement. They easily swept away the Baltic republics and advanced deep into Russia and Ukraine.

Operation Barbarossa had paradoxical effects. The old anticommunist imperialist Churchill now found himself on the same side as Stalin. As he declared in one of his famous asides, "If Hitler invaded hell, I would at least make a favorable reference to the devil in the House of Commons." For their part, the Japanese felt emboldened by the German offensive. Their protectorate Manchukuo (the Chinese province of Manchuria) shared a long border with the Soviet Union, which led them to worry about Soviet intentions. Now that the Nazis were keeping the Soviets occupied, the Japanese felt free to look to the south. A month later, they moved into French Indochina, putting themselves in a better position to threaten European and American territories: Hong Kong, Singapore, the Dutch East Indies, and the Philippines. The United States remained the only major power that was not a combatant. But every minute seemed to matter as the poorly prepared democracy—that up to that point was without a viable foreign intelligence agency—seemed to slip inexorably closer to war.

As soon as the president signed the COI directive, Donovan set up shop in three rooms in the old State, War, and Navy Building next door to the White House. The building delighted one wartime hire with "its old-fashioned, fussy architecture, its marble floors with their fossilized shells," and its waist-high swinging louvered doors, like those common in bars in the Old West.[19] Though its name had not changed, in 1941 the State, War, and Navy Building mostly housed the Department of State and had a small amount of room left over for the newcomers. Before long COI expanded into the Apex Building a few blocks away, a triangular structure that fit neatly into the intersection of Pennsylvania and Constitution Avenues. As the summer drew to a close, the young agency found a bigger home on a few acres of land known as Navy Hill in the somewhat run-down part of town known as Foggy Bottom, so named because of the fog that settled in the area from the nearby Potomac.

The new headquarters overlooked Washington's own Heurich Brewery, a rambling brick enterprise with a "copper roof green with age" that emitted yeasty but not unpleasant odors—and would be replaced by the Kennedy Center later in the century.[20] The COI occupied the three-story brick and limestone office buildings clustered around what one recruit called "a delightful little quadrangle," like a private world with its own flowering trees and shrubs.[21] Adorned with Greek columns that made it look like a county courthouse in the South, the building in the center became Donovan's headquarters. He himself occupied a modest (by Washington standards) ground-floor corner suite, Room 109. All in all, Navy Hill was fine real estate for COI. It was near the Departments of the Navy and the Army on the Mall and not far from the White House and the Department of State off Seventeenth Street.

Donovan relied on his own network and force of personality to recruit staff. Edgar Mowrer, the reporter from Chicago who had traveled with Donovan to London in 1940, got the treatment in August 1941. Through Secretary Knox, he was bidden to appear at 1647 Thirtieth Street. Over a hearty breakfast, Donovan launched into his breathless pitch:[22]

> I want you to go to Southeast Asia. You will size up the situation in various countries; judge . . . the possibilities of disseminating American information; and . . . bring back to me whatever you may learn concerning Japan's intentions. . . . You will pass as a newspaper correspondent as you did in England. When you come back, I want you to enter the information service as my personal advisor. How soon can you leave?

Mowrer agreed and went out to Asia on a grand tour, returning in the fall with his impressions only to find that Donovan no longer needed him. This was Donovan's way of taking the initiative, appealing to one of America's best and brightest, sending him off without much planning on another fact-finding trip that was neither reporting nor spying, but something in between, and then dispensing with his services.

By Labor Day 1941, the staff had grown to about forty members by dint of Donovan's recruitment techniques. At its core was an assortment of New York lawyers, many of them from the firm of Donovan, Leisure, along with the occasional old comrade from World War I, like the aging, philosophical G. Edward Buxton. Among them was a referral from Roosevelt, Bob Sherwood, the impossibly tall—six foot eight—speechwriter who was also a playwright, equally at home on Pennsylvania Avenue and Broadway. In World War I, he had volunteered to serve with the Canadian forces in the trenches and been wounded; after World War II he would write one of the best memoirs of life in the Roosevelt White House, *Roosevelt and Hopkins: An Intimate History*. As head of the COI's Foreign Information Service, Sherwood would take charge of propaganda, heading something like a public relations firm for the government.[23] The columnist James Reston characterized its mission as "rebutting Dr. Goebbels"; the FIS would broadcast news to Europe that the Axis did not want anyone to hear.[24] It would be a form of psychological warfare.

To conduct strategic research and analysis, Donovan attracted a dazzling cast of academics—mostly from the east coast and largely from Ivy League universities. At their head was James Phinney Baxter III, historian and president of Williams College, who in turn called on his good friend and former colleague, the chairman of the History Department at Harvard, William L. Langer, to assist him.[25] The future Pulitzer Prize winner Baxter evoked an absent-minded professor with super-charged energy; one of his former students at Harvard remembered him as "bald, mustached, bespectacled, moon-faced," the pockets of his three-button suits bulging "with a miscellany of notes."[26] Of German origin, Langer was solid and tidy, more conventional but no less brilliant or energetic.

There were few formalities. COI was allowed to circumvent civil service regulations. Many of the newcomers came to work for low or—like Donovan himself—no pay.[27] Some of them did not cut their ties to their old employers—again, like Donovan himself—and split their time between jobs. Roosevelt's son James joined to serve as liaison between COI and other parts of the government.[28] This was a promising development since the younger Roosevelt, a Marine Corps reserve officer, already knew his way around Washington and of course had the ear of Donovan's one and only superior officer; only he could send COI memos to Roosevelt's secretary, Miss Tully, that he thought "Father" should read.[29]

Elaborating on the mission was as important as hiring good people and finding a place for them to work. It was clear that COI would have two main business areas: "correlating" information and psychological warfare. But, as early as July 16, only five days after the president signed the COI order, the Bureau of the Budget was frustrated. Administrative analyst and budget examiner William O. Hall, who was assigned to the COI account, noted that, with the exception of Sherwood, "Donovan's staff did not seem particularly able."[30] Donovan's own management style did not help. He had "a tendency to commit himself too quickly on personnel and financial arrangements and probably lack[ed] the general background which should be present in the person directing . . . [such] activities." A month later, Hall was taken aback by Donovan's declaration that he was using "the oil slick principle of organization," giving his subordinates general guidance and then encouraging them to grow their offices.[31] It was a Darwinian approach to bureaucracy that did not sit well with Hall. He surmised that Roosevelt's written order had been too general, and that the president might have said something to Donovan that encouraged the colonel to think he could amass ever greater powers.

Setting up a research and analysis branch should have been straightforward. Donovan met with the Librarian of Congress, the famous man of letters Archibald MacLeish, to discuss cooperation (and how much that would cost—MacLeish charged what amounted to rent). More brilliant academics began to arrive in the nation's capital to continue their life's work—reading, thinking, and writing—now for their country's benefit. But the first few months were frustrating. No one inside or outside Donovan's agency "would or could enlighten us as to what we should do and how to go about it."[32] They then decided, more or less on their own, to create a news digest. *The War This Week* would meld information from newspapers, radio broadcasts, and various State Department or Department of War sources. It would be classified secret and have a limited distribution. Copy "A" would go to the president.[33] It would literally meet the requirement for which the Coordinator of Information was established.

Preparing to conduct psychological warfare—the second mission that Donovan had consistently promoted, an offshoot of his obsession with the fifth column—was more nettlesome. Would COI broadcast news or propaganda? Would it beam its broadcasts only to foreign countries? How would it mesh with other government enterprises? Bob Sherwood began energetically working through the issues. Donovan and Sherwood, the two New Yorkers with Irish blood and enviable combat records, seemed to be on the same wavelength.

Then came a dizzying series of distractions. Donovan the imaginative innovator was generating or endorsing a new initiative every day: preparing plans for the new world order—for use at the peace talks that would one day follow the war; opening an office in London; hiring Hollywood talent to produce propaganda films; planning to build a modern war room, with elaborate maps and projectors; reporting on American readiness, not just on developments overseas; standing up a psychoanalytical unit; and even hiring economists to help wage economic warfare. Verging on the otherworldly was a proposal for a Center for Arctic Studies within COI.[34]

Donovan received a late-summer windfall from the Department of the Navy and the Department of the Army. Not part of the mainstream, the few spies here and there that each department ran were considered a distraction from what the departments saw as more important and sensitive work.[35] Why not turn the spies over to COI? That would reduce the potential for needless competition and leave spying to specialists. The idea appears to have originally come from the navy, where Donovan's friend Knox held sway.[36] By October, the navy's spies, about thirteen in all, came under COI.[37]

The colonel happily accepted the additional work, which was an important part of the British vision that he had espoused. The budgeteers endorsed the plan, finding that "Colonel Donovan undoubtedly has a flare [*sic*] for the direction of subversive and counterespionage activities" and, since the US had historically neglected them, they could make "a tremendous contribution to our national defense."[38] Donovan proudly reported the fact to Roosevelt, adding that he was hatching plans to send officers to North Africa to create a network of "agents of information."[39]

The Bureau of the Budget concluded that apart from psychological warfare and espionage, much of what Donovan wanted to do was already being done by other parts of the government. Its officers talked matters over with Donovan at least twice more, letting him plead his case and vent his frustrations.[40] They finally decided there was no alternative but to appeal to Roosevelt for guidance. In October, they started writing a memorandum for the president that went through multiple drafts. The original concept for COI was for a small, high-powered staff, which would have required a budget of about $1,000,000. Donovan now wanted to hire some 1,300 employees to perform many more functions at more than ten times the cost. Was this what the president had had in mind?

FDR reviewed the Bureau's recommendations line by line, handwriting comments and approvals.[41] First on the list was "intelligence activities," described as "counterespionage and secret activities in Europe . . . of high strategic importance." Roosevelt authorized the full amount. He did the same for his friend Sherwood's international propaganda work, described as "broadcasting." Work on domestic morale was completely cut. The president slashed almost half of the budget for the colonel's state-of-the-art "war theater," leaving $2,000,000 for this needless luxury. This was still more than twice what Research & Analysis, the branch charged with COI's core mission, would receive. FDR concluded mildly that "the whole program was a little over expanded"; he wanted to support the new agency while limiting its size and power. He would stick to his policy of maintaining a dynamic balance among his security organs. He would not forget his commitments to the likes of Vincent Astor, John Franklin Carter, and of course J. Edgar Hoover.

On the Road to War

All things considered, Roosevelt liked Donovan well enough, and met with him regularly in the Oval Office from July to December 1941, one of the clearest signs of presidential favor.[42] Donovan prepared diligently for the meetings, gathering background information, and often worked from a typed agenda with as many as ten to twelve subjects ranging from the trivial to the substantive and occasionally to the fantastic. Since their meetings averaged thirty minutes to an hour each time, Donovan must have galloped hard to cover as much ground as possible.

In October, Donovan used a few minutes of his time to tell the president about the turf wars that had enmeshed COI from the first day. Other topics included a British account of the death throes of the *Bismarck* and two special requests from British friends that had little to do with intelligence. One topic that made a difference was German subversion in the Western Hemisphere. At his meeting with the president on the twenty-first, Donovan produced a map of South America that he had received from British Security Coordination.[43] It purported to show the German plans for dividing up the continent after the Third Reich won the war.

Roosevelt was so taken with the BSC map that he talked about it in his Navy Day Speech on October 27 that began, "Colonel Donovan, Ladies and Gentlemen."[44] Speaking after dinner in a ballroom at the historic Mayflower Hotel a few blocks from the White House, he claimed to have in his possession an intelligence coup:

> a secret map made in Germany by Hitler's government—by the planners of the new world order. It is a map of South America and a part of Central America, as Hitler proposes to reorganize it. Today in this area there are fourteen separate countries. But the geographical experts of Berlin have ruthlessly obliterated all existing boundary lines. They have divided South America into five vassal states, bringing the whole continent under their domination.

The White House estimated that fifty million Americans listened to the president's speech. In that light, he was not particularly worried about Nazi counterclaims that they had had no hand in the matter. On the contrary, he told reporters that the denunciations from Berlin were "a scream"—music to his ears.[45] It was also music to Stephenson's ears. His goal was to generate anti-German propaganda that would embarrass isolationists in the United States and propel the country into war—like the Zimmermann Telegram that the British government had leaked to the Wilson Administration in 1917. Without a doubt, the map fit Roosevelt's agenda; his most recent Fireside Chat on September 11 had denounced Hitler's "intrigues, his plots, his machinations" in the New World.[46] This was a natural follow-on topic. But the map looked like it had been sketched by a child—and it turned out to have been forged by BSC acting largely on its own.[47]

The BSC manipulation happened at a time when Churchill's intelligence advisor Morton noted smugly that "to all intents and purposes US security is being run . . . by the British," meaning Stephenson.[48] Assistant Secretary of State Berle was coming to the same conclusion, complaining to Undersecretary Sumner Welles that the BSC resident in Washington, Ellis, was in fact "running Donovan's intelligence service."[49] Berle remembered how Wiseman had manipulated the Wilson administration, and he wanted to protect FDR from another Wiseman.[50]

THE LAST WEEKS OF THE prewar Donovan-Roosevelt honeymoon came in November when the president saw the president at least four more times, apparently one-on-one. The topics they covered were less dramatic than the secret map: a potpourri of newsy tidbits and unusual ideas. For example, there was a proposal to fly an American bomber from Canada to Berlin and back, not to drop bombs, but leaflets.[51] The president was "delighted" at the thought of dropping leaflets on Berlin but stopped well short of ordering the mission.

On November 28, a day after the army and navy had sent a loosely worded "war warning" to their commands scattered across the Pacific, FDR invited Donovan to breakfast at the White House—one of the most sought-after slots on his schedule—to expound on the deepening crisis.[52] During the forty-minute meeting, the president spoke as if war was well nigh inevitable. He had been unable "to find a formula for dealing with Japan." Secretary of State Hull had decided that it was "impossible"; further negotiations were next to pointless. Japan was trying to make the United States commit an overt act that would plunge the two countries into war—something that he wanted to avoid, especially in the Pacific.[53]

A little more than a week later, on the chilly afternoon of Sunday, December 7, 1941, Donovan joined fifty-five thousand other football fans at the Polo Grounds, the aging but well-loved bathtub-shaped stadium off West 158th Street in Upper Manhattan. It was so cold that at least one of the players wore gloves on the field. The National Football League's Brooklyn Dodgers (the franchise would not survive the war) stole the spotlight from the formidable New York Giants, winning 21–7. Fans would soon forget the results. Instead they would remember that this was where they were when they heard the news about Pearl Harbor. Their first inkling of bad news was an announcement over the loudspeaker that "Colonel William J. Donovan . . . was being paged by Washington," followed immediately by a call to "all officers and men of the Army and Navy" to report for duty.[54] Before long, the "ominous buzzing" that greeted the announcements started to give way to news: there had been an incident, one that involved shooting, in the Pacific. That turned into the flash that the Japanese had attacked Pearl Harbor. War had finally come for the United States.

Even in Washington no one had much hard news. Details came slowly, in maddeningly small fragments. First came the famous one-line message off a navy teleprinter: "Air raid on Pearl Harbor. This is no drill." Then came inquiries from offices around town, and urgent attempts by Main Navy to telephone Hawaii to find out more. The rudimentary links could not handle the volume and dropped many calls. But enough got through for the scope of the devastation to emerge as the afternoon wore on. Staying outwardly calm, the president grimly shook his head in shock and disbelief at what he was hearing about his navy—the service he had helped to run in World War I. He called a cabinet meeting for 8:30 p.m., then started work on what would become the "day of infamy" speech, his call for a declaration of war against Japan on December 8. While he was working, the White House staff made dozens of calls to find and summon cabinet members, congressmen, and senators.[55]

On their way into the White House, the summoned officials passed a small crowd lining the fence on Pennsylvania Avenue, singing patriotic songs like "My Country, 'Tis of Thee."[56] Secretary Knox was one of the last to arrive. He and an aide shared as much of the painful news as they could, then had to put up with a different kind of attack: "uncharted hell" from Senator Tom Connally of Texas, chairman of the Foreign Relations Committee, who wanted to know how the navy could have been caught so off guard.[57] Connally's attack agitated Knox even more. Face red, voice rising, he vowed to fly to Hawaii to uncover the truth.

It was probably James Roosevelt who had known where Donovan was and placed the call to the Polo Grounds. Donovan got to Washington as quickly as he could. At LaGuardia Field he caught a plane to DC—Eastern and American Airlines added additional flights to accommodate the sudden surge in demand on what had begun as just another Sunday.[58] But it still took an hour and a half to two hours for the twin-engine DC-3 workhorses to fly the 250-mile stretch, and Donovan did not land until after 8 p.m. He went first to the COI to touch base with his staff. Focused mostly on Europe, they had little to add to the reports that were coming in from Hawaii. One staffer would remember that, on that fateful evening, it was *Donovan* who was telling *them* what had happened.[59] Around 11 p.m., Donovan proceeded to the White House.

By midnight, Donovan had found the president in the Oval Office chatting with reporter Ed Murrow, just back from London, over the beer and sandwiches that waiters had served after the larger meetings had broken up.[60] Roosevelt asked Murrow about mutual friends in England and how the country was bearing up, then rehashed what he knew about the attack. When Donovan joined in, the three men talked about how the American people would view the attack and speculated about the future of the war in the Pacific. It was not a long conversation—just "a few minutes' talk," in the words of FDR's secretary.[61] Donovan had little to say. Nothing was decided. Still, the president had included the colonel when he gathered his official family at a pivotal moment, and that was not a bad sign for the future of COI.

6

SPYING OR RIDING TO THE SOUND OF THE GUNS?

The Months After Pearl Harbor

At 8 a.m. on December 9, two days after the surprise attack, Secretary Knox boarded his official US Navy two-engine, fourteen-passenger Lockheed "Lodestar" at Anacostia, the small naval air station across the river from downtown Washington. He was determined to fulfill his vow to fly to Hawaii and uncover the truth about Pearl Harbor. In 1941 this was no easy flight, especially in winter. After making their way first to Memphis, then to El Paso and Douglas, Arizona, the Knox entourage encountered turbulent weather over the continental divide but survived to land in San Diego. There they transferred to a four-engine PB2Y "Coronado" flying boat—not unlike the Clippers that Donovan had taken across the Atlantic, but far more austere. Before taking off for Hawaii, the secretary consulted a weather officer who predicted smooth flying. Knox, who was taking no chances after the rough flight across the country, acknowledged the prediction with a kindly, "That's just fine, son," then told the young man to guarantee his work with his life by coming along on the long flight over the ocean.[1] The new member of the team had to shift for himself among the many boxes of plasma and serum, for the burn victims at Pearl, that filled the Coronado almost to overflowing. He wound up standing for almost sixteen hours. Otherwise, the flight was uneventful.[2]

On December 11, three days after taking off from Anacostia, a cold and tired Knox alighted from the poorly insulated flying boat at Naval Air Station, Kaneohe Bay on Oahu, fifteen miles from Pearl Harbor by car, but close enough by air to have been a target for the Japanese fighters and bombers on December 7. He encountered a landscape of wrecked seaplanes, burned-out hangars, and cars mottled with large bullet holes. One-third of the main runway at Kaneohe was charred black. This was just a foretaste of what awaited him at Pearl—eight battleships sunk, capsized, or holed; lifeless bodies still being hauled from the harbor; horribly burned survivors at Naval Hospital Pearl Harbor.[3] Seeing the wounded made him "as angry as I have ever been in my life," and drove him to work even harder than usual.[4] The secretary conducted lengthy interviews with Adm. Husband E. Kimmel, commander in chief of the Pacific Fleet, and Gen. Walter C. Short, his army counterpart in the Hawaiian Department, and spoke briefly to a number of others, including then-commander Edwin T. Layton, USN, Kimmel's fleet intelligence officer.[5] Knox was glad to find that no one tried to "alibi" the general lack of readiness for the kind of attack that had occurred.[6]

After some thirty-two breathless hours on the ground, the secretary began to retrace his route to Washington. Like an old newsman on deadline, he sat down at the navigator's desk on the Coronado and wrote his story out in longhand, then discussed parts of it with members of his staff, who borrowed a typewriter from the air station at San Diego and put a young lieutenant to work transcribing Knox's handwriting on the two remaining legs of the trip. After landing in Washington on the night of December 14, Knox hurried over to the

White House to deliver the nineteen-page report. The president, even more willing to work at odd hours than before December 7, met with the secretary from 9:45 to 10:30 p.m.

Knox described how the attack from 150 to 300 Japanese aircraft from three to six carriers had come as a complete surprise.[7] More concerned about other dangers, the army and navy were not prepared for an air attack. American planes had been grouped together on the ground to make them easier to guard from saboteurs, and the results had been catastrophic. While the "officers and men of the fleet [had] exhibited magnificent courage and resourcefulness," the Japanese had also been resourceful and meticulous in their preparations. Without citing evidence, Knox could not stress enough that a "most efficient" Japanese fifth column of spies had laid the groundwork for the attack (supposedly by spying on the fleet), then spread misleading rumors afterward. This should, he concluded, "serve as a mighty incentive to our defense forces to spare no effort to achieve a final victory."

The next day—presumably after a few hours of sleep in his own bed—Knox conferred again with the president and then received some two hundred reporters for a press conference.[8] They crowded into his suite at Main Navy, one of the more comfortable spaces—it even had a fireplace and a mantel—in the otherwise utilitarian building. Eagerly accepting the printed handout that the navy had prepared for the press conference, the reporters asked questions of their onetime colleague. In the news photo, Knox sits at his desk surrounded by surprisingly well-dressed gentlemen of the press, themselves sitting, squatting, and standing in every available space, including on the cork floor. Leaning back in his chair, the secretary looks at ease. He covered much of the same ground that he had for the president but today he was more upbeat, downplaying losses and recounting many instances of valor under fire. He concluded that if the Japanese thought they could knock the navy out with one blow, they had another thing coming. Indeed, most of the sailors at Pearl hoped "the b—s" (meaning the word "bastards," politely shortened for publication) would come back for another round.

Knox announced that the president intended to convene a more formal investigation to uncover any "error of judgment which had contributed to the surprise" or "any dereliction of duty prior to the attack." This occurred on December 18 when Roosevelt appointed a body known as the Roberts Commission, named after its chairman, Supreme Court Justice Owen J. Roberts, to conduct the first of many Pearl Harbor investigations between 1941 and 1946.[9] Working fast, the Roberts Commission produced its report by January 24, 1942, and found fault with the commanders on the ground, Kimmel and Short, both of whom had by then been relieved of their duties.[10]

Especially early on, the Pearl Harbor investigations were more about personal culpability than anything else. Had commanders like Kimmel and Short been prepared for an unlikely but not impossible contingency? Hardly anyone in the army or navy thought the Japanese would strike at Pearl Harbor from the air. Most American commanders underestimated their enemy, sharing Kimmel's attitude, which he expressed in a pithy aside to one of the investigators: he had "never thought those little yellow bastards could pull off such an attack, so far from Japan."[11] But shouldn't a prudent commander have been on guard against both more and less likely dangers? Washington might not have shared everything it knew, but it had shared enough to put most commanders on alert, even using the words "war warning" in the November 27 message. That was the crux of the case against Kimmel and Short.[12] What neither the Knox Report nor the Roberts Commission stated was that Pearl Harbor was an intelligence failure. That would come, but only with time.

The enemy would not wait for the United States to run through all its Pearl Harbor investigations and rethink the structure of American intelligence. Passed by nearly unanimous majorities in the House and Senate, the original US declaration of war on December 8 was only against Japan. But Hitler obligingly declared war on the

US on December 11, and Italy followed suit. The US was finally in the European war that had preoccupied Roosevelt and Donovan since 1940.

The advent of war had a paradoxical effect on Donovan. He did not spend time contemplating intelligence failure. Instead he did two other things: he ordered the broadcast of Allied propaganda to Europe, and sped up the flow of information to Roosevelt's office. A broadcast in late December compared Hitler's invasion of Russia, which had been bogged down in the snow and ice of a brutal winter, to Napoleon's misadventure in 1812.[13] While the COI took to the airwaves, Donovan filled the president's inbox with his signature mix of partially processed and marginally useful reports, along with the occasional policy prescription. Donovan's input ranged from a claim by one French source "in touch with personalities in Vichy" that Vichy had put its naval and air bases at the Germans' disposition; to a revelation that the Germans were using the wrong kinds of lubricants in their mechanical equipment; and finally to a tip that the US should make repeated use of the World War I song "Over There" in its broadcasts to Europe.[14]

Still unsure that running an intelligence agency was right for him, Donovan hedged his bets. As December was drawing to a close, he contemplated what the United States could do in the short term in the Pacific. Immediately after Pearl Harbor, powerful Japanese forces continued to spread through Southeast Asia, threatening the British and Dutch colonies of Hong Kong, Singapore, Burma, and the Dutch East Indies. The Philippines, a commonwealth under the aegis of the United States, was learning that their comparatively weak American and Philippine Army was no match for the invaders. Its small air force had been largely destroyed on the ground on December 8 much like its counterpart at Pearl, and the Philippine capital, Manila, would prove to be easy prey. To Donovan's way of thinking, this showed that the US needed a new kind of force to counter the enemy.

Accordingly, Donovan began promoting what he was now calling "the British commando principle."[15] Once again, he worked from the top down, hoping that the president would decide in his favor and rule by decree. In a lengthy memorandum dated December 22, Donovan argued for an independent corps of commandos, one distinct from the army and the navy.[16] Small units of commandos could operate independently behind enemy lines, cultivating relations with locals and supporting major military operations. The Imperial General Staff had adopted the commando principle after Britain's conventional forces were defeated on the continent; it was their way of compensating for weakness. The implication was that the US now faced a similar dilemma. Roosevelt replied the next day.[17] He liked the idea enough to suggest that Donovan proceed by discussing his ideas with Churchill, who had come from London and made himself at home in the White House for an extended stay—he seemed to take Roosevelt's hospitality for granted now that the two countries were fully committed to each other in the war.

Donovan's proposal attracted attention. Once lit, the fuse smoldered and sputtered for weeks, even throwing off occasional sparks. The record is silent on whether Donovan discussed commandos with Churchill. It could have happened at one of the functions Donovan attended in December or January when the after-dinner talk drifted to strategy over brandy.[18] Donovan did send an officer to London to benchmark best practices.[19] Knox and FDR considered the idea, and Knox told Adm. Ernest J. King, the newly appointed commander in chief of the navy, that "the President is much interested in the development and use of the equivalent of British 'commandos.'"[20] After learning that Knox had told Roosevelt that the Marines—a corps of expeditionary light infantry—had "such groups in training," King asked the Marines for proposals "as to such use."[21]

No part of Donovan's initiative found favor with the Marine Corps commandant, then major general Thomas Holcomb, who feared that the president would direct the Corps to give Donovan a

Marine commission and Marines to command. This would be unfair to long-serving Marine officers, and smack of the kind of favoritism common in monarchies. At best, it would drain Marine resources desperately needed for other purposes. The usually even-tempered Holcomb—a veteran of trench warfare in World War I as well as in the corridors of power in Washington—had a visceral reaction to "the Donovan affair." In January 1942, Holcomb wrote that it was "uppermost in [my] mind. I am terrified that I may be forced to take this man. I feel that it will be the worst slap in the face that the Marine Corps ever was given."[22]

Holcomb personally expressed his "utter disapproval" to FDR and decided to preempt Donovan by quickly setting up commando units in the Marine Corps.[23] The threat to the Corps receded. But the colonel did not give up, relentlessly probing for other openings. On February 6, 1942, Donovan wrote to Knox with a reminder of his vision: an American guerrilla corps could be called "Special Service Troops" and unofficially something snappier like the "Yankee Raiders."[24] It could conduct independent raids and coups de main or support large conventional operations. The Raiders would accept volunteers from the army and navy and comprise an independent command reporting to the president through the secretary of the navy. Perhaps on Donovan's orders, someone in the COI drafted an order from Roosevelt to Donovan directing him immediately to organize the corps. All it lacked was a date and a signature that never would never materialize.

On February 21, another memorandum from Donovan crossed the president's desk. This was "an appeal from a soldier to his commander in chief."[25] By now, US and Philippine forces were retreating to the Bataan Peninsula, where they would endure a monthslong siege. Their situation was tenuous, given the US inability to reinforce or even resupply them. Donovan told Roosevelt that he saw both "an obligation and an opportunity." He called for a comprehensive, long-term plan: support the Resistance in the islands and prepare the ground for an eventual attack on Japan; establish a force of heavy bombers in India; send a naval task force to attack targets in Japanese waters. In the meantime, using blockade runners, the US should land infantrymen on the island of Mindanao, some seven hundred miles to the south of Bataan. This would not be a large-scale operation, but more like guerrilla warfare, a strategy that was worth a try. Donovan closed his appeal by asking to "be permitted to serve with this force in any combat capacity."

Roosevelt acknowledged the appeal with good humor—he, too, would ride to the sound of the guns if he could—and gently admonished the colonel to consult General Marshall who, after all, was the arbiter of strategy and army personnel assignments. Marshall would have been within his rights to issue one of his icy rebukes to Donovan, but instead took his cue from the president, commending the colonel for his desire to serve on the front lines. That was "typical" of him.[26] He, Marshall, would keep Donovan in mind and promised to "watch for a suitable assignment, in that area, and . . . [to] call . . . as soon as it develops." Not one of the three men wrote or said anything about the work of the COI, or how important it was for Donovan to coordinate American intelligence instead of becoming a guerrilla chieftain.

While he was fighting his war by memorandum, Donovan also tried a flanking attack: organizing his own guerrilla units within the COI. He saw a possible opening after meeting in early January 1942 with the senior American officer in the China-Burma-India (CBI) theater, Lt. Gen. Joseph W. Stilwell. Stilwell's was a near-impossible job. With few American forces of his own, he had to rely on Nationalist Chinese forces to fight the Japanese. Their army was his to train and equip, and occasionally to use—but only under the constraints imposed by the Nationalist Chinese generalissimo Chiang Kai-shek, who wanted to husband its strength for the near-inevitable postwar struggle with the communists under Mao Tse-tung.

When Donovan and Stilwell met, they discussed the various ways the COI could support the general's difficult mission overseas—perhaps by sending a small American task force with a variety of skills, one that would be able to both undermine the enemy's will to fight through propaganda and deploy commandos at Stilwell's discretion.[27] Stilwell seemed generally receptive, but stipulated that his approval would depend on the officer Donovan chose to head the task force. Stilwell mentioned a suitable officer named Carl F. Eifler, a career employee of the Customs Service who was also an army reservist. Stilwell and Eifler were kindred spirits: infantry officers who had trained together in the 1930s in California, where Stilwell had noted Eifler's endurance and iron sense of duty.[28] In late 1941, Eifler was on active duty in Hawaii, serving as a strict—some have said abusive—commander of an internment camp for Japanese nationals who happened to be on US territory when the war started.[29] One senior noncommissioned officer spoke for many when he praised Eifler for running his unit with a needed dose of "extreme discipline."[30]

Donovan's staff made inquiries and decided that Eifler was right for the job. In February 1942, roughly six weeks after the meeting between Donovan and Stilwell, Eifler received a mysterious official inquiry asking if he was available for assignment in the Far East. On the heels of his reply, he received a set of orders directing him to report to an organization and a man he had never heard of.[31] Eifler soon learned that his would be the first deployment of the COI's irregular warriors, and began to assemble his team, roughly twenty strong, drawing heavily on soldiers that he had served with before. These men then received basic training in guerrilla warfare, some of it provided by the British SOE at their Camp X on the shores of Lake Ontario in Canada; drew the weapons and explosives they thought they would need; and set out on an odyssey halfway around the world to represent an organization that almost ceased to exist while they were en route.

His experience in the first months of 1942 taught Donovan that if he wanted to mount operations like the Eifler mission, the White House was not the best home for COI. There, he had to work through the Bureau of the Budget to eke out dollars from the president's relatively modest Emergency Fund, and his only superior, Roosevelt, now had even less time to spare for a small office with growing pains and many enemies. To get his men overseas he had to importune the uniformed officers who would support them. Perhaps it would be better for him to work directly for the military, to be at a high level in its chain of command.[32] By March, he was sending more memos to the president with his ideas.[33]

Did FDR read and digest the memoranda? Probably not. He simply did not have the time. But he did hear from others about Donovan, and he did discuss how best to organize this small but not insignificant slice of the war effort. On February 19, Roosevelt intimate Bob Sherwood, Donovan's ostensible collaborator, wrote the president that Bill was "most anxious" to find a way forward, hoping that the military would take the COI under its wing. Donovan (who was still serving as a civilian) himself would be overjoyed to be recalled to active duty—as the head of the COI or perhaps in another job that was of a "special, secret, even mysterious nature."[34] By now more of an enemy than a friend, Sherwood added his own opinion—the president should simply dissolve the COI.

Two days later, the budget referent William O. Hall found that Donovan was ready to lead the COI into an alliance with the military.[35] Hall's guess was that the colonel would prefer active duty in combat over another Washington assignment. The secretary of the Joint Chiefs of Staff, then brigadier general Walter Bedell Smith, had mixed feelings about taking the COI into the Joint Staff, but decided in the end to endorse a draft order to make that happen: "Colonel Donovan has pioneered this planning and he will be most effective if working from the inside rather than outside the JCS organization."[36] Smith noted that Donovan had included provisions for "special service units" in one of his proposals, a reference to commandos, the subject "nearest Donovan's heart."[37]

Then came April 1, 1942. Donovan was sitting in the back seat of his staff car in Washington traffic on his way to Union Station when the collision (most likely with a taxicab) occurred. He was thrown forward, wrenching his knee.[38] The old warrior refused treatment in DC, saying the specialists were better in New York, and let his driver, James Freeman, carry him onto the train with the help of a porter. Doctors in New York set the leg in plaster and kept an eye on their patient while he worked from a suite in the luxurious St. Regis Hotel on Fifth Avenue that happened to belong to Roosevelt's friend, the amateur spy Vincent Astor, and was a few blocks away from the offices of the British spies in Stephenson's BSC at Rockefeller Center (where both the OSS and the FBI also had offices at various times during the war). Stephenson himself stayed at the St. Regis when he was in town. It was not a coincidence that Ian Fleming would one day have the fictional spy James Bond patronize the hotel.

Some days the pain was bad enough that the usually abstemious Donovan nursed a whiskey and soda or two.[39] A clot traveled from his legs to his lungs, a possibly fatal complication, but one that passed. COI staff came with papers for him. Political friends called. Knox brought the unwelcome news that the president thought Donovan's accident now made him unfit for active duty.[40] No less disturbing were the rumors that appeared in the press in the first few days of April: Roosevelt wanted to consolidate the various offices for official news and propaganda, but Donovan was standing in the way.[41] On April 9, Mrs. Roosevelt wrote to offer "sympathy and best wishes for a speedy recovery," and on April 13, the president followed suit.[42] FDR was friendly, but his words were equivocal. He was "terribly sorry" to hear about the accident, and wanted Donovan to "take the proper rest at this time" and not to hurry back. He closed with the hope that he would see Donovan when he returned to the capital.

Sent the following day, Donovan's answer barely concealed his exasperation.[43] After thanking the president for his thoughtful note,

Donovan was quick to assure him that "it will not be long before I shall be completely well." His tone verged on disrespect when he scolded Roosevelt for having yet to sign the order aligning the COI with the Joint Chiefs. He was "disturbed as well as surprised" by the FDR's inaction, recapping his arguments for the move and adding a lecture on modern warfare.

Roosevelt worked hard to be upbeat in his relations with his official family; balancing them against each other, he tried to keep them as happy as possible. But there was only so much that even FDR could take. In January, he had heard twice from his personal spy, John Franklin Carter, that Donovan's work was not up to par. COI was, Carter reported, "very hazy on actual operations" and "not doing its job efficiently."[44] Two days later, Roosevelt commented to Assistant Secretary of State Berle, the New Deal insider, that "Bill was doing a pretty good job on propaganda" but only "something of a job in terms of intelligence."[45] By April, the president's attitude was more negative. Caught up in the dispute over COI prerogatives, FDR heard and read the bitter complaints about Donovan from Undersecretary of State Welles and FBI Director Hoover.[46] Then Donovan had scolded him for inaction on the fourteenth. A little more than two weeks later, on April 30, Roosevelt let slip his exasperation during a White House lunch with Berle. The president was in excellent spirits. Conversation did not lag as they worked their way through their meal of trout and eggs Benedict. When their talk turned to political warfare, Berle asked FDR if he had decided what to do with the COI. Roosevelt said he had been trying to get the colonel promoted to brigadier general. Perhaps when the promotion came through, he would send Donovan to establish a beachhead on an isolated island held by the Japanese, "where he could have a scrap with some Japs [*sic*] every morning before breakfast." Then, he thought, Donovan would be "out of trouble and . . . entirely happy."

On May 9, after he finally returned to Washington, Donovan sent another testy memorandum to the White House, this time to refute the allegations against him, point by legalistic point. He had nothing but contempt "for the people that retail such deliberate falsehoods." (These people happened to include men like Welles, whom the president had trusted for decades. The undersecretary had even been a page at the Roosevelts' wedding in 1905.) The president's real concern, Donovan preached, "must be with those who bring such stories to you."[47]

When he found time for Donovan six days later, FDR granted him a mere fifteen minutes.[48] Nevertheless, after a few more rounds of memos and meetings, Roosevelt was ready to agree to Donovan's basic request. A military order would transfer the COI to the jurisdiction of the Joint Chiefs of Staff and rename it the Office of Strategic Services (OSS).[49] It would be charged with collecting and analyzing strategic information, as well as planning and running special activities, all as required or directed by the Joint Chiefs. One important difference was that Donovan would now have little to do with "radio propaganda," the function that Roosevelt decided to transfer to the new Office of War Information along with almost half of the 1,800 members of the COI.

In this way Donovan lost exclusive control of one of the functions that made up his vision of modern warfare.[50] He would no longer be working directly for the president or charged with collating information for the first customer. Research and Analysis would continue to exist, but, at least in theory, it would meet the military's needs first. Nor would Donovan be able to tell any other organizations what to do. On the contrary, OSS would operate only in certain parts of the world after a two-step approval process, asking permission from the Joint Staff and the theater commander. But if the Joint Staff approved a plan, the theater commander would be more likely to follow suit. And now that the United States was at war, America would finally have an agency that was officially charged with conducting special operations and espionage. Donovan would have his beloved commandos and whatever spies OSS could recruit.

A few weeks later, Donovan would write to one of his British friends that he had succeeded in bringing about "something which has never been done in our military history."[51] Previously neglected secret activities would now have pride of place under his command. Though he was proud of what he had accomplished, he hoped that, "after a few months," the startup would take root and he could "get away."

Roosevelt and the Joint Chiefs knew that Donovan would not be in Washington on June 13, 1942, the day that the president actually signed the orders creating the OWI and OSS.[52] He might have waited until Donovan was out of the country so that the troublesome colonel would not be able to reroute the train that was about to leave the station.[53] The timing took Donovan by surprise. But he recovered quickly. One of his first meetings in London was with Sir Charles J. Hambro, the head of SOE, Britain's own special operations startup.[54] Though both Hambro and Donovan were newcomers to the field of special operations, they talked as if they were old hands at it, almost literally carving up the map of the world between them: American guerrillas would operate here, British commandos there.[55]

The next day, June 16, Donovan sat in on a meeting of the British Chiefs of Staff and explained the changes to some of his best official friends: British flag officers and senior civil servants. The president had, he proclaimed, been "anxious for some time that the intelligence arrangements in the United States should be organized under one hand"; the American Chiefs of Staff had requested him to set up an organization under their auspices.[56] It was as if the British had combined the spies in their SIS with the commandos in SOE.[57] Donovan was right. This was an unprecedented step in the American way of war. The country had never had an agency quite like the OSS, with spies, commandos, and analysts under one roof.

7

ARMY CIPHER BRAINS

Two Different Worlds

Washington's lukewarm welcome for the Office of Strategic Services came a little more than six months after the United States was caught unprepared by the Imperial Japanese Navy (IJN) at Pearl Harbor. Now, in the summer of 1942, America was embroiled in the greatest war the world had ever known. The fighting engulfed three continents, touched three more, and raged on the oceans in between. The war was not only about power politics, but also a clash of values: ultranationalism on one side and antifascism on the other. The country was more united than it had been in years; there had been only one vote against the declaration of war in Congress—even the diehard isolationists had voted with the president. Given the epic scale and intensity of the struggle, to say nothing of the stunning surprise at Pearl Harbor, it was the right time to build a mechanism to gather and coordinate information about the enemy. But neither the administration nor the armed services had mustered much enthusiasm for the idea.

There is no one explanation. Traditionally, the country had been ambivalent about intelligence, let alone the machine like the one that now seemed to be a natural part of the British government. This was especially true in peacetime. Like his predecessors, Roosevelt had little interest in a professional foreign intelligence service. He did agree to some military and FBI initiatives, but his disposition to divide rather than unite limited their prospects. When he came on the scene, Donovan had to rely on a president who had other things on his mind, then proceeded to make things more difficult for himself than they had to be. When he preached unity of effort, he seemed to mean "control by Donovan," but talked about picking up a rifle and decamping for foreign shores. The generals had their reservations about Donovan, but they also had a grudging admiration for him. And so the joint chiefs halfheartedly acquiesced when Roosevelt transformed COI into OSS, an agency that would be more or less under their control for the duration.

Even so, given the failures of the traditional approach, why wasn't there a greater cry for something better now that the country was at war? Perhaps Roosevelt's war cabinet felt there already was something better, something working quietly in the background: signals intelligence. It did not figure much in the raucous discussion of Donovan's ideas nor the early Pearl Harbor investigations, partly because it was so secret and partly because the War Department knew its value. It was as if "Magic"—as the intelligence on Japan gathered from codebreaking was known—existed in a world of its own.[1]

The military wanted very much to nurture and protect Magic. For General Miles, director of Military Intelligence in 1941, it was "essential that sources of certain highly confidential information received by the War Department be guarded."[2] The information Magic produced in 1941 was "some of the finest intelligence . . . in our history."[3] Admiral King was even more emphatic, citing "the extreme importance" of what the navy called radio intelligence: "From no other form of intelligence, can the enemy's intentions be so positively determined."[4]

The Pioneers

Modern American codebreaking had its start in World War I, the first major war since the advent of radio. Armies and navies could now communicate wirelessly over long distances. The problem was the enemy could also intercept radio transmissions—that is, unless they were encoded. This was generally done by hand from codebooks that were something like dictionaries, listing words or numbers to replace the original words.[5] A commander would write out his message in plaintext, then a clerk would look up the equivalents and encode the message before sending it out over the airwaves. The reaction to codemaking was codebreaking; one development called forth the other.

The pioneers of American codebreaking, Herbert O. Yardley and William F. Friedman, each got his start in an unusual way. The Midwesterner Yardley first learned about telegraphy from his father, a railroad station master, and in 1913 went to work as a telegraph clerk at the Department of State. To occupy himself on slow days, he deciphered messages not meant for his eyes—among them the amateur codes that Colonel House and President Wilson used for their private messages. It was, he said, child's play; Wilson's and House's intentions were transparent to most anyone who intercepted their messages. This was Yardley: breaking rules and breaking ground at the same time. He would never become a great codebreaker, but he had a firm grasp of the basics and an even better grasp of the potential of codebreaking. He became a leader and entrepreneur in a very new field.

Despite the skill they shared, Yardley and Friedman had only so much in common. Yardley was a free-spirited risk-taker, a man who could take off his shirt and relax outdoors or write explicit letters about his sexual exploits—things the proper Jewish immigrant Friedman, born in the Russian Empire but raised in Pittsburgh, would never do.[6] In photographs he usually appears dressed in a suit, hair carefully combed, bow tie in place, gaze focused and serious, sometimes with a courteous smile. He was an intellectual, a cerebral man who might have been a college professor under different circumstances.

Friedman started out as an agricultural geneticist, and spent the prewar years working for an eccentric millionaire named George Fabyan at the so-called Riverbank Laboratories on his estate some forty miles from downtown Chicago. Fabyan used Riverbank to conduct research on topics that struck his fancy: from chemistry and genetics to acoustics and cryptology. In the grandly named Departments of Genetics and Ciphers, for example, he ordered one small team of researchers to investigate what crops might grow better at night, and another to search for codes supposedly embedded in Shakespeare's works that would prove their author had actually been Sir Francis Bacon. This was an idle pursuit; Bacon did not ghostwrite Shakespeare. But the work awakened two passions in Friedman: one for codebreaking and another for Elizebeth Smith, an English major Fabyan hired to work on the Bacon/Shakespeare project. Friedman decided to dedicate his life to the study of codes and, on May 21, 1917, married Smith, who would blossom into a codebreaker in her own right.[7]

When the United States entered World War I in 1917, the US government had only the meagerest ability to make its own codes, let alone break enemy codes. Fabyan generously put his laboratories at the Allies' disposal "as a patriotic enterprise at his own expense."[8] The Departments of the Navy, War, State, and Justice all forwarded material they could not decode to Riverbank. Friedman wound up testing a machine for the British, decoding letters written by Indians plotting against the British Raj, and teaching cryptology to US Army officers at the recommendation of Joseph O. Mauborgne, the big, handsome, friendly young officer who was also a trailblazer in the field of American codemaking. (He has been credited with promoting the one-time pad, sets of codes that are used only once, then discarded, making them virtually unbreakable.[9]) In the years to come, Mauborgne would prove himself time and again as a professional and personal friend to Friedman, who with his help donned the uniform of an army officer in 1918 and traveled overseas. Upon his return to the United States, Friedman completed a series of groundbreaking technical manuals, then among the earliest and best American introductions to the field.

To do his part in the war, Yardley transferred from the Department of State to the army, where he became chief of the Cryptologic Bureau of the Military Intelligence Division (MID).[10] This bureau, designated MI8, grew quickly into a large organization—roughly 150 strong—with subsections for making code, breaking code, and working with secret inks.[11] Yardley's war included a trip to Europe to liaise with both the British Army and the Royal Navy, overcoming Admiral Hall's resistance to letting Americans learn from the Admiralty's Code and Cipher Bureau. After London, he went on to Paris, where despite a warmer welcome he found that the French would not allow him into their black chamber, which was becoming the term for the secret space where the codebreakers practiced their arcane craft.

At the end of the war, both Friedman and Yardley stood at the starting gates of a new kind of arms race. Citing his experience in Europe, Yardley argued for the creation of an American black chamber. With the support of MID and the Department of State, he obtained the financial means to establish a small enterprise officially known as the War-State Cipher Bureau. Its mission was "to read the secret code and cipher diplomatic telegrams of foreign governments."[12] Army and State jointly funded the enterprise, committing a yearly total of $100,000. Located in the four-story brownstone at 141 East Thirty-Seventh Street in New York City for most of its existence, the American Black Chamber appeared to have no connection with the US government. The number of staff members ran from one to two dozen, but that did not preclude outsize results. The Chamber's greatest contribution came in 1921 and 1922, when it deciphered some five thousand messages during the monthslong Washington Naval Conference. The codebreakers focused on Japan, technically an ally in World War I but now in a position to challenge America in the Pacific; its expansionist tendencies made it a natural rival.

First the codebreakers had to overcome the challenge of obtaining intercepts: the Japanese, like other governments, were sending encoded telegrams through companies like Western Union, whose primary responsibility by law was to their customers. Once the intercepts were obtained, codebreakers were able to decrypt the traffic between Tokyo and its diplomats.[13] Yardley sent the results verbatim to Washington with little to no embellishment or analysis; the War Department and the State Department typically saw the same text as the intended recipients, including policy guidance like the November 28, 1921, admonition "to avoid any clash with Great Britain and America, particularly America."[14]

With this priceless advantage, American negotiators were able to beat their Japanese counterparts back to their last-ditch negotiating position, and set shipbuilding ratios that favored the US and British navies over the Japanese Navy. Christmas 1921 was a happy time at the brownstone, "brightened by handsome presents to all . . . from officials in the State and War Departments."[15] But success came at a cost. By the time the conference ended in February 1922, Yardley and his staff had turned into "nervous wrecks" from overwork, one of the occupational hazards of codebreaking.[16] Yardley himself was too exhausted to even get out of bed for a month, and then only to go to Arizona to convalesce. On his return from the Southwest, Yardley learned that he was to receive the Distinguished Service Medal from the Department of War. The vague citation focused on his wartime development of "a science . . . to translate the most secret messages . . . of vital importance." It did not hint at his postwar accomplishments.[17]

Despite budget cuts, the Black Chamber continued to operate for most of the decade, making a solid if not remarkable contribution to the work of the State Department. It was still in existence when Herbert Hoover took office in March 1929 and made Henry Stimson his secretary of state. Having already held high office, including that of secretary of war, Stimson came to Washington with a well-formed belief system that might be described as principled internationalism. The officers at State and War who supported the Black Chamber allowed the new secretary time to acclimatize before approaching him. It was only in May that they placed "a few translations of Japanese code messages" on his desk in the venerable State, War, and Navy Building across from the White House.[18] His reaction was not long in coming. Stimson discussed the matter with the undersecretary of state, fellow patrician and Harvard Law School graduate Joseph P. Cotton. He told Cotton he needed to know how State had acquired the documents. Quickly establishing that they both felt it was "unethical . . . for this government . . . to be reading the messages coming to our . . . guests," Cotton and Stimson did not hesitate: State would no longer support the Chamber, which meant that it must close.[19] Looking back a few years later, Stimson did not regret his decision, which he justified with his famous declaration that "gentlemen do not read each other's mail."[20]

For Yardley, what Stimson had done was naïve, even unpatriotic. By closing the Chamber, he had "shut off the government's source of authentic information in critical situations."[21] Great powers spied on each other; it was the way of the world. They did not fret about gentlemanly behavior but did whatever necessary to further their interests. Instead of closing the Black Chamber, Stimson should have helped it thrive and grow. The great historian of cryptology David Kahn considered the secretary's action to be an expression of America's long tradition of holding itself apart from other countries, especially the great powers of Europe. "The foreign ministers of France, Germany, Italy, [and] the Soviet Union read intercepts without a moral quiver. Why did only the United States ban codebreaking?"[22] Kahn's answer was to invoke Puritanism: the idea that America should be held to a higher standard, that it was the City upon the Hill.

Stimson may have harkened back to an era that had never existed, but he was not a naïf or a Puritan, let alone an isolationist. When he declaimed about gentlemen's mail, Stimson was speaking as a peacetime secretary of state who, like so many other American statesmen since the founding of the republic, did not see a compelling need for a robust intelligence service except in time of war. In the 1920s, he nourished the hope that the great powers would be above board in their dealings with each other. His attitude would change a few years later after world leaders who were demonstrably not gentlemen had come onto the scene.

The US Army Signal Corps would fill the vacuum that Stimson had created. After finally cutting his ties to Riverbank, Friedman had gone to work for the Corps in Washington, starting a decades-long career. The army first wanted to give the wartime officer a regular commission, but he could not pass the physical examination. This became a blessing in disguise. The Corps then offered Friedman a civil service appointment, setting a precedent for civilian hires that would guarantee more stability and expertise than relying on officers in uniform, who were traditionally rotated from assignment to assignment. By 1922 Friedman was chief cryptographer of the Signal Corps, in charge of the Code and Cipher Section, part of the Research and Development Division in the Office of the Chief Signal Officer. In the nation's capital, away from the domineering Fabyan, the Friedmans blossomed, establishing their own salon for the arts. The couple especially enjoyed hosting private concerts in their Maryland home. The spirited Mauborgne, on his way to the top of the Signal Corps' ranks, came often with his violin.

Friedman was making codes in Washington while Yardley was breaking codes in New York. When Stimson withdrew his support for the Chamber, the Signal Corps sifted the fallout. What would happen to Yardley and his employees, and to the Chamber's records? The army decided to make Yardley an offer to work in Washington for $3,750 a year, exactly half of his previous (rather generous) salary, and less than Friedman's $5,600. At least two of the Chamber's other employees received smaller offers. All were rejected. What the Corps was able to save, however, were the files. In October 1929, as the Chamber was closing its doors, Friedman himself traveled to Manhattan to supervise their move to Washington.[23]

Few would miss Yardley. His ambivalent relationship with Friedman—they had been both colleagues and competitors—would sour, especially after Yardley published a devastating tell-all memoir, *The American Black Chamber.* It was a bestseller in Japan in the early 1930s, its title transliterated as "Burakku Chienba," that showed how the United States had manipulated Japan during the 1921–1922 naval disarmament conference. Yardley claimed that he wrote the book only to pay the bills, not for revenge. But the undercurrent of anger and frustration was impossible to hide. He had given sixteen of the best years of his working life to a government that had suddenly shown him the door and left him without resources.[24] He regretted that he had not worked for a "government, such as the Soviet Government, that understood and practised espionage in . . . [a] ruthless and intelligent manner."[25]

For the rest of his life, Yardley would ricochet from one job to another: moving to Hollywood to write screenplays, then to China to start a black chamber for the Nationalists; running a restaurant, the Rideau, at 1308 H Street NW in Washington that served the best bean soup in town (or so he claimed); playing high-stakes poker toward the end of his life and writing a book about it that was good enough to be published by Simon and Schuster, *The Education of a Poker Player.* At his death, he would be buried in Arlington Cemetery—an unusual end for a man who, Friedman claimed, had betrayed his country by revealing its secrets.[26]

Searching for Magic

With Yardley gone, the Corps quietly ventured deeper into forbidden territory, breaking foreign codes, while officially claiming that its mission was "code compilation," or codemaking.[27] For both missions, the

Corps set up a new office, the Signal Intelligence Service (abbreviated SIS), and put Friedman at its head as chief cryptanalyst.[28] With the savings from the army's share of the Black Chamber's expenses, he proceeded to hire a small cadre of employees. By the end of 1930, the initial staff was a miserly seven, the most the budget would allow—but Friedman had an eye for quality, and his hires were exceptional. Among them were four young men with backgrounds in foreign languages and mathematics who would become stellar codebreakers: Frank Rowlett, Abraham Sinkov, Solomon Kullback, and John Hurt. The New Yorkers Sinkov and Kullback would complete PhDs in mathematics at George Washington University. Rowlett liked to claim, disingenuously, that he was just a country boy from the mountains of Virginia, but he had an innate talent and passion for the work. Hurt, a somewhat eccentric man with a gift for languages—he learned Japanese from a college roommate at Roanoke College—was also from the Old Dominion.

Rowlett had almost no idea what he was signing up for—the field was so new that, even after Friedman described it, it made little sense to him. Things would only become clear once he started to immerse himself in the work.[29] The novices were introduced to the world headquarters of SIS on the third floor of the Munitions Building, the army's end of the block of "temporary" buildings between the Washington and Lincoln Memorials, with large, open bays like factory floors except that they were filled with rows of desks instead of machinery. SIS was originally in rooms 3416 and 3418, which included a makeshift vault, a walled-off space complete with a locking steel door. Barred windows let in light but only a little air during the hot, humid Washington summers. Air-conditioning had not yet come to most government offices. A sergeant roved the halls with a thermometer and something called a hygrometer to calculate the relationship between the heat and the humidity. When it reached a certain point, everyone was excused from work. Some went home; Kullback and Rowlett, who were becoming good friends, would call their wives and go for a round of golf at the nearby East Potomac Golf Course.

As an initiation exercise, Friedman turned the four loose on Yardley's files, now secreted away on the second floor in another vault, one without windows. The files themselves were dusty; years later Kullback and Rowlett both remembered how, as new hires, they had come to work in their Sunday best only to ruin their clothes in the file room. Rowlett looked over at "Kully [and saw that he] had a dirty streak across his forehead. Abe's shirt was spotted with smudges of dust and soaked with perspiration. My brand-new pin-striped trousers were no longer white."[30] But they did not care. They lost track of time while they worked, and when Friedman finally came for them, they emerged "starry-eyed" from the vault, in love with their strange new profession.[31]

Friedman had little competition anywhere in the world for the unofficial title of professor of cryptology. He would train and mentor his new hires for the next two years, giving them texts to study coupled with short lectures and impromptu exercises. Yardley's files on German and Japanese codes were useful. He had laid a respectable foundation for how the Japanese encoded their diplomatic messages. But methods change and improve. Starting in the 1920s, machines of various sorts played an increasingly central role.[32] As they came to the fore, codebooks yielded to algorithms, and the word "cipher" gained more and more currency.[33] Mathematicians became as important as linguists. The young American cryptanalysts' growing skill soon counted for more than anything in the files from New York.

When their apprenticeships ended, the novices were eager to test themselves on current messages. It is hard for any layman to comprehend just why and how they went about their work in the 1930s. Codebreakers have likened the challenge to mountain climbing: you climb a mountain because it is there—likewise, you attack a cipher because the unsolved puzzle is there, almost taunting you. You also want to defeat the codemaker. You keep at it until you succeed, sometimes for months or even years. It is not unlike an addiction or an obsession. One wartime codebreaker said it was like sex, but in a roundabout way. The process may be frustrating, but finally achieving a breakthrough was like a long-delayed climax. "Physiologically it's not the same," Thomas H. Dyer would admit in a remarkable semiofficial interview, "but the emotional feeling is pretty much the same."[34]

The work was a mixture of science and art. One way to break a system was to entertain a hypothesis about its structure, and then test that hypothesis by looking for patterns in the rows of numbers in intercepted messages. The more intercepts, the better; patterns emerged more clearly from a large stack of intercepts than from one or two messages. Sometimes a codebreaker would remember a message that he had seen months earlier that might hold the key to a newer message. If one hypothesis did not yield results, the codebreaker would move on to another hypothesis. What set the greatest codebreakers apart was an intuitive sense for how systems were structured, and where they might be vulnerable. This was something that the individual either had or did not have, no matter his or her qualifications. Yardley called it "cipher brains."[35] He estimated that out of a field of thousands, only a handful—perhaps a dozen—would excel.

Most days the workplace in the Munitions Building was as silent as the library in a monastery. The most common sound was from pencils marking groups of numbers on cross-hatched graph paper. Though Friedman formed his staff into small teams, the SIS codebreakers worked on their own most of the time. To break into a system, Dyer explained, an analyst might start with "five or six big files with maybe 500 messages in each." He would "sit there and thumb through the book" until a subconscious process "decide[d] itself": "You look at it until you see something that attracts your attention, your curiosity" because "that ought to mean so-and-so." Then you look for other occurrences, find a pattern, then a solution—which might at first be only a single letter of the alphabet. If not, "you go on to something else. The next day you come back and look at it again."[36]

One day, in 1933 or 1934, Kullback broke the afternoon stillness: "suddenly and without any warning, [he] . . . started to beat on the top of the table at which he was working and loudly exclaimed, 'That's it!'"[37] He justified this rare outpouring of emotion by showing his initially skeptical colleagues how he had figured out the structure of a system. Kullback's breakthrough signaled that the cryptanalysts were reaching the point where they could routinely solve relatively low-level Japanese diplomatic messages.

Loath to distribute its solutions—especially to the scrupulous Department of State—SIS kept the news mostly to itself.[38] If asked, the codebreakers would have claimed they were not spying, but training for the next war. That was fine with everyone, including G-2, the War Department's military intelligence office, which was (at least in theory) SIS's main customer. G-2 showed little interest in their work until the mid-1930s, when the codebreakers started to solve more sophisticated messages that concealed vital information. By 1937, they had broken into a Japanese diplomatic system they would call "Red," from which they gleaned information about secret negotiations among Germany, Japan, and Italy. This was too good not to share, as senior army intelligence officers quickly appreciated.[39]

In late 1938, the Japanese Foreign Ministry announced to its embassies overseas that it would transition from Red to a new system, labeled "Purple" by SIS, over the next few months. The first intercept was a message from the Japanese embassy in Warsaw to Tokyo in March 1939. Purple was far more complex than Red, and the shift meant that the army would be unable to read Japanese diplomatic traffic for more than a year. It was a terrible time to be blind: Japan continued to threaten the balance of power in Asia, while her ally Germany fomented crisis after crisis in Europe. G-2 pressed the cryptanalysts for results. But pressure from one part of the chain of command did not automatically translate into resources from another. The total strength of SIS was a mere nineteen when Germany attacked Poland in September 1939. Before the month was done the army authorized SIS to grow by twenty-six, which would bring the total to forty-five once the right personnel had been identified, cleared, and hired—a lengthy process.[40] Funding increased. Budgets rose to the neighborhood of $174,000, a welcome but not dramatic increase over the $100,000 allocated for Yardley's Black Chamber in 1919.[41]

Even with reinforcements on the way, the experienced SIS cryptanalysts could do little more than they were already doing, no matter how much anyone pushed. The chief signal officer himself, who was now Mauborgne, called Friedman in, and, in Rowlett's telling, told him to "drop everything and concentrate on this Purple. . . . You senior cryptanalysts go down there and solve [it]. . . . Friedman . . . tried to explain . . . that he had a good team working for him. They said . . . , 'Billy, drop it.'"[42] Rowlett felt that the pressure to work harder was tantamount to an insult; they were already working harder than almost everyone in peacetime Washington.[43] Nevertheless, Friedman went back onto the shop floor.

The pressure continued to mount as the months passed. The cryptanalysts lost their appetites and cycled through fitful sleep. Rowlett lay awake at night, rehashing the day's work.[44] Others had nightmares.[45] The occasional hours of hope—when testing a promising solution—gave way to despair upon its failure. The inability to share compounded the effect. Even Friedman, whose wife was herself a codebreaker (but not for the army), told his family nothing about his work.

Only in August or September 1940 did things change.[46] One afternoon, the quiet, attractive twenty-six-year-old mathematician named Genevieve Grotjan entered Rowlett's office. A summa cum laude graduate of the State Teachers College at Buffalo, she had had to content herself with a job as a statistical clerk for the Railroad Retirement Board until Friedman noticed her scores on a civil service examination in 1939. Since being hired by the army, she had worked unobtrusively. Today her eyes were uncharacteristically "beaming behind her rimless glasses." She asked politely if she could show Rowlett her worksheets. She had circled probable equivalents of known plaintext and unbroken ciphertext. Rowlett and his colleagues immediately grasped the significance of her work. Grotjan had broken into the system. Rowlett could not contain himself, waving his arms and, like Kullback, exclaiming, "That's it! That's it!" Friedman came over to see what the excitement was about. After acknowledging that "without a doubt we are experiencing one of the greatest moments of the Signal Intelligence Service," he famously authorized his staff to celebrate with a round of Coca-Colas—and returned to his office.[47] Grotjan, too, went back to work. Though she felt like the team member who happened to score the winning goal, her modesty could not hide her brilliant achievement.[48]

In the following weeks, Rowlett and his team worked to build their own replica—or analog—of the Japanese machine even though they had no idea what the original looked like. Rowlett, who had a talent for scrounging and inventing, discovered that he could sometimes make more progress through trial and error than by trying to puzzle out the design on paper first. More than once he took parts home with him and tried different arrangements while "relaxing."[49] It helped that Rowlett had worked with Friedman, one of the pioneers of machine ciphers, to invent strange-looking contraptions that randomized input and would be well-nigh unbreakable.[50] Knowing how to make your own machine made it that much easier to intuit how your Japanese counterparts might be thinking. Rowlett just seemed to know how to set the switches and solder the wires. One evening, at about eight o'clock, he and his colleague Leo Rosen, a young MIT-educated codebreaker, were ready to breathe life into the black box they had cobbled together. To their immense relief the contraption came alive when they flipped the switch. After a few modifications and improvements, it would look like two industrial-strength typewriters connected by a telephone switchboard. This Purple replica could perform magic, taking raw intercepts—pages of seemingly random letters—and turning them in to plaintext.

With some help from their counterparts at Main Navy, a handful of codebreakers in the Munitions Building in Washington had opened a window onto the forbidden city of Japanese foreign policy, where secretive men were making decisions about war and peace. Rowlett and Rosen's achievement was truly stunning. The army had broken the cipher first and then created a replica of the machine that had generated it. At roughly the same time the British were breaking German messages, but they had started with an Enigma machine in hand. Separately, they had tried to break Purple and had given up—as had the Germans.

Friedman took pride in breaking Purple but, like Grotjan, did not claim glory for himself, only for the team. A delayed reaction to the stressful months—coupled with the pressures of codemaking and the never-ending fight for resources—brought him down in December 1940. Like Yardley at the end of the Washington Naval Conference, Friedman collapsed one day at work.[51] It was just, Rowlett mused, "that he didn't have the physical stamina . . . to do all the things he thought he ought to be doing."[52] But unlike Yardley, who had recuperated in a warm, pleasant climate, Friedman checked into Walter Reed, the army's premier general hospital at the northern end of Georgia Avenue in Washington. When he presented himself, he said he was having a nervous breakdown but did not go into any detail about his work. Nonplussed, the doctors consigned him to a mental ward and spent the next three and a half months observing as much as treating him. He returned to work on April 1, 1941, but had to content himself with a lesser and somewhat amorphous position—director of Communications Research.[53] Plagued by insomnia and depression, he would never again be the dynamic leader he had been.

Putting Magic to Work

General Mauborgne worried that SIS's marvelous work could be for naught and moved to protect it. When Secretary Stimson was about to join the cabinet as secretary of war in July 1940, his reputation preceded him. Would he now close down SIS just as he had closed down the Black Chamber in 1929, curtailing "certain . . . projects in the War Department" just as he had ordered "the complete destruction and discrediting of the Black Chamber"?[54] Even before Stimson took up his duties, Mauborgne decided on an extraordinary preemptive strike. Circumventing the chain of command, he went to the White House to speak privately to Maj. Gen. Edwin M. Watson, longtime military advisor and personal friend to FDR, about protecting the once-in-a-lifetime source of intelligence. Mauborgne—or more likely Watson—spoke to the president, who "indicated that he desired the army to continue its program" and "also instructed that the War Department not let Secretary Stimson know that this was being done."[55] Foregoing the opportunity to take a stand for good intelligence and sensible procedures, Roosevelt once again took the middle ground, trying to please everyone.

With the Battle of Britain intensifying in the skies over England, and Europe's future hanging in the balance, Mauborgne decided not to follow the president's guidance. It was impossible to hide the breakthrough from the secretary and to realize its amazing potential at the same time. On September 25, Mauborgne got on Stimson's calendar to brief him on SIS, including "the new inventions that were coming along."[56] Far from admonishing Mauborgne about impropriety, Stimson recorded in his diary that it was a long but "very interesting talk." Mauborgne followed up by inviting the army's ranking officer, Chief of Staff Marshall, to watch a demonstration of the Purple machine in action. In what was, for him, an over-the-top outpouring of enthusiasm, Marshall thanked the cryptanalysts for the most interesting and informative briefing on codes that he had ever received. At Marshall's behest, the secretary himself and his two closest assistants, John J. McCloy and William P. Bundy, appeared some two weeks later at SIS's door. Rowlett conducted another demonstration.

Everything proceeded smoothly until Stimson picked up a sheaf of "exhibits"—some of the more interesting messages between Tokyo and its ambassador in Washington. Sifting through the papers, he came to a long report of a conversation between Secretary Hull and the Japanese ambassador, and bade McCloy and Bundy to pay attention. The floor was his; no one knew what would come next. Stimson then drew himself up, adjusted his eyeglasses, and, holding the message in his right hand, read it out loud as if making a proclamation. When he had finished, he declared that the nation was indeed "fortunate to have access to such important information" that was "vital" to the success of American diplomacy. Rowlett was stunned—could this be the same Stimson who had shut down the Black Chamber?—but also relieved.[57]

Stimson soon confessed how dramatically his attitude had changed since 1929, when he had been secretary of state and nourished hopes for a better world order. Early in the 1930s, he had become an outspoken critic of Japanese aggression in China, and then, as the decade progressed, he watched German expansion with growing concern. By the time he had returned to Washington, ready to oversee the country's preparations for war, he had no qualms about spying. On October 23, 1940, after a meeting with Knox about intelligence, he fairly gushed that "the Army had made some wonderful progress"—so wonderful that he could not, "go into some of the things they have done," even in his diary.[58] From now on, SIS no longer needed to fear Stimson's wrath.

A little bit of success bred a lot more. The army's flagship project was now unquestionably Japanese diplomatic traffic, and the numbers kept improving.[59] In 1937, on any given day, SIS could count the number of solutions on one hand; by the fall of 1940 the same tally could reach as high as fifty.[60] The army was doing so well that the navy was jealous. Op-20-G's primary mission was to intercept, analyze, and break IJN messages, but it was unwilling to cede Purple completely to the army. It wanted to share in the glory.

The result was a push-pull relationship in the summer and fall of 1940 between the two services. Joint committees met over and over again to hash out the details.[61] A system slowly began to emerge. General Mauborgne and his navy counterpart, Rear Adm. Leigh Noyes, reached an informal agreement for sharing the workload that was fair but not efficient: the navy would process messages that originated in Tokyo on odd-numbered days, while the army would take the even-numbered days. In January 1941, the two services arrived at a comprehensive written agreement that spelled out the procedures for distributing Magic according to a flowchart that was split down the middle: one side for the army and one for the navy.[62] Each service would distribute intercepts to its own chain of command, but only to the highest echelon. Alternating month by month, the army and the navy would share the duty of carrying the intercepts to the White House, where the aide from that service would hand the messages to the president. The originators (and no one else) would keep file copies, and then only one each.[63]

For any one message, the process typically started with interception by an army or a navy station. The navy, for example, had what David Kahn called "a big ear on a small base" on Bainbridge Island near Seattle that "reached up and snared [the coded Japanese message] . . . as it flashed overhead," then sent it to Washington by teletype, where, depending on the day, either the army or the navy would process it. A clerk typed the cipher into the Purple machine and watched it turn into Japanese plaintext, ready for the translator on duty. Once translated, the message went back to the originator along with carbon copies on flimsy paper—in as little as two hours from the time that the originator had consigned it to the airwaves. If it was an army message in 1941, it went to G-2, where Col. Rufus S. Bratton looked it over along with other messages and evaluated its importance. A tall, heavyset man with prominent jowls that made him look like a bloodhound, the infantry officer from South Carolina seemed an unlikely choice for the job. But he had served in Japan and learned Japanese before coming to Washington to serve at the War Department. He had a reputation for being "reserved and militarily correct" as well as knowledgeable and conscientious.[64] A navy codebreaker remembered him as "wooden" but referred to him, almost affectionately, as "Good old Ruffy."[65]

A day's haul might include a number of messages about the administration of the Japanese embassy in Washington, and far more about personnel matters than anyone in the US government cared to know—for example, who was due for transfer or eligible for a particular allowance, and so forth. Bratton relegated most of the administrative messages to the burn bag, selecting the 10 to 20 percent of messages with policy implications. (Even then, the reader might have to wade through a few pages of minutiae to find the passages that made a difference to Washington.) Bratton would then put the original or, more likely, one of the carbon copies that he called "flimsies" into a red-brown government-issue folder, and padlock the folder into a zippered leather briefcase. He himself would set out from his office with several briefcases under his arm to deliver them to the distribution list that started with the president and the secretary of state. In 1941, he could walk this route. The White House and the State Department sat side by side a few blocks away. If the principal was busy, Bratton left the case with an aide who had one of the two keys for each case and assumed responsibility for documents against a receipt. The aide—for Stimson it could be Maxwell Taylor or Walter Bedell Smith, both of whom would themselves rise to high ranks—extracted the intercept for the principal.[66] Once read, it went back into the briefcase to await Bratton, who would reappear no later than the following day to carry it back to the Munitions Building, where he would consign all but one copy to the incinerator.[67]

Most of what the customer saw in 1941 was raw or close to raw, a decrypt standing alone on a sheet of a paper.[68] Decrypts seldom came with any kind of notes or analysis, which was not surprising since no one was systematically analyzing the story that Magic was telling.[69] With the read-and-return regimen, customers did not have the luxury of pondering the meaning of intercepts. Sometimes they did not even have the time to read everything in the folder. Hull remembered handing the flimsies right back after looking at them.[70] Friedman recorded that the customers "had the messages . . . for so short a time that each message represented only a single frame, so to speak, in a long motion picture . . . a film which [*sic*] should have been shown and . . . intently studied as a continuous series of pictures."[71] It was hard to figure out how the frames fit together. Nor was it easy for customers to discuss what they were reading. Only rarely did the customer see the courier (who in Bratton's case happened to be an expert on Japan), or say much to the aide who might not know anything about the subject matter. Similarly, most of the cryptanalysts were just that; they were not intelligence officers. Friedman might have been able to assemble a figurative album, something that he could flip through, or go back to later. But he was working for the chief signal officer of the army, and his job was to break codes rather than to understand the enemy.

This system meant that customers were, more often than not, their own analysts. Rowlett characterized their attitude as give me "the word of God and . . . nothing less."[72] In Washington in 1941, General Marshall, Secretary Hull, and President Roosevelt all declared, more than once, that they wanted to read the actual words of the intercepts, and showed little interest in secondhand reports or analyses. If they wanted to address a particular issue, they would seek out another principal who had access to Magic rather than one of the US government's few experts.

Hull was probably the most engaged customer, testifying later that he was "at all times intensely interested in the contents of the intercepts."[73] After all, the main subject was foreign policy, specifically his negotiations with the Japanese ambassador Kichisaburō Nomura and special envoy Saburō Kurusu. Hull could use the decrypts to prepare for meetings. He could see if Nomura's and Kurusu's reports were accurate and gauge the difference between what Tokyo was thinking and what its ambassadors were saying. As the onetime lawyer and judge famously commented, it was like being able to read the mind of a hostile witness and know when he was dissembling and when he was telling the truth.[74] Over time, the decrypts were a key factor in his conclusion that the negotiations between the two countries were unlikely to bear fruit.[75]

Secretary of War Stimson was as astute as Hull when it came to Magic. His mind also disciplined by the practice of law and longtime government service, Stimson was a particularly quick study. He easily followed the trends in the decrypts and focused on significant passages. In early 1941, a remarkable series of reports seized his attention. Midmorning on January 2, Stimson briefed the president (who was working in bed), telling him that "he should read the important reports that had come in from Berlin" which offered a rare look inside the Chancellery. The president confessed that he had not read them.[76] That was Roosevelt's loss; Stimson knew that their source, Baron Hiroshi Ōshima, was exceptionally well connected.

A senior Imperial Japanese Army officer who was literally a samurai, Ōshima benefited from a close relationship with Hitler that dated to his arrival in Berlin, where he had served as military attaché before becoming ambassador. An extrovert who liked to drink, Ōshima spoke excellent German and socialized with Wehrmacht officers and Nazi officials; he was said to be more Nazi than many Nazis. When Ōshima and Hitler met, they spoke freely. Written within hours of each meeting, Ōshima's reports were on their way to Tokyo, filled with detail and nuance. One of Stimson's assistants found "the material . . . so orderly and well-presented that [it could simply be] . . . bound together . . . and supplied with a subject index."[77] This made him a wonderful virtual spy for Washington. It was even better than having an actual spy in Hitler's chancellery. No American had to cultivate Ōshima or puzzle out how to meet him without attracting attention.

FDR generally followed the narrative thread that emerged from the intercepts. But it is difficult to pinpoint specific intercepts that led him to take specific actions.[78] Nor did he change his casual approach to intelligence as the crisis with Japan deepened. For an unconscionably long time he tolerated the dysfunctional army-navy briefing system that assigned odd-numbered months to the army and even-numbered months to the navy.[79] Thus, in January, March, and May, Bratton delivered the locked briefcase to the president's army aide, the amiable General Watson, who went by the nickname "Pa." Watson took custody of the briefcase and delivered the flimsies to Roosevelt. One day Bratton came up short after a delivery cycle—until he found the missing document in Pa Watson's wastebasket. And so, after May, the army decided not to send any more decrypts to 1600 Pennsylvania Avenue because White House security was too lax![80]

Roosevelt eventually noticed the change, and pressed his navy aide, the sometimes-taciturn Capt. John R. Beardall, USN, for information that he did not have. It was only in September that FDR let it be known that "he wanted to see the [raw] material."[81] The navy stepped into the breach with a work-around—but, since it was an army month, would only do so after getting Bratton's "O.K." Beardall was to brief the commander in chief on the content of messages without showing him the actual decrypts. The stream of information started to dry up again in November, another army month, despite the ever-deepening crisis with Japan. FDR again asked Beardall for "original material." Beardall explained the situation. Roosevelt said he understood but finally exercised the power of his office and "directed [Beardall] to bring it anyway."[82]

As November gave way to December, individual decrypts stood out as signposts on the road to war without specifically pointing to Pearl Harbor. Tokyo urged an ever-tighter schedule for winding up the negotiations between Japan and the United States. Previously trivial administrative messages sounded more and more ominous: the Japanese Cultural Center in New York was preparing to close; employees of the Bank of Japan were set to evacuate with embassy personnel; and Japanese ships were being summoned home.[83] On November 29, Secretary Hull telephoned the president to warn him that negotiations might collapse at any time, which could presage a Japanese attack in the Pacific. On December 1, the flimsies included a secret report from Berlin to Tokyo with electrifying news from Hitler's foreign minister, Joachim von Ribbentrop: Germany would "join the war immediately" if Japan and the United States went to war.[84] At almost exactly the same time, a message from Tokyo to Berlin directed Ambassador Ōshima to tell Ribbentrop that "there is extreme danger that war may suddenly break out between the Anglo-Saxon nations and Japan through some clash of arms."[85] This message gave Roosevelt pause; he took the unusual step of requesting his own copy. It was at this point that he opted for a last-ditch effort to avert war.

The backstory is yet another example of the president's "do-it-yourself" approach to intelligence and foreign policy. On November 28, Roosevelt had traveled by train from Washington to Warm Springs, Georgia, where he regularly took the waters to soothe his paralyzed legs and relieve the pressures of office. After hearing Hull's gloomy prognosis the following day, he decided to return to Washington. On December 1, at Union Station, he found a letter from one Dr. E. Stanley Jones waiting for him. Jones was not a Roosevelt intimate, but his standing made him difficult to ignore. He was perhaps the leading evangelist of his day, a household name for many Americans; he had met the president at gatherings of church leaders at the White House. In the preceding months, he had been campaigning for peace, trying to forestall war between Japan and the United States.[86]

On the ride from Union Station to the White House, Roosevelt read the letter importuning him to appeal to the emperor. It was, Jones pleaded, the last chance to prevent war. Willing to explore alternatives to war, Roosevelt was receptive. On December 2, he surprised Stimson, Knox, and Undersecretary of State Welles (sitting in for Hull) by telling them that he was drafting a personal appeal to the emperor. Stimson was against the idea, but State's reservations softened now that the eleventh hour was approaching. On December 3, the president saw the reverend, off the record, at the White House. In the forty-five-minute meeting that followed, Jones explained how a diplomat at the Japanese embassy had proposed the initiative to him. One Hidenari Terasaki was fervently promoting the idea—and said to be risking his future by passing this message to the president. Terasaki's role must be protected. The president agreed, reportedly saying that Terasaki was a brave man and that he, Roosevelt, would never reveal his secret—even to his own intelligence officers.[87]

Intriguing in a good cause suited the president but was pointless.[88] An intercept soon revealed that the Japanese embassy had reported Terasaki's initiative to Tokyo; it was apparently not as unilateral or as risky as Jones had suggested.[89] By December 6, MID and ONI noted on another intercept that Terasaki was "the head of Japanese espionage in [the] Western Hemisphere."[90] Unaware, Roosevelt prepared and dispatched his appeal to the emperor.[91]

On December 2, Tokyo ordered the Japanese embassy in Washington to burn many of its codes and destroy one code machine completely, a telling indicator of war.[92] When presenting the message, decrypted on December 3 and likely distributed on December 4, Captain Beardall called FDR's attention to this "very significant" event. After reading the message, the president asked, "Well, when do you think it will happen?"—meaning the outbreak of war. The taciturn officer replied, "Most any time."

The next series of intercepts upset the weekend routine in Washington. In the early morning hours of Saturday December 6, army and navy codebreakers began to process messages from Tokyo to the Japanese embassy in Washington that started with a "pilot" message—a guide to the following fourteen messages. The army and navy worked together to process the resulting jumble that arrived out of order. The first thirteen messages rehashed arguments familiar from previous traffic, including justifications for Japanese aggression. The navy courier on that day, Lt. Cdr. Alwin D. Kramer, carried the decrypts to the White House between 9 and 10 p.m., arriving shortly after the president had dispatched his appeal to the emperor. Lt. Lester R. Schulz, USN, Beardall's brand-new assistant—his first day of work had been December 5—presented them to Roosevelt. They were the first ever delivered to him outside normal working hours.

Meanwhile, back at the Munitions Building, Rowlett and his colleagues waited impatiently for the fourteenth and final transmission to see what it would bring. On the night of December 6–7, they sat through what he called "the wee smallies of the morning . . . waiting and waiting and waiting."[93] It was only after breakfast on Sunday, December 7 that they had the message decoded, typed, and locked in attaché cases—well before the clerks at the Japanese embassy on Massachusetts Avenue could break it out. An addendum to the fourteenth message arrived separately and set a very specific deadline—1:00 p.m. Washington time—for the Japanese ambassadors to submit Tokyo's reply to Hull. Kramer, who was on duty again after less than seven hours off, realized the addendum was an ultimatum of some sort. The fastidious thirty-eight-year-old, wearing the heavy, dark blue wool uniform suited for winter but not for moving fast, made the rounds twice within ninety minutes: once for the concluding message and once for the addendum, sometimes breaking into a jog on his delivery route down Constitution Avenue from Main Navy to the State Department. Kramer passed the fourteenth message to Beardall, who carried it to the White House. Again the president barely reacted, concluding simply that the Japanese seemed to be breaking off negotiations.

Thanks to Kramer, Hull knew what the Japanese ambassadors Nomura and Kurusu were going to say at their early afternoon meeting before they did. Since the Japanese code clerks worked slower than the American codebreakers, it was about 2:20 p.m. when the envoys were shown into the secretary's second-floor office at State. Minutes earlier, the president had called with an unconfirmed report of the attack on Pearl Harbor. Nomura and Kurusu, who apparently knew nothing about it, started by apologizing for arriving late. Hull kept them standing. He pretended to read through the document that he had already seen. Stilted and formal, its ending had chilling implications: "The Japanese Government regrets to notify hereby the American Government that . . . it cannot but consider that is impossible to reach an agreement through further negotiations." Then Hull pronounced his verdict: never in his fifty years of public service had he read anything "more crowded with infamous falsehoods and distortions."[94] Heads down and speechless, Nomura and Kurusu walked out of Hull's office.

Anyone comparing the two versions of the fourteen-part message knew that American codebreaking had passed a major test. Putting them side by side and seeing that they were identical in almost every respect was irrefutable proof that the system worked. Wrought by a handful of introverts working on their own under less-than-optimal conditions with limited resources, the astonishing thing, in the words of Military Intelligence's General Miles, was not that it took the codebreakers as long as it did, but that they were able to do it all. "It was," he concluded in a rare bit of praise, "a marvelous piece of work."[95]

This was America's best—and close to only—insight into Japanese secrets in 1941. There was little else in the days before Pearl Harbor. Unlike the Soviets, America did not have a network of well-placed spies to tell them what Tokyo was thinking. Unlike the British, she did not have a mechanism for synthesizing information. The few American intelligence officers qualified to ponder the meaning of the Magic intercepts were not weaving them into reports—and were barely cooperating with each other. Breaking Purple might have been a triumph of codebreaking, but it was not a triumph of American intelligence.

8

MORE WALL STREET LAWYERS

Transforming Army Intelligence

Secretary Knox's reaction to the attack on Pearl Harbor was to dash off to Hawaii to conduct his whirlwind investigation. It was the reflex of a seasoned journalist: get the story while the embers of the fire are still glowing. His breathless report started the process of apportioning blame. A few years older than Knox, Secretary of War Stimson reacted in his more measured, lawyerly way, and took a longer view. On December 7, his "first feeling was of relief that the indecision was over and that a crisis had come in a way that would unite all our people."[1] The bitter conflict between the interventionists and the isolationists that had divided the country had finally ended; everyone could now focus on the task at hand. One of his first wartime priorities was to look at Magic and ponder its potential. It was, he confided in his diary at the end of the month, "a matter I have had on my mind for some time."[2] Unlike Knox, he was more interested in looking forward than backward. He decided to call in Raymond Lee, the former military attaché at the American embassy in London who was now the G-2, as well as his troubleshooter, John J. McCloy. On the last day of 1941, the three devoted time to

Magic. Before Pearl Harbor, when the responsibility for intercepts was divided between the army and the navy, no one had been able to pay "sufficiently close attention" to them, a situation they wanted to remedy.[3]

Stimson had lured McCloy to Washington, first on an ad hoc basis and then to be assistant secretary of war. In a rare display of affection and humor, he called the cheerful, tireless McCloy his "imp of Satan."[4] Even more than Lee, McCloy was the right man to engage on the issue of Magic. The New York lawyer had a strong interest in intelligence, having dealt with the legal fallout from German sabotage in America during World War I. Like Stimson, he believed lawyers were best suited to take on the most difficult tasks of administration and government. Their training had both sharpened their minds and taught them to see both sides of every issue; they were, he believed, both objective and incorruptible. The best lawyers worked at McCloy's firm, Cravath, Swaine & Moore, one of the oldest in the country. Thanks to the complicated suits they had litigated on behalf of the most powerful corporations in America, Cravath lawyers were not afraid of hard work, willing to devote up to seventy hours a week to sift through rivers of data to find the nuggets that mattered. Magic had already generated thousands of pages of files. A Cravath lawyer could find patterns that others had missed.

McCloy recommended one lawyer in particular, Alfred T. McCormack, who would have the "organizing ability" that they wanted. By the end of the meeting on December 31, Stimson had "authorized [McCloy] . . . to send for him."[5] McCormack was another inspired choice. The Brooklynite had received a Phi Beta Kappa key from Princeton and emerged from Columbia Law School as a budding member of the country's legal elite. After clerking for Supreme Court Justice Harlan F. Stone, he went on to practice law on Wall Street, becoming a partner at Cravath in 1935 and then a managing partner, one of the men who ran the firm. Even during the Depression, a partner at Cravath could not avoid becoming wealthy, and McCormack was no exception. In 1936 he was able to purchase a 250-acre estate in Connecticut. But, while hardly an incubator for the New Deal, Cravath also had a tradition of public service. When Stimson and McCloy called from Washington early in January 1942, McCormack hurried to Washington to hear their unusual job proposal. Their exact words are lost to history, but it amounted to: study the problem, come up with a solution, then implement it.

McCormack's partners had long suspected that he was as interested in military history as the law, and they were not surprised when he resigned from Cravath to do war work that he could not discuss.[6] He wasted no time, returning to Washington on the nineteenth. On his first day of work, McCloy escorted McCormack to the secretary's suite in the Munitions Building, where Stimson took the time to welcome "the man whom we have selected to brief the Magic papers and cross index them so that they will form a really useful basis for inferring what the enemy are going to do."[7] With the title of special assistant to the secretary, McCormack agreed to stand up a small Special Section in Colonel Bratton's office. Though vague, the title implied that he answered only to the secretary himself, a fact that seasoned bureaucrats would grasp. If he had any second thoughts, or doubts about his ability to do the job, he did not express them.

McCormack had the physical presence to match his legal brilliance; somewhat overweight at about 220 pounds and six feet, he filled out the dark suits he favored. With a full head of brown hair conservatively parted close to the middle of his scalp, he looked out intently from round, horn-rimmed classes, often with a hint of a smile. Those who matched his competence liked him, no matter what their own status. Those who did not would be shown the door; McCormack did not tolerate inferior work. Like McCloy, who seemed never to age, the forty-one-year-old had boundless energy. For the next two months he would put in long hours surveying the situation: studying back materials, investigating production facilities, and conferring regularly with McCloy and the leadership of MID.[8] It was not unlike preparing a complicated legal case.

McCormack's training made him a formidable researcher and a compelling, fastidious writer. He loved words, especially his own words, and enjoyed writing. (He even liked to read his compositions back to himself and to his colleagues.) Three weeks after starting work, he reflected in a thoughtful essay that Magic enabled Washington to "look behind the enemy's eyes, into his emotions and his brains," and see "that he is a formidable adversary, cunning, patient, infinitely painstaking, highly intelligent and wholly un-moral."[9] In contrast, McCormack wrote, the United States was "without a clear idea of either its ultimate purpose or its immediate objectives," and unprepared to translate "its verbal thunderbolts" into action—biting criticism of US policy from the new hire, who was clearly not a Roosevelt devotee.[10]

When McCormack wrote about Pearl Harbor, it was not to point fingers, but to puzzle out how to make the best use of Magic going forward. His goals, he said, were to "examine and study the past . . . for any light it may throw on current and future problems" and "to make sure that the material is used with maximum effectiveness."[11] He noted the problematic separation between the intelligence officers and the codebreakers. G-2 simply received whatever SIS sent over, trusting it to decide what to catch and process. Nor did G-2 send feedback to SIS. Next, McCormack focused on the folly of circulating raw intercepts to busy decision-makers. "The daily reporting of current messages was only one part of the job; the real job was to dig into the material, study it in light of outside information, follow up leads that it gave, and bring out of it the intelligence that did not appear on the surface."[12] He knew from years of experience with "masses of material" that you had to comb through things "over and over" to glean their true value.[13] He saw a way to produce the kind of reports that decision-makers needed by drawing on other sources, carefully checking facts, and identifying trends.[14]

McCormack thought that the War Department was the right place for codebreakers as long as the war lasted, no matter if the subject matter was diplomacy. On account of its "piety," the State Department had already proven that it could not be trusted with this important work, and now that the country was at war, military considerations must come first.[15] Nevertheless, he came close to stating that military intelligence was too important to be left to the military. He thought that regular officers seemed to have "a certain supine attitude toward intelligence [work]" and "disagreed with the notion that any reserve officer, or any civilian who had been graduated from college was qualified to handle cryptanalytic intelligence."[16] He wanted men with stronger qualifications. Like McCloy, McCormack believed the ideal candidates were "top-flight young lawyers, trained in research and the preparation of cases."[17] Lawyers, he wrote, "are better fitted for intelligence work of the type that must be done in the War Department, i.e., what the army calls 'strategic intelligence,' than is any other group in the community."[18] They could be leavened by "a couple of good economists and people trained in historical research, together with some language specialists." McCormack's vision was to bypass the old intelligence bureaucracy. Instead, he would use his elite to process raw intercepts, molding them into a far more useful product.

McCormack completed his initial survey in March 1942 and with Assistant Secretary McCloy's and General Lee's approval, prepared to assemble the staff he needed for his Special Section. The outsider started by acknowledging that he needed help from an insider in order to succeed. He was, an associate would say, "beautifully qualified" to stand up and run an elite bureau for analysis, but he was not a good fit for the army.[19]

Early on he met the officer who would offset that shortcoming. For some twenty-five years, Col. Carter W. Clarke had been a regular officer in the Signal Corps. Not a West Point man, he had originally enlisted in the army and worked his way up from the bottom. All his life he was trim and fit, and at six foot two, slightly taller than McCormack—but only a bit less energetic. He did not have McCormack's intuitive knack for grasping the theory of a case, or specialized knowledge of signals intelligence. He occupied a point on the far right of the political spectrum, unusual even among regular officers, who tended to keep their conservative attitudes in check. Clarke, on the other hand, was openly critical of the supposedly over-the-top liberal Mrs. Roosevelt, and carried a pistol around Washington during the war—just in case, he said, he ever got caught in a race riot.[20]

He was not afraid to use colorful language if it would help get results. Using a polite euphemism for the actual words, McCormack would never forget how Clarke referred to one general's morning meetings as "rodent intercourse." Above all, Clarke understood how the army worked and could bridge the gap between bureaucracy and McCormack. In January 1942, he was working in Washington as chief of the MID section for Safeguarding Military Information. By May, the new G-2, Maj. Gen. George V. Strong, had placed him at the head of the Special Service Branch (SSB), the successor to McCormack's Special Section, now expanded "to study, evaluate, disseminate, index, classify, and become custodians of . . . [the] material produced by the Signal Intelligence Service."[21] Clarke would report directly to Strong, who told Clarke that his job was "to go in there and get along with that fellow," meaning McCormack, and make the unusual arrangement work.[22] Strong also thought that McCormack should be in uniform and, perhaps unwisely, arranged a direct commission for him in June 1942. As a junior colonel, he became Clarke's nominal deputy. But for once seniority did not matter; the two men worked as a team to implement McCormack's ideas.[23]

Even side by side with Clarke, McCormack did not mesh with the army bureaucracy; he remained the foreign object that the military body would keep trying to reject. Not only had he vaulted into a colonelcy from civilian life, but he did not hide his attitude toward regular officers. McCormack thought they did not even seem to realize there was a war on. SSB was open for business from 7 a.m. (and sometimes earlier) until about 11 p.m.; its officers did their best to finish the task at hand before leaving for the day—and worked thirteen out of every fourteen days.[24] Other parts of G-2 kept peacetime hours. One afternoon at 4:40 p.m., McCormack went in search of an important bit of time-sensitive information. The colonel in charge of the European Branch was sorry that he would not be able to help until the next day; his officers were busy putting their papers away so that they would have clean desks before the 5:00 p.m. security inspection.[25]

As they staffed the SSB over the course of 1942, McCormack and Clarke struggled with military and civil service regulations. It was almost easier to find and commission the lawyers that McCormack wanted than it was to come up with low-ranking specialists to support them. By painful stages over the next year and a half, the staff grew to a total of some one hundred officers and three hundred civilians. One of the officers, Thomas E. Ervin, later estimated that an astounding 85 percent of the officers were lawyers, while the remaining 15 percent were academics, roughly the mix that McCormack had originally envisioned. There were so many lawyers that Ervin felt that it was like being back at law school and working on the law review.[26] Even so, morale was reportedly high: Japanese speaker Edwin O. Reischauer could not imagine "a wartime job more fascinating."[27]

McCormack and Clarke changed the way the army produced intelligence.[28] SIS codebreakers still received, decrypted, and processed Magic intercepts, the messages to and from Tokyo and Japanese embassies around the world. But an intelligence officer like Colonel Bratton no longer sorted through the daily haul. Instead, the codebreakers sent the decrypts to McCormack's office. Initially, every message flowed through his hands. Over time, as the volume increased from a few hundred to a few thousand a week, he delegated more and more to his staff, the newly minted officers who had just arrived from Wall Street law practices. They selected the most important intercepts for processing and added value by checking references, looking up other messages on the same subject, and generally supplying context. Once they had drafted a report, it went to McCormack for editing and quality control.

Most days the result was a product known as the Magic Summary, a double-spaced report printed on heavy eight-and-a-half-by-eleven-inch US government stock, written in plain English, up to ten pages long, and bound in a spiral notebook with a hard cover.[29] Each summary contained a number of reports. A report might begin by identifying the intercept, say, a detailed message from Ambassador Ōshima in Berlin about a meeting with Hitler to discuss German strategy. (Until the end of the war, Ōshima remained the Americans' only—and very productive—source in direct contact with the Führer.[30]) A lengthy quote might let him speak for himself—sometimes McCormack would include every word of a particularly important message—but now there were references to other messages on the same subject, as well as useful background notes like short biographies. It was the difference between reading a telegram that seemed important when it was ripped from the teleprinter—which is basically what the reader had to settle for before Pearl Harbor—and reading a curated newspaper article on the subject of the telegram.

The summaries went to many of the same addresses that Bratton and Kramer had serviced before Pearl Harbor: some eleven offices at the War Department, at least two at State, then ten to Navy.[31] The list still did not include Donovan at OSS or Hoover at FBI. Paradoxically, even though this was now more than ever an army operation, the navy officers detailed to the White House continued to actually carry the reports to Roosevelt. He might read them in bed while the aide waited, or ask the aide to read to him while he shaved, or while his troublesome sinuses were being packed, a gruesome procedure whereby his navy doctor used forceps to push pieces of gauze into the nasal cavity. Some days the president did not read them at all. He seldom commented on what he did read.[32] Like all other customers, he still had to surrender his copy to the courier, who might have waited in an outer office to retrieve the pouch. SSB kept careful notes of who saw what and ensured that most copies went into the incinerator.

THIS WAS THE COUNTRY'S BEST strategic intelligence during World War II. Based in the War Department in Washington and run by Wall Street lawyers, SSB served the most senior members of Roosevelt's cabinet well. It was a wartime expedient, a work-around of the old G-2 system. It was not the result of some grand plan, or the vision of the Bureau of the Budget's experts on management, or even that of a visiting spy from Britain. It was simply the way things had evolved in Washington—how the Signal Corps had become the custodian of Yardley's legacy; how the brilliant pioneer Friedman had hired other brilliant trailblazers to take up where Yardley had left off; how his largely civilian codebreakers had flourished outside the mainstream; how Stimson and McCloy had sensed an opportunity and brought in just the right outside help. The Magic Summaries were almost exclusively based on one kind of reporting, signals intelligence, that could dry up if the Japanese changed their codes. They were not yet fully finished intelligence—a coordinated synthesis of information from signals and other kinds of reporting. But they were still enormously valuable. No one in Washington had ever known anything better.

9

NAVY CIPHER BRAINS

Early Days

Until the 1920s, navy cryptology lagged well behind army cryptology. During World War I, the navy became adept at radio direction finding—locating the enemy at sea by tracking his transmissions—but showed far less interest in codes. By comparison, Royal Navy codemakers and codebreakers, led by "Blinker" Hall, were extremely active, venturing into mischief like intercepting American diplomatic communications and trolling the Zimmermann Telegram in an attempt to propel the United States into that war.[1] An official US history written in 1935 confessed that, before 1917, "we had no means of secret communication" apart from simple codes like the Larrabee Cipher, which was for sale in bookstores for ten cents.[2] Once the US entered the war, the navy edged into the world of modern codes with the help of the Royal Navy, more to guarantee the security of its own communications than to attack the enemy's communications.[3] After the war, the navy did not support the cutting-edge Black Chamber.

It was not until January 1924, "more by accident than by design," that the director of Naval Communications created an intelligence organization of sorts in the Code and Signal Section, and bestowed on it an innocuous name: "Research Desk."[4] The initial staff was just one officer, then lieutenant Laurance F. Safford, and four civilians, one of whom was a woman named Agnes Meyer Driscoll. Safford was not a typical Naval Academy graduate.[5] His uniform was often disheveled, his hair slightly too long, and his eyes darted this way and that while he spoke in short, unconnected bursts. Even as a young officer, he did not belong on the quarterdeck of a battleship, the spit-and-polish epitome of naval service in the 1920s, and had wound up instead in command of the unprepossessing 187-foot USS *Finch*, a minesweeper on the China Station. His command ended when one of his brother officers remembered his academic brilliance, especially in mathematics, and recommended him for the Research Desk.

The daughter of a German immigrant doctor, the rail-thin, intense, and mostly unsmiling Driscoll had also struggled to find her place in life.[6] Before the United States entered World War I, she had been the director of music at a military school in Texas and the head of a high school math department, as well as a stenographer, bookkeeper, and typist who spoke French and German. Although at twenty-nine she was years older than most recruits, in June 1918 she enlisted in the US Naval Reserve and became a yeoman in Naval Communications in Washington, DC, eventually landing in a section dedicated to codes. The day after her discharge from active duty in 1919, the navy hired her as a civilian employee in the same field. After a brief stint in private industry at the Hebern Electric Code Company, she returned to the navy and, though taking a significant pay cut, joined Safford on the Research Desk in 1924. She, too, had finally found her niche.

Safford viewed his tiny staff as the cadre for navy cryptology and planned for future expansion. He wrote textbooks, ran training courses, and advocated for a string of intercept stations in the Pacific.[7] He was as dynamic as any manager in this strange new field, while Driscoll proved to have one of its greatest minds, said to have been comparable to that of Friedman, the army's trailblazer.[8] Together Stafford and Driscoll created modern navy cryptology.

By the mid-1930s their office evolved into the Twentieth Office of the Office of Naval Communications, G Section / Communications Security, better known by the acronym Op-20-G.[9]

Part of Safford's plan was to attract naval officers with an interest in cryptology by inserting "puzzles and problems" in the official monthly Communication Division newsletter.[10] The response was underwhelming. The way to advance in the navy was to go to sea and stay there as long as possible, not to migrate to a new and obscure shore-based activity. Nevertheless, he did pique the interest of one lieutenant junior grade, Joseph J. Rochefort, a tall, slim officer born into a large, light-blue-collar Irish Catholic family at the turn of the century. His father was a rug salesman for an upscale furniture store in Los Angeles. Lying about his age, Rochefort had enlisted in the navy in 1918. Noting his potential, the navy sent him to an abbreviated engineering course at the Stevens Institute of Technology in New Jersey, a quick route to a commission in the Naval Auxiliary Reserve.

Rochefort felt that some Naval Academy graduates looked down on him because he was not a member of their brotherhood.[11] It did not take much to get him to push back. He bucked authority more than once over the course of his career, stopping just sort of disrespect or disobedience. (In retirement, he would go further, freely expressing colorful opinions about the "clowns" who had undermined his career.[12]) He made enemies among senior officers who would turn out to have long memories. But he was almost always able to connect with and lead enlisted sailors in this, the most formal branch of the American armed forces.[13]

In the 1920s and '30s, he had a variety of postings. He served as a junior officer on the oiler USS *Cuyama* for almost five years. After responding to one of Safford's newsletter inserts, he served at Op-20-G from 1926 to 1927, briefly sitting in for Safford after taking his introductory training course. Along with Driscoll he proceeded, successfully, to tackle one of the Japanese naval codes. Then he took advantage of an opportunity to go to Japan for three years to study the language, a program sponsored by the Office of Naval Intelligence (ONI) as part of its responsibility to keep track of foreign navies through the attaché service. Rochefort was in Japan when Yardley's tell-all book about American codebreaking became a bestseller in that country, an event that, he noted, caused quite a stir. (The book included Yardley's explanation of how he had broken Japanese codes. It would contribute to the hardening of their government's attitude toward the United States.) Ashore in San Pedro, California, from 1936 to 1938, Rochefort served as an intelligence officer for the Eleventh Naval District, and then went to sea again from 1939 to 1941, mostly on heavy cruisers. He enjoyed life on the larger warships—even on a cruiser: "You were still in the Grand Fleet, traveling with the high class, not being around the dregs or [in] the backwaters."[14] By the end of the 1930s he possessed a unique mix of skills. He was a seagoing navy officer—he always said that came first for him—but he was also a Japanese linguist who had worked in the fields of conventional intelligence and cryptanalysis.

The navy did not prepare itself for war in those two fields as well Rochefort did. Like the army, the navy continued to treat cryptology as an adjunct to communications, not intelligence. While the Army Signal Corps was not overly generous to Friedman's SIS, the Office of Naval Communications did even less for its cryptologists, resisting appeals for resources and personnel. A 1937 study noted that there was only one fully trained cryptanalyst on the rolls with a permanent appointment. This was Driscoll, the civilian who did not have to interrupt her work to go to sea every two or three years.[15] Op-20-G's relationship with ONI was poor, sometimes close to the breaking point, with codebreakers refusing to share the fruits of their labor with ONI for long stretches of time.[16] (Codebreaking has little purpose if its results are not disseminated to customers who can put it to use.) The fractures within the navy were compounded by the highly ambivalent relationship between army and navy codebreakers working less than a city block apart in Washington.

The army's Friedman was generally more willing to help his colleagues in the navy than the other way around. His subordinate Rowlett took the navy's half-hearted cooperation as a challenge. While SIS was working on Red, the then head of Op-20-G, an officer named Joseph N. Wenger, told Friedman that he had discovered a Japanese naval cipher machine using *kana* (phonetic renderings of Japanese characters).[17] But when asked, he would not provide the details that SIS needed—perhaps because, like other navy officers, he felt there were too many civilians working for the army, and as "everyone knew," civilians were inherently untrustworthy![18] The navy's refusal to share information spurred the army civilians to work that much harder. By the time the army broke Purple a few years later, Safford was once in again in charge of Op-20-G. More cooperative than Wenger, he put a machine shop in the Washington Navy Yard at the army's disposal to build another Purple analog,[19] and the two services came to the better-than-ever (but still imperfect) work-sharing agreements on Magic.

The Fateful Year

By 1941, the navy had intercept- and direction-finding stations covering the Pacific. Stations like "the Big Ears" on Bainbridge Island outside Seattle existed to capture coded transmissions, while the direction-finding stations, at locations like Samoa (in the Polynesian region of the Pacific), Dutch Harbor (on Amaknak Island in Unalaska, Alaska), and Midway Atoll (in the North Pacific), were built around strange-looking loop antennae that could fix the direction to enemy transmitters. Op-20-G had three locations to process the raw material that, ideally, worked with each other and the local chain of command. Code names were used for various important locations: "Negat" for the head office at Main Navy, and "Cast" for the small subsidiary at Cavite in the Philippines. Located in the Navy Yard at Pearl Harbor, "Hypo" was the largest and most important office outside

Washington. In June 1941, then lieutenant commander Rochefort took charge of Hypo.

Rochefort discovered that Hypo was about to be rehoused in the basement of the plain concrete administrative building that was headquarters for the Fourteenth Naval District, the shore-based command to which it belonged.[20] The new space—some five thousand square feet—gave his staff room to grow. But there was little by way of luxury. The windowless concrete walls and air handlers kept the place too cold for comfort, and the fact that most of the twenty-three officers and men used tobacco in some form made the air murky and hard to breathe. (Rochefort himself was a pipe and cigarette smoker.) The men at Hypo called it their "Dungeon."

Privacy was virtually guaranteed behind the nondescript steel door at street level that led down a flight of stairs to another steel door, guarded by a chief petty officer at a nearby desk, Durwood G. "Tex" Rorie. Rorie was old navy—he had enlisted in 1927 and planned to stay in uniform as long as the service would allow. He was a sailor who knew his way around a pool hall. More than once, he had transgressed against the good order and discipline of the navy, but, thanks to his abilities, continued to advance. Quite simply, he got things done. One officer mused that, if ordered to move the Washington Monument, Rorie would respond, "Well, it will take some time, some manpower, some material, but how far and in what direction?"[21]

Rorie was intensely loyal to Rochefort, in whom he sensed a kindred spirit. There was little that he would not do for Hypo and Rochefort. Among other duties, he ensured that no unauthorized personnel of any rank entered the Dungeon. To add another layer of security, Rochefort adopted the title "Combat Intelligence Unit" (CIU), which suggested that the mission was keeping track of the location of Japanese ships, a routine navy function that steered the curious away from the truth that the work included codebreaking.

What the CIU did was different from solving Japanese diplomatic codes. Thanks to Rowlett's hard work, operators in Washington could feed coded messages into the US-built analog machine, and it would render a plaintext message. Focused on the IJN, the officers at Hypo knew little or nothing about Purple. No one in the US Navy had come close to solving JN-25, the IJN's principal code, in the way that the army had solved Purple.[22] While others, especially Negat, worked on JN-25, Hypo tackled the IJN flag officers' code, but to little avail. This meant that Rochefort had to rely on other methods to track the enemy, like traffic analysis: looking for patterns of communication and drawing conclusions from those patterns.

Hypo's analysts—some of the best were senior petty officers with years of experience—learned to recognize individual enemy ships and shore commands by listening to their coded transmissions. Each operator had his own "fist," a distinctive way of working his Morse key; some were light and subtle, like some jazz pianists, others came down hard, like a big band drummer. The operator on the carrier *Akagi*, for example, had such a heavy touch that it was "as if he . . . [were] kicking [the key] . . . with his foot."[23] Combining data from at least two radio direction-finding stations could yield the target's location, or at least narrow it to a given area at the time of transmission—a significant achievement given the vast expanse of the Pacific. The volume of traffic could be important; high volume might suggest increased activity, and low or no volume could be an ominous sign. Actually breaking into a message and figuring out a few groups—perhaps a call sign or a salutation, or even the text—was an unexpected but well-deserved bonus.[24] Any messages or parts of messages that the CIU decoded would then have to be translated into English by one of the Japanese linguists in the unit, who were mostly navy or marine corps officers.

What Hypo produced was usually known as "radio intelligence." Rochefort's daily summaries read something like shipping news, the lists of arrivals and departures that newspapers in port cities printed in the days before the internet: "The carriers are mostly in the Kure-Sasebo area with the exception of a few which are operating in the Kyushu area," or "No movement from home waters has been detected."[25] Rochefort forwarded his reports to the fleet intelligence officer, who came under a different chain of command, that of the seagoing fleet.[26] In 1941 that officer happened to be a good friend, then lieutenant commander Edwin T. Layton, whom he first met on the liner SS *President Adams* in 1929. They were both on their way to Japan to learn the language. It was an intense, lonely experience; the students were, for the most part, left to their own devices to use the time well. Never particularly welcoming to foreigners, Japan was already somewhat hostile to Americans. These two US Navy officers came to rely on each other for support.

Layton had gone on to serve three tours with ONI: two in Washington and one as assistant naval attaché in Tokyo. Nevertheless, like Rochefort, his loyalties were torn between intelligence work on land and serving at sea. In 1941, ONI was still woefully weak.[27] Although Layton worked for the admiral at Pearl, he looked to ONI in Washington for support that seldom came fast. At Fleet Headquarters, with only one officer to assist him, Layton had fewer resources than Hypo.[28] But he compensated for the lack of manpower. Described as hardworking and confident, sometimes to the point of brashness, Layton took Rochefort's input and combined it with data from other sources to produce something like finished intelligence. For most of 1941 this was a good system. Rochefort's biographer concluded that, thanks in large part to Rochefort, Layton was able to present Pacific Fleet Commander Kimmel "with a remarkably accurate picture of Japanese moves in the Pacific, which were those of a nation preparing for war."[29]

Rochefort and Layton's system worked well until late November 1941, when Rochefort lost track of the IJN's heavy carriers.[30] This was mostly the result of the imposition of strict radio silence on the fleet that was sailing toward Hawaii.[31] Since the Japanese warships were not broadcasting to each other, there was nothing for the

Americans to hear. Matters took a turn for the worse on December 1 when the IJN changed the fifteen thousand call signs that its ships used, an ominous development that came only thirty days after the last change. This made Rochefort uneasy, and he had his men—already on something like a war footing—work even harder. Without data from Rochefort, Layton, too, drew a blank and, on December 2, had little to tell Kimmel about the Japanese fleet. His report contained nothing about the missing carriers. Kimmel focused on the shortfall, and then famously "looked at [Layton] . . . with a stern countenance and an icy twinkle in his penetrating blue eyes. 'Do you mean to say they could be rounding Diamond Head [on Oahu] and you wouldn't know it?' "[32]

The answer to Kimmel's question was yes. But that did not lead him to order additional reconnaissance or send the fleet to general quarters. On December 7, the enemy carriers did not round Diamond Head; undetected, they reached a point some 230 miles to the north of Oahu before launching their planes. But their high-level bombers did fly over Diamond Head on their way to making Kimmel's nightmare come true.

Few men were more devastated by the attack than Rochefort and Layton, but neither of them took time to reflect on what others would call "intelligence failure." Layton asked to be transferred; he wanted to command another destroyer and fight the Japanese on the high seas. But Kimmel's successor, Adm. Chester W. Nimitz, told him that he could do more for the navy by staying on at Pearl as fleet intelligence officer. It would be 1945 before Layton would go on to another job.[33] For his part, Rochefort felt personally responsible for the attack. Speaking about Station Hypo, he would declare after the war that "it was . . . my job, my task, my assignment . . . to tell the commander in chief today what the Japanese were going to do tomorrow"—something they had not been able to do on December 6.[34]

The Recovery

Rochefort told his men that they needed to "forget Pearl Harbor and get on with the war."[35] They would now work even harder. He and then lieutenant commander Thomas H. Dyer, the supremely talented navy cryptanalyst who had run the unit before Rochefort but stayed on as his subordinate, started what they called "a watch and watch proposition," meaning that the two would rotate, each on station for twenty-four hours then twenty-four off, at least in theory.[36] In fact, they overlapped for at least a few hours every day, which made their watches even longer. To keep from lapsing into sleep, Rochefort might nap for an hour or two on a cot during his watch. Dyer tended to work marathon shifts, at one point putting in ninety hours at his desk.[37] Working on the nearly impenetrable IJN flag officers' code before December 7 had been so stressful that Dyer developed inflammatory colitis, for which he was prescribed drugs that slowed him down, and then another set of drugs to speed him up so that he could keep working. This would eventually lead to the half-truth that a jar of "goofballs"—drugs—powered the men at Hypo through the impossibly long hours.[38]

Rochefort had always been a conscientious, productive officer, but now, in the Dungeon, at war, he outdid himself. Supported by Safford at Op-20-G, he had assembled an exceptional team, which he led by example with a light touch and a little humor. The Japanese attack was motivation enough for most sailors at Pearl Harbor, especially those assigned to Hypo. They did not need much by way of conventional navy discipline. Rochefort did not enforce uniform regulations on men working as hard as they could; he himself would pull an unauthorized red smoking jacket over his uniform in order to keep warm.[39] Everyone benefited from Rochefort's common-sense approach, and from seeing that he cared for his men. (There were no women at Hypo early in the war.) One anecdote has him dispatching

Rorie to the brig to retrieve sailors who had been locked up for misbehaving on liberty. They were not, he said, helping to win the war while behind bars.[40]

Not only were Rochefort and his men more motivated than ever before, but they were reinforced by sailors who had lost their ships. One large contingent—"recruited" by Rorie in the time-honored way of senior NCOs foraging for men and equipment—came from the ship's band on USS *California*.[41] The musicians turned out to be adept at running Hypo's rudimentary IBM data processors, machines that were especially useful now that Op-20-G had directed Hypo to join other stations, including Cast and Negat, in collaborative work on the current version of JN-25, the IJN code that was now surrendering a few of its secrets.

Dyer thought highly of Rochefort as a leader and analyst. As outstanding as Rochefort's leadership was, Dyer respected him even more for his uncanny ability to take "fragmentary information and arriv[e] . . . at a correct analysis of . . . what it meant."[42] That information might come from codebreaking, traffic analysis, direction finding—or Rochefort's phenomenal memory of something he had seen months earlier. He would blend the fragments with what he knew about the Japanese language and the country's culture, as well as the IJN, to come up with an educated guess about the enemy's intentions. Even though he might have thought of himself first and foremost as a seagoing officer and only secondarily as a linguist or an intelligence officer, this was his unique gift, one that he now had a unique opportunity to bring to bear.

By mid-April 1942, Hypo's hard work was starting to pay off. Navy codebreakers began to see a pattern emerge from fragments of traffic that pointed to IJN movements toward Port Moresby, New Guinea, in the Southwest Pacific, which could threaten the sea lanes from America to Australia.[43] First they noticed the massive buildup of enemy forces at Rabaul in nearby New Britain, and sensed that the enemy was getting ready to shift into a higher gear. Even at Op-20-G in Washington, officers "began to eat lunch at their desks" and "no one needed to be told to go on working long after the rest of the building had emptied."[44] Eventually the pattern was clear enough to convince Admiral Nimitz to send two carrier task forces to the Coral Sea to fight the Japanese in early May. Neither side came away with a decisive victory. Japan was arguably more successful at the tactical level, while the United States won an operational victory of sorts. Despite losing the carrier *Lexington* and sustaining damage to another carrier, *Yorktown,* the Americans sank the light carrier *Shoho* and damaged the heavy carrier *Shokaku.* Her sister carrier *Zuikaku* sustained so many loses to aircraft and aircrew that she had to sail back to Japan—which meant that two carriers that had been part of the six-carrier attack on Pearl Harbor would be out of commission for months.

Even before the Battle of the Coral Sea had begun, intercepts were pointing to yet another major Japanese offensive in the Pacific. Patterns of communication were shifting; new task forces were taking shape; supplies were being staged. The kinds of ships and supplies suggested that this was not just a raid, like Pearl Harbor, but an invasion. But what was the target? By mid-May, Rochefort was able to argue from the evidence at his disposal that the principal target was Midway Atoll, the US territory in the middle of the North Pacific about 1,300 miles northwest of Oahu, roughly halfway between Japan and the United States. Though tiny, with only a little more than two square miles of rocks and sand and birds, its location and airstrip lent it outsize strategic importance. If the Japanese seized Midway, it could serve as the fulcrum of a defensive line on their eastern flank, and perhaps as a base for operations against Hawaii. If the Americans held the atoll, they would have the reverse advantages: it would remain a vital part of their western defensive line and facilitate future operations against Japanese-held islands in the central Pacific.

The Battle of Midway

After three months, Nimitz was already proving to be an inspired choice for one of the most difficult jobs of the war: assuming command and restoring the morale of the Pacific Fleet after the disaster at Pearl Harbor. He possessed a rare combination of experience, judgment, and temperament, along with a talent for meshing the operations of diverse groups. A Naval Academy graduate who had qualified in submarines—the "Silent Service"—he had a reputation for being a careful listener who weighed his options before making sound decisions. Rochefort and Layton were two of the officers that Nimitz listened to.

Sometime between May 9 and 11, Rochefort called Layton on the private, sound-powered field telephone that connected their offices through a concrete conduit to tell him that he had "bits and pieces [of evidence about Japanese intentions] . . . so hot they were burning [holes in his] . . . desk."[45] Layton passed the news on to Nimitz, telling the admiral that he needed to go down to the Dungeon for a firsthand look at the evidence. Pleading the pressure of work, Nimitz delegated the task to his war plans officer and fellow submariner, Capt. Lynde D. McCormick.

On the morning of May 14, 1942, McCormick walked down the steps, past the chief guarding the big steel door, and onto the shop floor. Under the artificial light, sheets of plywood resting on sawhorses were covered with partially decrypted Japanese messages in some kind of order that was not immediately obvious. Over the hours that followed, with Layton looking on, Rochefort painstakingly explained how the messages related to one another, and how the whole came together to indicate that the target was Midway. "We went over the papers one by one," Layton remembered. "We went through the whole compilation of traffic analysis, how each command . . . became associated with others . . . how these associations continued [or changed] . . . how . . . all the ships of a division [were] brought into [a] common traffic association."[46] It was, Layton mused later, not unlike the Virginia Reel. If you saw the ships as dancers, you could imagine them starting off lined up in rows, then meeting other partners as they went through the prescribed steps, forming a new set of partnerships, while occasionally returning, even if only back to back, to the partners they started with.[47]

But even after the fourteenth, Washington had its doubts. At first it was hard to convince the new commander in chief, Adm. Ernest J. King, who worried that the target might be somewhere else in the Pacific.[48] But he slowly came around to Nimitz's way of thinking. Not so, though, Op-20-G, which was no longer in friendly hands now that Safford had been shifted from its direction. Reading the same intercepts and solutions as Hypo, and aware that Nimitz did not have the forces to defend multiple targets, Negat was unwilling to support an assessment that could lead to unaffordable losses.

In May 1942, five months since the battle line of the American Fleet had settled into the mud at the bottom of Pearl Harbor, Japan enjoyed a considerable advantage in the variety and number of their ships. The one exception was between enemy and friendly aircraft carriers in the central Pacific.[49] With two Japanese carriers in the yard at home and one American carrier, *Yorktown*, about to emerge from a round of emergency repairs, the ratio was four to three. Counting Midway with its airstrip, it was four to four. Nimitz could offer battle at or near parity if Layton and Rochefort could tell him where the enemy carriers would be. Like Midway itself, Rochefort's masterfully assembled puzzle assumed outsize importance.

Rochefort settled the argument about Japanese intentions through a clever ruse devised by one of his subordinates and approved by Nimitz.[50] Communicating via underwater cable, Pearl Harbor directed Midway to report over the air that Midway's desalination plant had broken down. It was the kind of administrative message that the navy would routinely send in the clear, one that the Japanese might intercept and then relay in a coded message that Hypo in turn might intercept and read. This is just what happened on May 20 in a Japanese message reporting that "AF" had water problems. Hypo already knew that the target was "AF," but had not been able to prove that "AF" was Midway. This was the confirmation Rochefort needed. A few days later Hypo made a further break into JN-25, allowing it to read the embedded date and time groups to predict the time of attack. On May 31, in a cable to the fleet drafted by Layton, Nimitz conveyed the enemy's plans in the kind of detail that stood out so far from run-of-the-mill intelligence forecasts that it made some officers wonder how CINCPAC (Commander in Chief Pacific Fleet) could know so much about the enemy. Was the English-speaking Japanese radio personality known as Tokyo Rose actually an American spy transmitting coded information in her otherwise shameless broadcasts? Or had the US perhaps broken the Japanese code?[51]

On the strength of what his intelligence officers told him, Nimitz ordered his three carriers to position themselves to the northeast of Midway and prepare to ambush the Japanese. The week that followed was one of nearly unbearable tension at Pearl Harbor, perhaps most intense in the Dungeon where Rochefort and his men worked around the clock to track the enemy and follow the flow of battle from their imperfect sources. On June 3, a PBY "Catalina" flying boat out of Midway sighted the enemy force, confirming Hypo's predictions. Its officers breathed a collective sigh of relief, releasing some of the pent-up tension. Then on June 4, the principal day of battle, the long stretches of air silence were deafening, broken only by fragmentary messages, some of them in plaintext, sent by Japanese operators under extreme pressure: "Inform us of position of enemy carriers!" "Attack, attack, attack!" "Attack enemy carriers!" By the end of the day, American information rounded the picture out: US Navy pilots had hit all four Japanese fleet carriers, three of them within the span of a few minutes around 10:30 a.m., leaving them all in flames.[52]

The attackers reported the hits and saw the fireworks—including a massive fireball from a fuel-air explosion—but they could not circle the enemy ships to see if they would sink. Sink they all did by the following morning, but no one at Pearl or in Washington could be sure.[53] Sunken ships and dead pilots do not transmit. Then an intercept showed that a flagship's call sign, "FIRST AIR 'COMMAND'," [had] shifted from the [carrier] AKAGI to the [cruiser] KAGA[,] indicating that the former is lost."[54] Even this was not conclusive. Finally, word came from an unusual source. An easy-going, Texas-born Douglas TBD Devastator pilot from *Hornet*, Ensign George H. Gay Jr. told an incredible story; shot down while attacking, he had spent hours bobbing in the water surrounded by Japanese ships. Wounded, hoping that no one would notice him, he hid under his seat cushion when they passed by close enough to spot him. (Sailors on one ship even pointed at the cushion.) From his ringside seat, Gay was able to see three enemy carriers sink. He survived the next thirty hours to be rescued by a passing Catalina on June 5. His testimony was one of the tipping points that persuaded CINCPAC that the US had arrested the Japanese advance to the west and set the stage for the US to fight its way back across the Pacific.[55]

On June 6, Admiral King sent his congratulations from Washington. Nimitz assembled his staff for champagne at his makeshift headquarters in Facility 661 on the Submarine Base at Quarry Point, about a mile from the Dungeon.[56] Perhaps at the loyal Layton's urging, the admiral sent a car for Rochefort—who took so long to change into a clean, regulation uniform that the champagne was all gone by the time he appeared. Nevertheless, Nimitz graciously introduced the lieutenant commander to a room full of captains and admirals as the officer who deserved a major share of the credit for the victory.[57]

This was not empty praise, but recognition of a monumental, clear-cut achievement. Intelligence did not win the battle, but it did make victory possible. "Oh naturally I was pleased," Rochefort would say years later about Nimitz's words. He added characteristically, "Even at this point I felt, and many of my people also felt over in

Station Hypo, that this December 7th thing was to a very great extent our responsibility and we had failed."[58] The victory was satisfying, but it did not entirely erase the stain of the earlier defeat. Nor did it lessen his sense of duty. Rochefort returned to the Dungeon from Nimitz's office, and kept his men on alert for at least another forty-eight hours, just in case of any surprises. He finally allowed them to stand down on June 8 or 9, the first real break in almost a year. He told them to go up out of the darkness into the tropical sunlight, to get away from the base and the pressures of work. They were not to come back for three or four days. His men did not exactly follow orders. Someone had the key to a house out near Diamond Head, invited the others—both officer and enlisted, a rare thing in a navy still governed by a strict social hierarchy—and started drinking. Normally abstemious, Rochefort himself joined in the celebration with twenty to thirty of his subordinates. It was, in his words, "a straight out-and-out drunken brawl"—but no one had to call the Shore Patrol—and after two days or so, the drinking tapered off, everyone sobered up, and they "drifted back" to work.[59]

After the Battle of Midway came the Battle over Midway. Even before the house party at Diamond Head had ended, Main Navy had directed the Pacific Fleet to prepare award citations for those who had performed so well during the crucial days.[60] The consensus at Pearl was that Rochefort merited the Distinguished Service Medal, at the time the navy's second-highest award, commonly given to senior officers for exceptionally meritorious service, especially when that service produced a result like a battle won.[61] By June 8, Nimitz had endorsed the citation and forwarded it to the next addressee. By June 22, the citation was on the desk of King's chief of staff at Main Navy in Washington, Rear Adm. Russell L. Willson. He had unhappy memories of working as a cryptanalyst—and of serving on the same ship as Rochefort. He was, he once said, happy to have left *that* field as soon as he could.[62] The proposal from Pearl irritated this busy man enough for him to take the time to write a page-long explanation for his recommendation against awarding the medal to Rochefort. Willson reasoned that while Rochefort had performed well, he was simply a member of a team who had made good use of the mechanism that he had found at Hypo when he took charge. Besides, Willson added in what read suspiciously like an appeal to his own commander's ego: "equal credit is due to the COMINCH Planning Section for the correct evaluation of enemy intentions."[63]

Willson's assertions were wrong. It was Rochefort who had assembled and led the winning team at Hypo; Nimitz had to use Rochefort and Layton's data to overcome Washington's reservations about Japanese intentions. Only then did King's staff come up with "the correct evaluation." But Willson's ploy worked, and King agreed with his recommendation. As long as he lived, Rochefort would not receive the Distinguished Service Medal that he deserved.

No matter what the men in the Dungeon accomplished, or how many battles they helped to win, the Naval Academy graduates who ran the service seldom seemed to notice. Before Midway, they had done little to create the conditions for Rochefort or Layton to succeed. This pair succeeded *despite* the system—not thanks to the system. They succeeded because Rochefort had a unique set of skills, because he partnered with Layton, and because Nimitz listened closely to his intelligence officers. It was the country's great good fortune that all three happened to be at Pearl in 1942.

10

REORGANIZING NAVAL INTELLIGENCE

Rochefort and the Fusion Center

While the Battle of Midway was raging in June 1942, the country was beginning to hit its stride in the long race for victory. The political divide between interventionists and isolationists had nearly closed. With few exceptions, that would not change for the duration of the war. War became America's business. The country built on initiatives that the administration had launched even before Pearl Harbor. As early as June 1941, the secretary of the navy reported that 697 ships were under construction, among them seventeen battleships and twelve aircraft carriers.[1] Many of these new warships were commissioned in 1942 as both civilian and naval shipyards laid new keels at rates that America's enemies would never be able to match.[2] Thanks in part to the draft, the number of sailors and soldiers was also growing at a staggering rate, the navy adding more than 878,000 men (and a few thousand women) to the 337,000 who were on duty on December 7, 1941, while the army ballooned from 1.5 million in mid-1941 to 5.4 million by the end of 1942.[3] The growth led to debates—sometimes heated—about the best way to organize for victory.

This was especially true for the navy. Before the war, the navy had two principal officers. First, the commander in chief of the US Fleet (abbreviated CINCUS and, amazingly, pronounced "sink us") traditionally embarked on a battleship, ready to steam off to direct the next great sea battle. Then there was the chief of naval operations, who actually had little to do with operations, but rather with the functions that enabled operations—like intelligence, communications, and personnel. Each had his own staff and presided over a set of subordinate commands. Those commands often had a remarkable amount of autonomy, a holdover from the days when it was hard to communicate over the horizon, and the senior officer present had to make decisions on his own.

King had replaced Kimmel as commander in chief of the navy in December 1941. Then in March 1942, he had also replaced his friend and mentor Admiral Stark as chief of naval operations, and would exercise both functions. Shedding the title CINCUS, he became known as COMINCH. Though the sixty-three-year-old was nearing retirement age, he was smart, demanding, and very energetic. King was described by a junior officer at Main Navy as "a tall bald eagle of a man" who looked so predatory that, instead of just chewing junior officers out, he simply ate them whole.[4] A sea story had it that his favorite dish was raw ensign.[5] Whatever his diet, there were things that even King found difficult to control—one of which was intelligence. Admiral Nimitz as CINCPAC answered to King as COMINCH. Nimitz's intelligence officer, Layton, answered to Nimitz as CINCPAC. But when Layton reached back to Washington, it was usually to the Office of Naval Intelligence (ONI), part of the CNO's staff, despite the fact that there was an increasingly powerful intelligence office (designated F-11) that reported directly to COMINCH. Rochefort's life remained similarly complicated. Nimitz and Layton desperately needed his services, and he granted them willingly. But he was still under the Fourteenth Naval District, part of the chain of command that led to the CNO through Op-20-G, which, crucially, struggled with ONI over the right to control communications intelligence.

The impending shift from defense to offense complicated matters still further. Even before the Japanese had been checked at Midway, American generals and admirals knew that they would have to fight their way across the Pacific, liberating US territories and ultimately invading the enemy's homeland. This called for new kinds of intelligence—not just about the Japanese fleet but also about the Japanese Army, as well as the islands the United States would have to assault.[6] Since Marines would conduct many of the landings, the forward-thinking commandant of the Marine Corps, Lt. Gen. Thomas Holcomb, USMC, called for the navy to create intelligence centers to support them.[7] As early as March 1942, he wrote to Admiral King to propose five centers strategically placed in the Pacific for the collection and dissemination of information from "all sources," which might include aerial and submarine reconnaissance, library research, and finding former residents to interview, in addition to puzzling out enemy strengths and intentions.[8]

Almost immediately, King endorsed the concept and called for a formal plan, which Holcomb produced by April 11.[9] Nimitz, who enjoyed excellent relations with Holcomb, saw the merit in the proposal but had reservations about large staffs.[10] The compromise solution was for the shore establishment to run the fusion center, to be known as the Intelligence Center, Pacific Ocean Areas (ICPOA).[11] The next step was for the Fourteenth Naval District to establish the new organization.[12] The District made Rochefort the temporary head of ICPOA without relieving him of his duties as officer-in-charge of Hypo.[13]

Though it was a vote of confidence, Rochefort had mixed feelings about the new responsibility. He delegated as much of it as he could to Lt. Wilfred J. "Jasper" Holmes, the medically retired submariner who had come back on active duty. Rochefort simply walked up to him one day and dropped the paperwork on his desk. What Rochefort really wanted was to continue to run Hypo. Besides, the relationship between his two titles was hazy. Was Rochefort of ICPOA now the boss of Rochefort of Hypo? Holmes, who would become one of the longest-serving officers in the Dungeon, found it almost impossible to sketch out the organization and flow of intelligence at Pearl Harbor.[14]

Despite the confusion, Rochefort felt that he was in a position of strength. Backed by Layton and Nimitz, perhaps buoyed by his new title, Rochefort asserted his independence from Op-20-G in Washington, that part of the Office of Naval Communications that believed in its right to control navy radio intelligence throughout the world. He would later remember trying to convey the message that he was not working "any longer for you clowns."[15] Unfortunately for Rochefort, the clowns had the upper hand. Cdr. John R. Redman had replaced Safford at the helm of Op-20-G.[16] Redman reported to the director of Naval Communications, who happened to be his elder brother, Capt. Joseph R. Redman. Both were smart, ambitious, and sometimes ruthless graduates of the Naval Academy, where John Redman had been such a good wrestler that he made the American team for the 1920 Olympics. Neither was particularly well-liked, at least not by the officers in Rochefort's camp. Dyer, Rochefort's second-in-command, had known Joseph Redman since the 1920s when they were both radio officers on the battleship *New Mexico*, and had found it difficult to deal with him then. Over time he had, Dyer thought, become even more difficult to deal with. Part of the problem was that Redman seemed to look down on cryptology, Dyer's emerging specialty. "Usually when he spoke to me," Dyer would declare, "it was with that sort of sneering smile on his lips that said, 'You poor ignorant bastard, you think you're doing something useful.'"[17] It did not help that Dyer and Rochefort had been right about Midway while the Redmans had questioned their analysis.

Within days of Rochefort's appointment to head ICPOA, John Redman had acknowledged "the definite need of organizing an intelligence service . . . [to] ensure the collection, evaluation, and dissemination of intelligence."[18] He did not assail Holcomb's proposal.

Nevertheless, Redman argued for radio intelligence—that is, the kind of work that Hypo did—to remain under the control of Op-20-G, not some other organization, and certainly not ICPOA. The reason was that Op-20-G had the resources and the expertise that run-of-the-mill intelligence officers lacked.

On June 20, Joseph Redman attached his brother's memorandum to a personal attack on Layton and Rochefort, dismissing them as a pair of "ex-Japanese language student[s] . . . not technically trained in Naval Communications."[19] As a result, he claimed, "Radio Traffic Analysis, Deception, and Tracking" at Pearl had suffered—an outrageous claim to make less than two weeks after they had proven themselves during the Battle of Midway. "Strong people," Redman concluded, "should be in strong places, and I do not believe the Pacific organization is strong." His recommendation: send John Redman to Hawaii to gather firsthand information and then propose "remedial action."

Over the next few months, the Redmans would deprive Rochefort of his independence.[20] To replace him, they groomed an amiable, combat-wounded communicator, Capt. William B. Goggins. In October 1942, Goggins and John Redman appeared in Pearl Harbor, Goggins ostensibly to work at ICPOA and Redman to serve as Nimitz's chief of communications.[21] Citing Rochefort's impolitic declarations of independence, Redman spread the story that he was not a team player, and even that he had not gotten along with his good friend Layton. Redman fine-tuned his moves by communicating with co-conspirators in Washington in a private code. He even shared a small cottage on base with Goggins, which made it easy to ensure close coordination.

Rochefort's countermoves were marked by the "just-try-and-make-me" obstinacy that he displayed from time to time during his career. Directed to proceed forthwith to Washington for consultations in October 1942, he took fifteen days of leave along the way to visit his family on the West Coast. After arriving at Main Navy, he learned that he would not be returning to Hypo; the Bureau of Personnel (BUPERS) was ready to issue him a set of orders for undefined "special work" with Op-20-G, a major step down from the responsibility he had exercised at Pearl Harbor. After the war he claimed that he "flatly refused" the orders, and pushed instead for a command at sea.[22] When BUPERS offered him an ammunition ship that was about to sail from San Francisco, he countered that he could only take the job after he had visited his son at West Point, which would literally mean missing the boat. He wound up spending the rest of war in a variety of assignments. The highlight was supervising the construction of a floating drydock, the USS *ABSD-2*, a repair vessel that came in ten sections.[23]

The career officer's willingness to accept an unusual assignment and make the best of it was still part of Rochefort's character. Even though it was hardly the best use of his skills—he had no particular qualifications for the *ABSD* job apart from previous service afloat—he found the work interesting and put his heart into it. It was, sadly for navy intelligence, a welcome change of pace for him. The field he had contributed so much to had worn him out.

Life without Rochefort

Rochefort's departure and his replacement by Goggins were heavy blows to the morale of a uniquely skilled workforce. Everyone was surprised when Goggins appeared one day and was introduced as Rochefort's replacement. But Rochefort himself did what he could to smooth the transition, writing to his former comrade Holmes on November 16 that Goggins was not to blame for the intrigues against him, and that he, Goggins, deserved the support of his officers, which they now granted more willingly.[24]

As the war progressed, Hypo and ICPOA continued to grow but maintained their separate identities. Compared to its complement on December 7, 1941—Rorie remembers the count coming to about thirty—the number of officers and men at Hypo in the spring of 1942 was something like 120.[25] Processing more messages called for more men, and the total continued to grow by the hundreds, eventually peaking around one thousand, most of whom would work in a new, purpose-built space at Makalapa on a nearby ridge on base.[26] The unit continued to distinguish itself through its codebreaking. Bit by bit, the principal IJN codes, all twelve in the JN-25 series, fell to the team led by Dyer.[27] This phenomenal achievement gave the US Navy a virtual seat on the bridges of most Japanese ships.

Deciding how to use the decrypts was a recurring issue. Midway had been a matter of life-or-death, the decisions important enough to have been made by Nimitz himself. Most messages did not reach that threshold. What should Hypo do with an intercept that gave up the location of an important Japanese warship, one that an American submarine could use to sink it? Throughout the war, the answer rested to an amazing degree on personal relationships rather than formal procedures. Having been a submariner, Jasper Holmes still knew members of the Commander, Submarine Force, US Pacific Fleet (COMSUBPAC) staff. From the Dungeon he would make his way over to their headquarters and pass information orally to his onetime shipmate, the chief of staff. If Hypo had learned that, say, a Japanese carrier—a high-value target—would be at a certain latitude and longitude at a given time, Holmes would write the information in ink on the palm of his hand and then scrub it off after the meeting. Holmes did not tell COMSUBPAC where the information came from, and the submariners never asked. For his part, Holmes apparently never asked how it would be used, let alone protected. No one kept any records; it was as if the meetings had never occurred.[28]

By early 1943, after five months of intense effort, Dyer's team defeated the so-called *maru* code used by the Japanese merchant marine. It was a four-digit system—somewhat less sophisticated than the five-digit JN-25 system—used to form convoys, which was a sensible response to the threat from American submarines even though it called for extensive coordination.[29] Convoys and escorts needed to synchronize their routes from many points of the compass to meet at a given time and place, then steam together to another rendezvous point and repeat the drill, losing some ships, adding others. Helpful to the Americans was a Japanese requirement for the merchantmen to report their positions twice a day, at 8 a.m. and 8 p.m.[30]

Hypo again chose to pass this information from person to person. Holmes worked out a procedure with another fellow submariner, Cdr. Richard G. Voge, an old friend who was the operations and combat intelligence officer for COMSUBPAC from September 1942 to June 1945. For the first few months of the war, Voge had commanded a submarine in combat; he now proved to be an exceptional staff officer. Every morning he would come to Holmes's office with a chart on tracing paper showing the approximate tracks of American submarines on patrol, and then review the previous day's *maru* decrypts in search of matches that indicated where submarines and *maru*s could cross paths. Voge would then craft orders for the submarines, using a code within a code to protect the source of the information. (Layton remembered one early code that used a simple substitution: "whales" were aircraft carriers, "bears" were cruisers.[31]) Holmes saw his daily conferences with Voge as "the basis for almost perfect coordination between operations and intelligence."[32]

This kind of cooperation worked for Holmes and Voge, but its extreme informality, unspoken assumptions, and lack of record-keeping were signs of an immature organization. It was impossible to know just how valuable their work was—in plain terms, how many sinkings resulted from intercepts. A curious fact emerged from a painstaking Op-20-G postwar study of ten months in 1943: American submarines sank roughly the same number of Japanese ships whether or not COMSUBPAC alerted them to the presence of specific targets.[33]

Nevertheless, intercepts contributed to the knowledge of the routes that the Japanese used, and made it possible to position submarines advantageously. Two conclusions emerge. One is that codebreaking, however valuable, was perhaps not *the* decisive factor in the war at sea in the Pacific theater.[34] The other is that, given what Holmes and Voge knew at the time, they had to use the information from codebreaking to the best of their ability. In the end, the war of attrition at sea in the Pacific *was* decisive; sinking most of her merchant fleet weakened Japan to the point of collapse. After the war, Holmes and Voge would both receive well-deserved Distinguished Service Medals, the senior award that had been denied to Rochefort.

Admiral Yamamoto

Late in the afternoon of April 13, 1943, Hypo spotted and broke out from JN-25 most of the itinerary for an inspection trip by Adm. Isoroku Yamamoto, the charismatic commander in chief of Japan's Combined Fleet, responsible for both the Pearl Harbor and Midway operations.[35] Marine major Alva B. Lasswell, a Japanese-language officer, worked through the night after the message landed on his desk.[36] The next morning he and Holmes carried the decrypt, with only a few remaining blanks, to Layton, who immediately set out for Nimitz's office.[37] The admiral weighed the merits of targeting Yamamoto. He asked his intelligence officer if Yamamoto was irreplaceable; was there another leader of equal stature who could take his place? After years of studying the Japanese Navy, Layton knew there was no one else like him. In that case, it might be worth the risk. At stake was the priceless secret that the US was able to read such Japanese messages.

Nimitz likely made an active decision to consult Washington. Hypo sent a copy of the partial decrypt to COMINCH—King—on April 14, and some twenty hours later both Hypo and Op-20-G circulated more complete versions of the same message.[38] This put Admiral King and

Secretary Knox on notice, giving them the opportunity to concur or dissent. No one knows whether they briefed the president before the fact.[39] A junior staff officer at Main Navy who handled the incoming decrypt remembered hearing that Knox had qualms about ordering a targeted killing, and asked his judge advocate general to study the legal pros and cons.[40] In the end, Nimitz decided to proceed, forwarding the information to Adm. William F. Halsey, the commander of the South Pacific Area, and authorizing him to intercept and kill Yamamoto. The aggressive Halsey did not hesitate. On Sunday, April 18, a beautiful clear day in the tropics, sixteen US Army Air Force P-38 "Lightning" fighter-bombers flew hundreds of miles to a pinpoint in space and time to intercept Yamamoto's flight over the island of Bougainville in the Solomon Islands.[41] Scattering the six escorting Zero fighters, the Americans shot down the two Betty bombers believed to be carrying Yamamoto and his staff. The next day a Japanese patrol found the crash site after a Solomon Islander excitedly called their attention to it. No one had survived the crash. According to a postwar interview, Yamamoto was still upright in the plane commander's seat, holding his samurai sword with both hands. If so, he died with more ceremony than most victims of the war he had played a key role in starting.[42]

On the American side, it was impossible to be sure that Yamamoto was dead even though Hypo soon noticed that Japanese messages were no longer addressed to him but to his chief of staff.[43] Tokyo waited over a month to announce his death. On May 21, the Japanese Navy sent a plaintext message to all of its stations, announcing that Yamamoto had "died a heroic death in April of this year in air combat with the enemy."[44] Three days later, President Roosevelt penned a joke that reflected his feelings about the enemy in the Pacific. Writing in his own hand, he directed "Bill," presumably his chief of staff Adm. William D. Leahy, to convey to "the old girl"—"Mrs. Admiral Yamamoto"—that "time was a great leveler" and that he was sorry he could not attend the funeral "because I approve of it."[45] At least after the fact, Roosevelt endorsed the targeted killing of an enemy commander, a rarity even in the brutal Pacific War. He would not have quarreled with the codename for the operation, "Vengeance," chosen to indicate that the US was evening the score for December 7.

For his part, Layton felt a twinge of regret that he had, in effect, signed Yamamoto's death warrant; before the war, he had known and liked the Japanese admiral. Now, in wartime, he concluded reluctantly, it was his duty to hurt the enemy in any way that he could.[46] His original assessment had not been wrong. The risk appeared to have been worth taking. Not only did Yamamoto prove to be irreplaceable, but the Japanese also chalked up the attack to the fortunes of war, not to a broken code.[47]

The Wave of the Future

While Hypo was thriving, the Intelligence Center, Pacific Ocean Areas (ICPOA) was off to a slow start. The Center opened its doors too late to support the Marines at Guadalcanal and Tulagi in August 1942, making the initial landings little more than "a stab in the dark," which made them an object lesson for better intelligence (and the kind of services that OSS might have provided).[48] At first, ICPOA's main stream of information was radio intelligence from Hypo—welcome and useful, but not enough.[49] Early on, the preexisting Photographic Reconnaissance and Interpretation Section Intelligence Center came under ICPOA, adding another key capability. Submarine reconnaissance—taking pictures of enemy islands through a periscope—became an art form. As time went on, US forces captured Japanese documents in ever-larger quantities. In shallow water off Guadalcanal, two New Zealand corvettes sank a Japanese submarine, the *I-1*, that turned out to have a treasure trove of "valuable secret documents," including code books, call signs, and charts.[50]

Notably lacking was human intelligence. Not only did the United States not have any spies in the emperor's government or his forces, but most Japanese soldiers and sailors chose death over surrender. Especially early in the war, there were few prisoners of war to interrogate.[51] In September 1943, the navy agreed to allow ICPOA to become a joint organization, making it JICPOA, with representatives from every branch of the armed services. By January 1945, JICPOA would have a complement of 544 officers and 1,223 enlisted men, all primed to cull the myriad bits of information that an army and a navy needed to seize increasingly larger islands and contemplate landings on the Japanese home islands.[52] Though imperfect, JICPOA was a good tactical and operational intelligence center, a far cry from the small, navy-only staffs in 1941. It might not have prevented the attack on Pearl Harbor, but it enabled the operations to drive Japanese forces back across the Pacific.

When ICPOA became JICPOA, Hypo took on the title of Fleet Radio Unit, Pacific Fleet (FRUPAC).[53] This seemed like the long-overdue official declaration that the function belonged to Nimitz at Pearl, not OP-20-G in Washington. Layton surmised that Rochefort would have been pleased.[54] But while Hypo might have won a battle, the Redman brothers were winning the war. The principle of central control would prevail. There would be no going back to the days of the independent operator, who, nearly starved of resources in a basement, nevertheless assembled and led a pick-up team to victory.

11

ARMY AND NAVY CODEBREAKERS IN WASHINGTON

The Navy

We'll never change our course, so Army you steer shy-y-y-y.

—"Anchors Aweigh"

The Redman brothers did not know as much about intelligence as they thought they did. Both skilled communications officers who had come of age in a much smaller peacetime navy, they understood how to set up and run complicated networks over long distances, enabling the different parts of the fleet to work together. They also had an instinctive feel for how the navy bureaucracy worked, and knew how to promote each other, if not literally, then certainly figuratively. While signing their names to the documents that sealed Rochefort's fate, they also signed others that went a good ways toward cementing the place of signals intelligence in the US military.

Many of those documents—including some of the screeds against Rochefort—were drafted by Joseph Numa Wenger, the officer who told Friedman that the navy had discovered the kind of characters that the Japanese Navy was using in its messages—then refused to share the details. Wenger comes across as a figure who seems to have been comfortable in the background. Photographs show a tall, thin, almost ghostly man, with big ears and thick, unsmiling lips. He wore a large ring on one of his long fingers, most likely his Naval Academy class ring. Born one hundred miles from New Orleans in the small city of Patterson, Louisiana, in 1901, he attended the Academy from 1919 to 1923, where his nickname was (unsurprisingly) "Skinny." Like most junior officers, he went to sea for the first few years after graduation, then alternated sea duty with shore duty, where he developed an affinity for cryptology. In 1924 he was the first officer to take Safford's informal course on cryptanalysis and then, not unlike Rochefort, served as a radio intelligence officer.[1] It was in the Asiatic Fleet from 1932 to 1934 that he explored, firsthand, the many ways that the US Navy could learn about the Japanese Navy over the airwaves.

In the mid-1930s, Wenger worked on the research desk at Op-20-G in Washington. He took some of the first steps toward mechanizing American cryptanalysis by studying how the navy could use business machines to break codes, and helped to defeat the first Japanese machine ciphers. Though he would later claim that the two services collaborated well before the war, his ambivalence about collaborating with the army was like that of other navy officers. He justified his position by invoking the peacetime fight for survival of the two codebreaking services; "each depended for its slim existence on the intelligence it was able to produce."[2]

Wenger returned to Op-20-G in February 1942, and though still only a lieutenant commander junior to both Redmans, took the lead in strategic planning. By now he had spent a great deal of time thinking about the best way to organize and conduct navy signals intelligence, perhaps more than anyone, even the trailblazer Safford.[3]

He had come to understand how important it was for the various streams to flow together, how it was better to have two or three stations working on a problem, continuously sharing raw intercepts and developing solutions, rather than allowing far-flung stations to work independently.[4] Optimizing the system called for a central authority, like traffic police who signaled drivers when to stop and when to go.[5] Someone had to decide which messages were worth the effort—a decision dependent on the message's code, the addressees, or earlier messages in the same series—and which station would take the lead. Not everyone would have the same skills or corporate memory, but stations could reinforce or complement one another's efforts. Not surprisingly, Wenger argued that Op-20-G should be the central authority for the navy because it would be able to muster the necessary resources. Located in Washington, it was also far from the threat of enemy action.

Even before the war began, navy codebreakers had been far-sighted in their own way. They understood that wartime codebreaking would call for a sharp increase in the number of hands on deck. The service would need people who had demonstrated ability in mathematics, computing, and languages. Women could fill those requirements, so long as they had the right intellectual, social, and political qualifications. As early as September 1941, Joseph Redman's predecessor as director of naval communications, Rear Adm. Leigh Noyes, had reached out to the president of Radcliffe College, Ada Comstock, asking for her help in identifying women who would join the navy "in the event of total war."[6] The best candidates would be from the Seven Sisters, women's colleges equivalent to the all-male Ivy League.[7] Safford had followed up with a list of detailed qualifications, many of which had elitist overtones: "no fifth columnists, nor those whose true allegiance may be to Moscow," and no refugees "from persecuted nations or races"—which then included many Europeans and Jews.[8] Friedman, who was foreign-born and Jewish, would not have qualified.

Now that the country was at war, and the American Navy was breaking more and more Japanese Navy codes, Op-20-G acquired "business machines," early computers that could perform "statistical processes" on intercepted traffic and help increase output.[9] Bringing women into the navy also increased output. They were especially welcome because they could replace men, who would deploy outside the continental United States. (The navy forbade women from serving overseas with the eventual exception of Hawaii.) The goal was ambitious. By mid-1942, the onboard strength of Op-20-G had increased from some 240 to 650.[10] In 1943, the upper limit would be set at a stunning 5,000. According to the plan endorsed by the CNO's office, "two-thirds of the total . . . will be enlisted WAVES."[11] This would upend the social order that had prevailed before the war, when few women worked outside the home; those in favor of bringing women into the service faced bitter opposition from conservatives who feared that the move would shred the fabric of society. Secretary Knox enabled the navy to proceed.

It would be impossible for Negat, as Op-20-G was still known, to remain at Main Navy with thousands of codebreakers, to say nothing of the codebreaking equipment it was acquiring, much of it from the National Cash Register Company of Dayton, Ohio.[12] Obeying the unwritten imperative for government agencies to expand, Negat went house hunting. Wartime authorities made it relatively easy to acquire property at bargain prices, which were often unfair to the owners. By the fall of 1942, the navy had settled on an extraordinary piece of real estate and forced a sale at a price so low, $800,000, that the government was later obliged to pay an additional $300,000.[13]

Despite its title, Mount Vernon Seminary and College was not anywhere near George Washington's plantation on the Virginia side of the Potomac, nor was it sectarian. Instead, it was a progressive, upscale boarding school for girls, its redbrick buildings covered with ivy, on some thirty acres at the intersection of Nebraska and Massachusetts Avenues NW in an upscale neighborhood. It was perfectly suited for a workforce that would include thousands of women doing ultrasecret work: a dedicated, secure facility complete with a fence that Marines could patrol, not too far from Main Navy. It would become known as the Naval Communications Annex.

Not surprisingly, Op-20-G was overwhelmed by the increasing volume—and value—of naval traffic as the war intensified. Upon his return to Washington in February 1942, Wenger had been distressed to find that the navy was still spending much of its time on Magic—Japanese diplomatic traffic—under the vestiges of the strange prewar work-sharing arrangement with the army. Wenger's idea was as compelling as it was simple and overdue. He suggested to John Redman that "we ask the Army to take over the entire job" of Magic.[14] By late June—about the time that both Redmans were downplaying Rochefort's accomplishments during the Battle of Midway—Redman had asked the CNO to approve the idea on the grounds that "the Navy has more Orange [Japanese] Naval [traffic] than it can handle."[15] Approval in hand, Redman convened a meeting with his counterparts, including Alfred McCormack, and eventually reached a cautious "gentleman's agreement," valid only for the duration of the war, for the navy to cede its share of Magic to the army.[16]

The strategically minded Wenger saw collateral benefits to the agreement. The two services could use it "to have all cryptanalytic activities confined to the Army, Navy, and FBI." If the services could "demonstrate . . . that they had taken positive steps to eliminate all unnecessary duplication, they would be in a much stronger position." Wenger's point of view resonated with other members of the small cryptanalytic committee and showed how much they had learned since 1941. Another round of bureaucratic maneuvers led to an omnibus agreement on June 30 that expanded the army-navy agreement: the navy would focus on naval traffic only; the army would continue to tackle foreign diplomatic traffic and, when possible, foreign military traffic; the FBI would take on clandestine and criminal transmitters.[17] In the tradition of American intelligence, the new agreement was not about collaboration—joining forces to fight the enemy—but about setting boundaries.

It turned out that Donovan had been buying equipment and hiring employees in order to start his own black chamber. When John Redman found out, he complained to the Bureau of the Budget that OSS was ruining the market for new hires, offering annual salaries of up to $10,000 when the best the army or the navy could do was something like $4,600.[18] The services believed, naively, that the best way to protect their domain would be to involve the president and have him issue some kind of order.[19] The result was a memorandum for the president stating that "in the interests of maximum security and efficiency, cryptanalytical activities should be limited to the Army, the Navy, and the Federal Bureau of Investigation."[20] Any self-appointed codebreakers would have to cease and desist.

Within two days the president acted in his own way. He directed the Bureau of the Budget to issue "the proper instructions . . . discontinuing the cryptanalytical units" outside the army, navy, and FBI.[21] This was vintage Roosevelt: more genial agreement than decisive action. His bureau of administrative experts proceeded to write more memoranda and make phone calls until OSS admitted defeat; Donovan agreed not to run his own black chamber.[22]

The army and the navy would have preferred something more definitive. In August, at the first meeting of the Standing Committee for Coordination of Cryptanalytic Work, a by-product of the new army-navy work sharing agreement, Wenger and McCormack lamented the absence of "formal notification . . . on the action of the President."[23] Then they went on to act as if he had issued the requisite order. From then on, the army and navy, with an occasional nod in the direction of the FBI, would make cryptanalytic policy and decide who would or would not be in the exclusive cryptanalytic club.

The Army

March along, sing our song, with the Army of the free.

—"The Army Goes Rolling Along"

When the navy ceded its share of Magic to the army, it was like the resolution of a paternity suit. Redman was, in effect, admitting that the pretty baby was not actually the navy's, but the army's. Of course the army agreed. The first breaks into the Japanese diplomatic system had occurred in the Munitions Building, not Main Navy. The army had gone on to design the first Purple analog machines that transformed ciphertext into plaintext as if by magic. No one could argue that now, more than ever, the baby was thriving, thanks to McCormack and Clarke's reforms. The navy had great codebreakers, but no one like McCormack, who was running the nation's best cadre of analysts and distributing the Magic Summaries.[24]

But Magic was no longer enough. The army was preparing to engage the enemy in ground combat on three continents and needed to try its hand at military traffic. After Pearl Harbor, it started tackling what it called "the Japanese Army Problem," meaning the various codes and ciphers that the enemy used to run its far-flung army.[25] When army codebreakers did not at first succeed, they redoubled their efforts, adding more machines and especially people, continuing on a trajectory from some 181 souls on December 7, 1941, to a wartime high of 7,848.[26] Like the navy, the army felt the need for more space. Codebreakers of any stripe would be welcome to space at the new Pentagon across the river. But, like the navy, the army coveted a campus of its own, in this case a girl's finishing school known as Arlington Hall Junior College.

On high ground at the end of a circular driveway, Arlington Hall itself was a stately colonial revival edifice with a touch of antebellum. Six Greek columns set off the formal entrance to the large, yellow brick building; dormers projected from the roof. Behind the Hall was a lush one-hundred-acre campus, complete with a pond, bridle paths, and a gymnasium. It offered the prospect of security without being off the grid. The army could fence off the small base and grant entry only to codebreakers. It was only about four miles south of the riverside flats where the Pentagon was taking shape. Invoking wartime powers and leaving the owners no choice, the army acquired Arlington Hall for the bargain price of $650,000 on June 14, 1942, and started to move in by the twenty-fifth.[27]

By the end of August, the entire Signal Intelligence Service—the army's codebreaking enterprise known as SIS—was ensconced at Arlington Hall Station.[28] The main building quickly filled up. Classrooms turned into offices where workers shared long tables stacked with papers and reference books. Dormitory bathtubs were transformed into makeshift filing cabinets. The hundreds of machines that the army was purchasing turned out to be too heavy for the upper floors and had to be relocated to the basement.[29] For the overflow, the army added various buildings around the campus, some of them flimsy temporary structures that were suspiciously like the "temporaries" on the Mall that many codebreakers had been happy to escape.

Even amid the crowding and the unrelenting pressure to produce, Arlington Hall was not a bad place to work. SIS acknowledged the stress of codebreaking and tried to be as accommodating as possible, providing amenities—like the cafeteria that served meals around the clock, a base theater, and health care providers who did not judge their patients. A sense of mission drove the unusually diverse workforce.

No American uniformed service was ready for diversity in 1942. Arlington Hall was not a test bed for gender (let alone racial) equality, but SIS learned how to remove enough barriers for the work to flow. This was mostly a matter of accepting women and civilians in the workplace, and placing a premium on skill rather than gender or status. Officers and enlisted men and women performed similar tasks alongside civilian men and women. With only a little reluctance, the army began to recognize the value of the first steps toward diversity:[30]

> Had it been possible to operate . . . entirely with military personnel, or entirely with civilians, some friction might have been avoided but an [agency] made up only of military, or only of civilians, would have lost immeasurably the contributions of the other group.

The brilliant civilians hired in the 1930s, many of them Jewish, few of them very military even after receiving reserve commissions, helped to set the casual but productive tone at a place where intellectual capital mattered more than military rank or social standing. As time went on, more women, both civilian and military, were brought on board, hired from all over the country for their skills, not just from the exclusive colleges that the navy favored. The early attitude was that women were better than men at certain repetitive tasks common to codebreaking—but that shortchanged brilliant women who rivaled their male counterparts.[31] *New York Times* reporter Sally Reston was close to the mark when she commented that "men may have started this war, but women are running it."[32] By the fall of 1942, the women—and men—of SIS and Op-20-G were poised to make ever greater contributions to the war effort as the Allies planned their return to the European continent.

12

JEEPING INTO ACTION*

Building Capacity

Just as the ranks of army and navy codebreakers were swelling a few miles away, OSS continued to grow in the second half of 1942, but in its own unique way. The first tranche of staffers had come from COI, automatically transferred from the old agency to the new, still conveniently centered on Navy Hill between the new Department of State building and the Heurich Brewery. To round out the staff, amateurs of many stripes now poured into Washington. They included many who were part of the "right" social and professional networks, and others who were artistic and unconventional—or simply looking for adventure—and happened to be in the right place at the right time.[1] Together they would create what one German immigrant who joined R&A called *"ein seltsames Gebild"*—a peculiar entity.[2]

The way that James Grafton Rogers came to OSS was typical of many senior leaders, and his service reflects the agency's growing

* "Jeeping," originating from the multipurpose Jeep (beloved by GIs), was World War II slang for finding informal, quick solutions.

pains in the first year of its life. With his varied interests and accomplishments, Rogers was something of a renaissance man.[3] During World War I he served as an artillery officer and tactics instructor. Like many OSS leaders, he had a strong connection to the Ivy League. At Yale he had been a student, a professor, and the head of one of the colleges that made up the university, Timothy Dwight College. A pragmatic Republican, he had been assistant secretary of state under Henry L. Stimson from 1931 to 1933, some two years after Stimson had fired Yardley. Later in life he moved out west, where he served as dean of the University of Colorado Law School, wrote and staged plays, and in his spare time, hiked and climbed with the hardcore mountaineers of the American Alpine Club.[4]

Like Stimson and Donovan, Rogers nourished a strong interest in foreign affairs. When the war broke out in Europe in September 1939, he happened to be in Germany on a self-financed fact-finding mission. After being repatriated, he wrote about his experiences, suggesting in a letter to President Roosevelt that the country needed to be able to produce strategic intelligence and to conduct irregular warfare. Roosevelt recognized Donovan and Rogers as kindred spirits—they seemed to be promoting the same ideas—and turned the letter over to the colonel, who invited Rogers to join OSS.

When he arrived in Washington on July 6, 1942, Rogers was a hardy fifty-nine-year-old, five feet, eleven inches tall, whose receding hairline and glasses belied his youthful athleticism—he thought nothing of walking six miles to a friend's house for lunch. Even a man of his standing had to content himself with a tiny room "only two or three times the size of the iron bed in it," which he counted himself lucky to find at Sixteenth and I Streets NW in overcrowded wartime Washington.[5] As soon as he settled into the city, he showed up for work at OSS headquarters.[6]

Rogers found Donovan impossible to dislike: a "charming Irishman, winning, daring, imaginative . . . a knight-errant of war," who loved a fight, whether on (preferably) or off the battlefield.[7] He was impressed that the colonel worked nights and Sundays, even if "always discursively, unsystematically," and that he read two or three books a week, looking for useful lessons that he could apply to this war. Donovan made Rogers a special assistant at a high pay grade and gave him "a roving commission." He was to make a survey of the organization in order "to inform the [Joint] Chiefs of Staff 'how best to employ' it."[8] Donovan was telling Rogers to figure out how to articulate OSS's purpose in life. "Grand! Now I can explore" was how he reacted.[9]

Rogers quickly learned that OSS had three main business areas: research and analysis; espionage; and subversion and sabotage. Running them was "the most interesting collection of people . . . I have ever . . . known."[10] He described James Phinney Baxter III, the historian at the head of Research and Analysis, as "truculent, rude, quick, energetic, and victorious." Still at Baxter's side was William L. Langer, the World War I veteran and distinguished historian from Harvard, rated by Rogers as "tough minded" and "hard working."[11] A historian from Yale, Sherman Kent, had joined them. All three were recruiting other men like themselves. Kent would remember that of the first dozen or so people he met in the OSS, five were from Yale, four from Harvard, two each from Columbia and Williams, one from the University of Virginia, and one from "one of the mid-west universities" whose name he could not recall, a priceless reflection of his East Coast outlook.[12]

The Secret Intelligence branch, responsible for espionage, was first under David K. E. Bruce, who has been called the last American aristocrat.[13] Handsome and elegant, touched by fabulous wealth and fame, he was a charming dilettante with roots in Virginia and Maryland who had married into the Mellon family of bankers and philanthropists. His father had served in the US Senate and won a Pulitzer Prize for a biography of Benjamin Franklin. When the war broke out, Bruce decided to dedicate himself to public service. As other Americans were streaming home in the fall of 1939, Bruce boarded the SS *Washington,* a luxury liner going the other way, and found he was her only passenger. In London he led the American Red Cross, providing services to civilians in need, and mixed with the same officials and socialites whom Donovan encountered on his trips across the water in 1940 and 1941. From the Red Cross he went on to the army air forces, but with his connections and interests, he was a far better fit for Donovan's startup.[14] When COI morphed into OSS, he took on the responsibility of defining the espionage mission and attracting like-minded men and women.

For one OSS employee, the Research and Analysis Branch seemed "the heartbeat of the organization," the apple of Donovan's eye.[15] This was perhaps true in the early days of COI, since its mission aligned most closely with the agency's stated purpose of collating and presenting information. By now anyone who knew Donovan at all understood that his heart was in the third branch that Rogers was exploring, the Special Operations branch charged with subversion and sabotage as well as morale operations.[16] Its first chief was M. Preston Goodfellow, a New Yorker with the right credentials: war correspondent on the Mexican border in the 1910s, army officer in World War I, editor and publisher of the *Brooklyn Eagle.* Not unlike Frank Knox, he had started his newspaper career in his teens—he was fifteen or sixteen on his first day on the job—and had worked his way up from the bottom.[17] Assessing him "as irresponsible as a blue-bottle fly," Rogers saw Goodfellow as a "promoter character" who meddled in everyone else's business.[18] Goodfellow's executive officer was William H. Vanderbilt III, one of the great-grandsons of the railroad magnate and a recent one-term governor of Rhode Island.

Its original makeup opened OSS up to taunts. One joke was that its initials stood for "Oh So Social." Another was that it was the "Bad Eyes Brigade" because so many of the Ivy League intellectuals in Research and Analysis wore glasses. Perceptions notwithstanding, most members of OSS were not off a social register or from an Ivy League university. OSS seldom hesitated to hire the foreign-born, left-leaning, and eccentric when it saw a need for their services.[19] The great German political philosopher Herbert Marcuse found a wartime home in R&A despite his radical views. The distinguished African American academic Ralph J. Bunche, who would earn a Nobel Peace Prize after the war, also joined the staff of R&A.

Veterans of the European labor movement were welcome on the assumption that they might be able to mobilize old contacts on the continent. Three members of the Hemingway family—Ernest himself, his brother Leicester, and his son John—each had his own relationship with the OSS.[20] Hollywood movie stars and directors—the likes of Sterling Hayden and John Ford—came on board, some because they wanted to appear to be doing their patriotic duty, others because they were willing to take risks and were attracted by the lure of spying. Like the navy and the army, OSS was hiring women in significant numbers, mostly for clerical duty in Washington or New York, but sometimes for duty overseas, where a number of them distinguished themselves.[21]

The OSS workforce would eventually peak around thirteen thousand.[22] Some two-thirds of that total were not direct hires like the first tranche of employees but came from the military. They had already been inducted into one of the services, usually the army, and found their way to OSS in response to official requests for manpower.[23] OSS told the army chief of staff how many officers and men it needed. He would put out calls down the line for volunteers for "overseas duty of a secret and highly hazardous nature . . . that is similar to commando operations."[24] There were few further details; even after they filled out the paperwork, applicants were unable to learn much about OSS or just what it did before they received orders to "Report to O.S.S. Wash." One officer with a PhD in psychology from Yale was happy that the army was finally going to make use of his skills by sending him to what he imagined to be the Office of Scientific Services.[25]

The question in the fall of 1942 was still what, exactly, these employees would do. It was one thing to realize that the US government needed a particular set of capabilities, but it was quite another to puzzle out how best to provide products and services, and especially challenging to create demand for them. The experiences of R&A had not been encouraging.

The branch had been trying to do the right thing by producing *The War This Week*, the government's only secret roundup of strategic news for busy decisionmakers.[26] Most weeks it comprised some thirty well-edited pages. The main focus was, not surprisingly, on the epic battles unfolding around the globe and their significance. But it also covered a mind-expanding range of topics, some of them decidedly arcane: from morale on the home front in Germany to Mahatma Gandhi's relationship with Chiang Kai-shek to "Salt and Smuggling in China." Like a good newspaper, *The War This Week* carried frequent updates on continuing stories.[27]

The breadth and nature of OSS reporting makes one omission all the more striking, especially in retrospect. Despite a growing awareness in Washington from mid-1942 on that Nazi Germany had shifted from the persecution of European Jews to their systematic extermination, the OSS devoted scant resources to the Holocaust.[28] The reasons were varied and complex—from latent anti-Semitism to a sense that the best way to end Nazi crimes was just to win the war. What is fair to say is that the Holocaust was a subject that, sadly, Washington preferred to overlook for much of the war.[29]

Even without any articles on the Holocaust, *The War This Week* elicited angry denunciations from the military for supposed indiscretions. R&A was alleged to be bandying sensitive information around town without adequate safeguards. In January 1943, the Joint Chiefs insisted that Donovan shut it down. He passed the order on to Langer with the wry comment that his professors with their knowledge were like chorus girls with wonderful legs who could not resist showing them off. Langer was left to conclude that the newsletter could not survive because it was just too good.[30] Whether or not he was right, it was clear that there was limited demand for R&A's services in Washington.

In the second half of 1942, the pipe-smoking, even-tempered Rogers tried his best to address the young agency's many problems, including its relationship with the military. His commission as a roving explorer morphed into the vice chairmanship and then the chairmanship of an ad hoc body, the OSS Planning Group.[31] Rogers optimistically recorded in his diary that the appointment meant that "Bill and I now run OSS."[32] Not afraid of long hours and hard work, he embraced the challenge.

The pressures on Rogers soon led his Catholic secretary to regularly "run the beads" (say the rosary) for him.[33] His day would start with the OSS staff meeting around 8:30 in the headquarters building at Twenty-Fifth and E.[34] At around 9:30 he would convene the Planning Group in his own office. Together they would review the latest developments around the world and brainstorm how OSS could best serve the war effort. Much of the work was, Rogers wrote, "trying to harness Bill's mustang plunges into pulling a wagon."[35]

A Planning Group meeting could last up to three hours. By then it would be close to midday, time to break for lunch, which might mean walking down the hill to the exclusive, Republican-dominated F Street Club housed in a nearly one-hundred-year-old mansion at 1925 F Street NW. In the afternoon, there would be informal conferences with Donovan and a steady stream of visitors, not to mention meetings of the Psychological Warfare Panel, which looked at the various ways to undermine the enemy's will to fight.[36] Around 7 p.m., Rogers would call it a day and walk about a mile to the Cosmos Club, then in buildings around Lafayette Square opposite the White House. There he would have dinner on his own or with one of his many distinguished personal and official friends, men like the president of

Harvard, James B. Conant, a chemist who was serving as the chairman of the National Defense Research Committee.

Before long, Rogers concluded that planning was not a core value of OSS—and that the problem started at the top. Donovan's official policy was to support coordination and strategic planning, but his actions were anything but planful. No matter what he said, he was reluctant to yield power to his subordinates; Rogers found him "grasping," snatching authority back after delegating it.[37] He saw how Donovan could not restrain himself when it came to a new idea, embracing it wholeheartedly for a few days, and then moving on to the next. "Bill's impulses, subtleties, and disregard of system" turned OSS into "an endless zoo of curious diversions."[38]

For Rogers, too much of the work of OSS fell into the "curious diversion" category. For example, the Research and Development (R&D) Branch under Stanley P. Lovell—who, according to Rogers, was "a vigorous, salty little Yankee inventor" from Massachusetts—was largely left to his own devices to produce what Rogers called "amusing murder nonsense we will never employ—poisons, bleaching devices" and such, along with some rare explosives and fuses that had more potential.[39] Rogers was not surprised to discover that the Special Operations and the Secret Intelligence branches proved equally averse to establishing goals and procedures. They wanted to deploy to a theater of war and improvise, exploiting whatever opportunities happened to present themselves—an attitude that was anathema not only to Rogers but also to many in the military.[40] Donovan allowed—and even encouraged—this attitude, insisting "on a free hand for his SO and SI."[41]

Working with General Strong

Just as challenging as bringing order to OSS was trying to harmonize with the military. The president might have signed an order placing OSS under the Joint Chiefs of Staff, but, as Rogers grasped almost immediately, that was hardly the end of the matter. He was frustrated that so many army officers thought mostly in terms of kinetic force and overlooked the new dimensions of modern warfare—the contributions that spies, saboteurs, and propagandists could make. This meant that OSS had to get through "a high hedge against . . . irregular warfare" because "the rough-shod boom-boom soldier has no faith or sympathy with total war."[42]

That was nothing compared to the attitude of one officer who actually understood modern warfare, the G-2, Maj. Gen. George V. Strong, who presided over army intelligence in Washington for much of the war. He could be self-important, even imperious—a common joke was that the "V" in his name stood not for his middle name, but for "the Fifth," like the British king and emperor of India George V, who had died in 1936. Strong spoke at a pace so slow that some listeners found it maddening, and even condescending, as if he were explaining the basics to a lesser being. But few, even his enemies, questioned Strong's competence.[43]

After graduation, the West Pointer became a cavalryman, but that designation alone did not define him. In World War I he served in combat and on a staff, returning home with two purple hearts and a Distinguished Service Medal.[44] Between the First and Second World Wars, he obtained a law degree from Northwestern University, worked as a military attaché in Tokyo, and prepared a Japanese-English dictionary of military expressions. From 1937 to 1938, he served as a branch chief for the G-2 in Washington. Considering the glacial pace of peacetime promotions, he rose much faster than his peers. He was relatively forward-thinking and tough-minded; even before Pearl Harbor, at a time when other senior officers hesitated, he was in favor of expanding cooperation with Britain, especially in the field of intelligence. After Pearl Harbor, he was receptive—even supportive—of the army's signals intelligence program. The reformers McCormack and Clarke had happily allied themselves with him.

Wartime photographs suggest an officer who was sure of himself and not afraid of a little affectation. Aging but trim, with thinning hair so short that he looks almost bald, in one photograph the sixty-two-year-old stands not at attention but leaning against a mantelpiece, smoke rising from a lit cigarette fixed in the black holder in his right hand. (The most prominent American at the time, Roosevelt, also smoked his cigarettes through a holder, often set at a jaunty angle in his teeth.) In another photograph he sits on a couch in his formal World War II coat-and-tie uniform with legs crossed while he tunes a shortwave radio. The implicit message: he was technically savvy and in touch with the world.

Rogers quickly learned that Strong simply could not abide Donovan, the amateur interloper whose agency's very existence was, the general claimed, undermining real intelligence professionals in the army and the navy.[45] Seeing him as a vicious, vigorous bully, Rogers came to believe that Strong's behavior merited a court-martial—and would eventually tell General Marshall as much.[46] Marshall simply smiled, commenting that OSS was on a different plane than G-2 or ONI. Each had its mission and purpose; everyone needed to work together. Marshall's wishes made little difference. From 1942 to 1944, Strong actively and passively blocked OSS.

There were two basic points of contention. One remained signals intelligence. OSS did not attempt to revive COI's foray into cryptanalysis, but Donovan continued to claim, not inappropriately, that OSS had a right to receive the best intelligence products possible. "We have," he wrote, "been handicapped by our inability to get adequate disclosure of military and naval intelligence."[47] In October 1942, Donovan complained that he had agreed not to work in the field of cryptanalysis on the assumption that "the proceeds resulting from the decoding by the Armed Forces would be made available."[48] Donovan went on to challenge Strong and his colleagues to own the reason for not putting OSS on the distribution list: namely that they doubted "the loyalty, discretion, or intelligence" of its members. This was true. In 1942, under Strong, G-2 started to set up its own highly secret corps of spies because its leader did not trust OSS.

This was Strong's way of reneging on the army's and navy's offers to leave spying to COI and OSS. The head of the new startup was John Grombach, the army officer whom Donovan had not wanted to trust with COI's secrets. Grombach called his group "the Pond." The G-2 supported the Pond by funding it, initially to the tune of $150,000. The startup would work harmoniously with Roosevelt intimates like Assistant Secretary of State Berle, but not with OSS or ONI. After the war, one of Grombach's supervisors would claim that Roosevelt had approved the initiative, and Grombach himself would claim significant wartime accomplishments.[49]

Preparing for Operation Torch

When not fending off internal and external challenges, Rogers tried to push OSS forward. Progress came slowly. OSS was, he told his co-workers, still "in the primitive stages of undercover work and military support."[50] It was the military's "bargain basement," where a shopper could find an assortment of "remnants and novelties." To move up to the next level, OSS had to embrace the mundane but traditional work of preparing for contingencies and operations.

Rogers and his team looked at specific areas where OSS might deploy and how it should operate if it did. The range was broad, including contingencies on the Iberian Peninsula, Italy, Germany, the Far East, and the Balkans. Some of these plans would be implemented, others would sit on the shelf against the day when they were needed. In the second half of 1942, Rogers spent much of his time supporting America and Britain's first major joint offensive.[51]

Operation Torch was the Allied plan to invade Northwest Africa. The origins of Torch are complicated. French Morocco, Algeria, and Tunisia were controlled by Vichy France, the rump state left after the French collapse in June 1940. Vichy collaborated with Germany to a large extent. In a photograph taken in October 1940, head of state Philippe Pétain, the eighty-four-year-old field marshal who had saved France from Germany on the battlefield in the First World War, is seen shaking hands with Hitler. But Vichy maintained a degree of independence from Germany, especially in France's overseas territories. The Roosevelt administration wanted to encourage such independence. Unlike Great Britain, the United States had diplomatic relations with Vichy, and the American ambassador, Adm. William D. Leahy, had numerous meetings with Pétain that were both civil and productive.

In the same vein, an American diplomat named Robert D. Murphy negotiated what amounted to an aid agreement with Vichy's proconsul in North Africa in February, 1941: the US was to supply his realm with badly needed basic goods like cotton, sugar, and petroleum.[52] Tall enough that he sometimes appeared slightly stooped, as if he wanted to come down to everyone else's level, the forty-seven-year-old Murphy was a charming, energetic Irish American and a Roosevelt favorite. The president told him to deliver the aid, bypassing Department of State channels and reporting directly to the White House. Twelve newly minted vice consuls—mostly successful middle-aged men from the East Coast elite who had served in World War I and stayed on as reserve officers—would help Murphy administer the program. Briefed by G-2 and ONI, the vice consuls would double as amateur intelligence officers.

Donovan sensed an opportunity. During his travels in early 1941, he had come to appreciate the strategic importance of North Africa. As it was taking shape, he grasped that Operation Torch could be a testing ground for his new form of warfare; Rogers knew that "if the political plans go well [as, he believed, they would] . . . it will be a good beginning for a long war."[53] Donovan persuaded Roosevelt "to turn him loose in French Africa . . . with plenty of money to create [a] . . . spy-subversion secret service."[54] Murphy gratefully accepted when Donovan offered his agency's services. He needed a worthy partner who, working behind the scenes, could complement his diplomatic overtures.

William A. Eddy was a warrior-scholar after Donovan's heart. Born in 1896 to American missionaries in Sidon, Lebanon, he was as comfortable in a Bedouin tent in the desert as he was in a drawing room in New York City. Within a month of graduating from Princeton in 1917, the athletic young man found his way to Officer Candidate School at Fort Myer, Virginia, but soon transferred to the Marine Corps. He eventually became the intelligence officer for the Sixth Marine Regiment in France.[55] His interpretation of the job was to personally venture into German lines to gather information and direct artillery fire—which he did from an exposed observation post during the Battle of Belleau Wood. His gallantry came at a cost. Wounded in the leg, then falling victim to what might have been the Spanish flu in September 1918, he came close to losing his life, but survived to return home with a set of medals that rivaled Donovan's: a Navy Cross, a Distinguished Service Cross, two Silver Stars, and two Purple Hearts. The difficult convalescence continued into the summer of 1919 and left him with a painful, frozen hip joint that caused a lifelong limp.[56] Once he was as healthy as he could be, Eddy returned to Princeton to obtain a PhD in English, writing his dissertation on *Gulliver's Travels*. Next he accepted a post as the chairman of the English Department at the American University of Cairo. There, in his spare time, he introduced baseball to the Nile Valley and translated the rules of baseball into Arabic. By 1936, he had returned home and was on his way to becoming the reform president of Hobart and William Smith Colleges in upstate New York. By mid-1941, exhausted by campus politics, he made a memorable declaration: "college presidency is a job with which I am definitely out of love, I [just] want to be a Marine [again]."[57]

With his record and his connections—his first wartime commander was Thomas Holcomb, now commandant of the Marine Corps—Eddy was soon back in uniform despite his physical condition, and en route to Cairo to serve as naval attaché. Donovan learned of Eddy's good work there and appealed to Frank Knox to detail the Marine to his agency.[58] In January 1942, Eddy reported for duty at the American Legation in Tangier to run one of Donovan's first overseas offices. Robert Murphy gave him a warm welcome. "We could," the diplomat would write, "have used a hundred like him."[59]

Tangier sits on the northwest shoulder of Africa at the straits of Gibraltar where the Atlantic and the Mediterranean meet. Though surrounded by Spanish Morocco and occupied by neutral Spain for the duration of the war, the city constituted a demilitarized "international zone" by treaty. It was an open city that belonged to no one and everyone at once, a hotbed of wartime intrigue where enemies, future enemies, and neutrals faced off against the backdrop of the age-old whitewashed houses in the casbah. Eddy saw it as "a compact musical comedy set peopled by as strange a crowd as ever gathered on an opera stage."[60]

Donovan sent Eddy out with a broad mandate to prepare for two contingencies: a possible invasion of North Africa by the Axis in the short term or, more likely, by the Allies in the medium term. In either case, the work was roughly the same: recruit a network of agents and set up clandestine radio stations. The agents were to provide intelligence and prepare to fight the enemy on D-day. Eddy and his men proceeded to spread American propaganda; reconnoiter potential landing areas; establish a resistance network among the French; and recruit local tribes, especially in the region known as the Rif.[61] For that purpose, Eddy relied on adventurous American academics who knew the language and the people.[62]

Eddy did his work in the background while Murphy, "the policy man," focused on French officialdom.[63] Both played important roles in shaping Torch. In late July, wearing his World War I combat ribbons that seldom failed to impress, Eddy flew to London to brief American military leaders on the challenges they faced. The audience included a skeptical George V. Strong, inclined to doubt the soundness of any

OSS operation. Nevertheless, the many details that Eddy coolly presented over dinner at Claridge's, the elegant, old-world hotel favored by Donovan, won Strong over.[64]

While in London, Eddy also met with then lieutenant general Dwight D. Eisenhower, who would command the operation, as well as Eisenhower's deputy, then major general Mark Clark. Clark welcomed the "diversion" from his routine work.[65] In writing, Eddy recommended to Clark that he "be authorized to arrange for the assassination of the members of the German Armistice Commission at Casablanca and for any members of the German or Italian Armistice Commissions who may then be in the city of Oran."[66] He had assigned men to specific targets; they had demolition materials "already in their hands" and were "impatiently awaiting permission" to act. Clark wrote, "O.K. Looks good to me" in the margin, but, as he would write in his memoirs, the proposal was "of course out of bounds" and "squashed at a higher level."

On Labor Day, Donovan managed to get an appointment with Admiral Leahy, who had returned from serving as the US ambassador at Vichy to become chief of staff to the commander in chief, which made him the president's principal military advisor as well as the de facto chairman of the Joint Chiefs of Staff. Donovan told Leahy that Murphy and Eddy were on track to persuade "fourteen poorly equipped French and colonial divisions . . . [to come] over to our side in the event of a successful invasion."[67] Donovan was talking about buying their support—a stratagem that was new to a traditionalist like Leahy. When Donovan told the admiral that purchasing their allegiance would cost no more than $2,000,000, he was taken aback, writing in his diary that Donovan "did not seem to be short of money."

Murphy was grateful when Eddy short-circuited another Donovan proposal to purchase the support of a local ruler for the unheard-of sum of $50,000.[68] By October, Murphy and Eddy had—without paying out large sums of money—found "secret sympathizers in every [French] military headquarters, in various governmental and police establishments, the youth organization, and a tight little group of ardent civilians."[69] The finishing touch was strategic deception. OSS helped to spread rumors that the Allies intended to resupply the besieged British island of Malta, which meshed with a complicated British deception campaign to draw German attention away from the North African beaches.[70]

One concrete example of what the agency could accomplish is the case of René Malavergne, a French patriot and marine pilot who was willing to share his priceless knowledge of the local waters that American forces would have to navigate. Malavergne had already been arrested and tried by the Vichy regime for his politics, and it was not difficult to persuade him to rally to the Americans. Eddy's assistant, Thomas Holcomb's son Frank, smuggled Malavergne by car through Spanish Morocco to Tangier.[71] In order to cross the borders from French Morocco to Spanish Morocco to Tangier, Holcomb confined Malavergne in a crate that he loaded onto a trailer filled with gas cans, carpets, and luggage. The Frenchman was left, immobile, to inhale exhaust fumes. "In this minuscule box," Malavergne would write, "the knees touch the chin, and if I move my feet even a little, I lose my equilibrium and fall over."[72] Holcomb stopped the car more than once to ask in French if Malavergne had inhaled too much carbon monoxide, and each time he was conscious enough to answer and continue the journey.

Joining Battle in North Africa

In early November 1942, Eddy moved across the straits to Gibraltar to sit with Eisenhower's staff in an airless dungeon burrowed deep into the rock on the tiny British base; the supreme commander insisted on close coordination between conventional and unconventional operations. The landings started on November 8, and succeeded in putting some 110,000 men ashore in French Morocco and Algeria, but not without confusion and bloodshed. In many places, Vichy French forces chose to fight, and did not hold back against their former World War I allies until local cease-fires began to take hold on November 10. Many of the OSS secret operations failed due to poor coordination and "French defections"—one key actor turned traitor; others simply did not report for duty to seize or sabotage installations.[73] On another level, the young agency did better. OSS intelligence about French forces was accurate, as was the plethora of information on natural features and man-made installations.[74] Some OSS affiliates literally helped guide the ships to shore, while others met troops as they waded through the surf. Malavergne covered himself with glory, piloting the antique four-stack destroyer USS *Dallas* twelve miles up the Sebou River under artillery, machine gun, and sniper fire through a channel littered with shipwrecks to a strategically located airfield. For this he would receive the Navy Cross, becoming one of the few foreign civilians ever to receive the navy's second highest award.[75]

The political dimension remained messy. Uncertain whether to remain true to Vichy or accommodate the invaders, senior French officers dithered, quarreling among themselves and with the American envoys who wanted a commitment to the Allies. The most senior of those officers was Adm. François Darlan, one of the pillars of the Vichy regime. Much to every American official's surprise, on November 8 he happened to be in Algiers to visit his son, who was gravely ill with polio. Described by one unsympathetic officer as "short, bald-headed, pink-faced, needle-nosed, [and] sharp-chinned," Darlan had collaborated with Germany over the past two years, going so far as to offer military support to Hitler in his war with Britain.[76] But Murphy found him approachable and even likable.[77] After tacking this way and that, Darlan more or less agreed to switch sides and bring at least part of the French military with him. In return the Allies would recognize him as the de facto head of state for French North Africa. On November 13, Eisenhower flew over from Gibraltar, met briefly with Darlan, and blessed the agreement. This sparked a firestorm of criticism in Britain and the United States. Joining forces with a man who had turned over political refugees and French hostages to the Germans compromised the moral tone of the western alliance; Ed Murrow was not alone in the outspoken denunciation that he broadcast to millions of listeners over CBS.[78]

From Navy Hill in Washington, Donovan had been following the progress of the operation. The first day of the invasion, November 8, fell on a Sunday, and many OSS seniors spent the weekend at work; one account has Donovan sleeping fitfully on an iron cot in his office.[79] They were waiting anxiously to see if their contribution to Torch would prove the worth of the young agency. Eddy's initial cables were upbeat, but then the tone shifted, reflecting the complicated situation on the ground in North Africa: the pathfinding that worked, the plots that failed. Donovan read the reports as they came in.[80]

Taken by surprise, OSS played no more than a supporting role in the deal that confirmed Darlan in power and made both Rogers and Donovan uneasy. In early December, Donovan drafted a memorandum in his own handwriting that lamented "the identification of Darlan with our operations" and went on to declare that "we have before us the very practical problem of eliminating [his] . . . political leadership."[81] He went on to suggest a convoluted solution whereby the United States would "stimulate the setting up in Occupied France of a national committee" to siphon off Darlan's legitimacy. Two weeks later, Rogers expressed his concern. "Darlan has bought our souls—a dreadful error," he wrote, and, like Donovan, wondered if OSS could somehow remedy the mistake.[82]

Later in the month, Donovan received official recognition for his agency's performance during Torch and could breathe a sigh of relief. He knew OSS had filled a niche. In the words of the OSS War Report, which sounded like one of Donovan's memoranda from 1940 and 1941, "the techniques developed during Torch for informing the invasion commanders of last[-]minute conditions . . . represented a new kind of efficiency in warfare."[83] It was a hybrid between tactical intelligence—trying to learn the enemy's plans and capabilities, usually through reconnaissance—and classical espionage—recruiting spies to steal secrets. Donovan would have known, too, that his R&A Branch had catalogued and disseminated reams of information for its military customers, operating much like a good reference service, and that Eddy's men had appeared at the beachhead, helping to get the troops ashore.[84]

On the twenty-third, General Marshall took the time to send Christmas greetings that came perilously close to an emotional gush. Marshall thanked Donovan for his personal support "in the trying times of the past year."[85] He went on to praise OSS for rendering "invaluable service, particularly with reference to the North African Campaign." Marshall even added a conciliatory note, regretting that OSS "has not had smoother sailing" after coming under the Joint Chiefs. A day later, on Christmas Eve, Donovan answered with his own expression of goodwill. He acknowledged Marshall's "very cordial note" and praised him for supporting the military's "revolutionary and courageous" arrangement with OSS.[86]

Donovan did not have long to bask in the warm glow of Marshall's message. On the same day that he drafted his note to Marshall, he learned that Darlan had been assassinated. On Christmas Eve, a twenty-year-old French Royalist named Fernand Bonnier de la Chapelle had ended the debate over Darlan with a handgun outside the admiral's office in Algiers. His departure from the scene, welcome though it was in some circles, left many unresolved issues. Bonnier de la Chapelle did not act on behalf of the American government.[87] But the circumstances remained murky. Had someone else manipulated the impressionable assassin?[88] Summarily tried before he could say much, the hapless youth was tied to a stake and shot by a French firing squad at 7:45 a.m. on the day after Christmas.

From the field, Eddy kept Donovan abreast of developments in the crisis, including an unexpected twist almost guaranteed to make a spy manager draw deep, anxious breaths. It turned out that an OSS man, the former journalist Edmond Taylor, had a relationship with leading Royalists. One day in December over a pleasant lunch of couscous and strong Algerian wine, the topic of conversation turned to Darlan and how to replace him "with something less objectionable to the democratic conscience and the ideals of the Resistance."[89] Taylor spoke vaguely about US policy goals. When the cleric Abbé Pierre-Marie Cordier passionately burst out that he could kill "that man" with his own hands, the conversation instantly stopped, as if Cordier had betrayed a dark secret.[90] Taylor reported his suspicions to Murphy, who had also cultivated the Royalists. He did not act on this report.

It also turned out that Bonnier de la Chapelle had recently received paramilitary training in communications, explosives, and firearms at an SOE camp some twenty miles outside Algiers. The camp's ostensible goal was to prepare members of the French Resistance to deploy to Southern Europe. One of his trainers at the camp was an OSS man, Carleton Coon, a forty-two-year-old Arabic-speaking anthropologist from Harvard with a weight problem. Coon himself had only minimal military training and, somewhat mysteriously, wore British battle dress but drove a late-model Studebaker with official American license plates.[91] Professor Coon happened to be driving his sedan through Algiers soon after the assassination and found himself caught in the traffic jam of first responders.

Empowered by the murder of Darlan, the Vichy police proceeded to arrest and interrogate members of the Resistance, including the French instructors who had worked with Coon at the camp. Eddy decided that, even though Coon was innocent of any wrongdoing, it would be best to make him scarce before anyone started in on him. On January 1, the two men drove east, away from Algiers and toward the battlefront, where Coon would run a project to confound the enemy with land mines that looked like mule turds.

Four thousand miles away on the same day, Rogers noted grimly that "the Africa business, since the assassination of Darlan, turns out to be a maelstrom of murder, intrigue, and jealousy."[92] OSS was caught up in what Taylor called the "extreme moral ambiguity" of the affair.[93] First Rogers and Donovan had worried that the United States had sold its soul to Darlan, then Donovan had considered a far-fetched political solution. When a third party imposed a violent solution, Rogers was not any happier. OSS was learning that there was only so much that secret operations could accomplish. Not only were the desired effects hard to achieve, but the unintended consequences were impossible to predict.

General Strong's Last Offensive

The same was true of the turf wars in Washington that, amazingly, started up again in February 1943. Donovan could be forgiven for thinking that Marshall's kind words at Christmas two months earlier signaled the end of the threats to his agency's existence. But the G-2 General Strong was not yet ready to concede defeat. By chance, he was almost able to deal OSS a mortal blow by taking advantage of FDR's ambivalence toward the young agency and leveraging a dispute between OSS and the Office of War Information (OWI), which had absorbed many of Donovan's original employees.[94] The director of OWI, a well-known journalist named Elmer Davis, had been protesting supposed incursions on his turf by Donovan. On February 18, Davis presented his case to the president over lunch at the White House. In search of reinforcements, Davis asked Roosevelt to summon Strong, who hurried over from the Pentagon.[95]

The three met in the Oval Office for a little more than half an hour. By the end of the meeting, Roosevelt had agreed with Davis and Strong that OWI, not OSS, should run propaganda for the government—and that the War Department should run OSS. Even though he had placed OSS under the Joint Chiefs within the past year, Roosevelt told Strong to draft an order for his signature that would shift OSS to the department where Strong held sway over intelligence matters.

Rogers and Donovan soon learned that the president was preparing to consign their agency to oblivion. Under the Joint Chiefs, it had lost some independence, but it had also gained some support. Coming under the army would have been a big step down. Especially galling to them was the implication that the only part of OSS worth saving was R&A, the analysts—not the spies or even the director's favored guerrillas.[96]

The two OSS leaders decided that they must resign if Roosevelt went ahead with the plan. But first, Donovan would write a letter of protest and, as a last resort, try to see the president in person, a far more difficult proposition now that the war had intensified and OSS was officially a dependency of JCS. Dated February 23, the letter was polite, almost beseeching: the charges that OSS had invaded the province of the OWI were false; transferring the OSS to the War Department would "disrupt our usefulness" and be "a valuable gift to the enemy."[97] As the president had acknowledged early on, Donovan continued, "this work could not live if it were buried in the machinery of a great department." And preserving OSS would have benefits on the battlefield: "Our connection with underground channels will . . . count heavily when invasions are ready."

Almost at the last minute, OSS was rescued by officers on the JCS staff who stepped in to save what they now viewed as "their" agency. Marshall waffled but finally came out against Strong's draft. Admiral Leahy worked the issue until the solution was acceptable to OSS. Figuratively changing horses in midstream, the president now professed to hate the idea of putting OSS directly under the army. In the end, Executive Order 9312 preserved OSS as a supporting agency of the JCS.[98]

13

TRAVELING THE WORLD

Seabiscuit and China-Burma-India

Operation Torch demonstrated what OSS could accomplish in the murky new world of strategic services. Eddy and his men helped create the conditions that enabled the Allied invasion of French North Africa to succeed. He did not meet all of his goals; many of his intrigues did not pan out. But thanks largely to him, the military knew what to expect—on the beach and after it moved inland.[1]

On the other side of the world, another remarkable man was making a name for OSS. This was Carl Eifler, who first met Donovan early in 1942 when the COI had been looking for ways to get his agency into the war as quickly as possible.[2] Standing at well over six feet and usually weighing in at about 250 pounds, as intelligent and forceful—sometimes brutal—as he was big, he had come recommended by Lt. Gen. Joseph Stilwell, the China-Burma-India (CBI) commander. The thirty-five-year-old army reservist had assembled some twenty officers and men, and led them through a short course on guerrilla warfare. Preston Goodfellow, the Special Operations branch chief, had supervised their preparations to deploy to China and, in May 1942, arranged a remarkable set of orders for them to proceed by the most expeditious means from Washington, DC, to Stilwell's headquarters in Chungking, China.[3] Carrying hundreds of pounds of military gear—some of it lethal—as carry-on and checked baggage, Eifler had set off for the other side of the world. By June 12, he found himself in front of Stilwell.

The general was surprised to see Eifler and wanted to know why he had come halfway around the world.[4] Stilwell might not have remembered that he had met with Donovan in January 1942 and agreed to accept an OSS team.[5] He had a great deal on his mind: he was running a newly established theater-level command and had not focused long on this, a lesser priority. But now he said he had no need for American "agents" like Eifler in China. It was not clear why; Eifler thought he detected general reservations about guerrilla warfare.[6] But Stilwell had not forgotten his prewar impressions of the reservist who would not give up, and was willing to give him a chance to make himself useful in a part of CBI where he could do little harm: Burma, the former British colony now occupied by the Japanese. Stilwell authorized Eifler to establish a base in northeast India and from there, run operations against the Japanese.[7] He reportedly told Eifler that he did not want to hear from him again until the Detachment 101 had started making "booms" in the Burmese jungle.[8] On a long leash from Stilwell, in touch with OSS mostly for administrative support, Eifler would be largely on his own.

Eifler was soon at work creating a viable guerrilla force virtually from scratch, the first for OSS. Starting with his small band of irregulars, the men who only a few months earlier had been army infantrymen, many of them reservists like Eifler on extended active duty, he established a camp on a tea plantation in Nazira in the state of Assam, India. With a little help from the British, Detachment 101 proceeded to recruit and train native forces—largely Kachin tribesmen—to operate in Burma. The camp grew quickly, covering some twenty-five square miles, and was soon split into subcamps.

The Detachment's broad mandate from Stilwell would allow it to run the gamut from conducting hit-and-run raids out of India to waging war from small bases in parts of Burma where there were few Japanese troops. The Det would send tribesmen out to observe the enemy, rescue downed Allied pilots, and harass the occupiers with small acts of sabotage or simple ambushes. One rationale for the harassment was to provoke the enemy into retaliating against the local population, which in turn would discourage collaboration.[9]

In December 1942, Eifler himself led the first foray into the remote areas of Burma to establish a base. Another base followed in February 1943. As his operations unfolded, Eifler was on a swift, upward trajectory. A captain when COI had come looking for him in early 1942, he was a lieutenant colonel now, a little more than a year later. His luck changed on the night of March 7–8, 1943, on a stretch of enemy-held coastline near Sandoway, Burma. Eifler and a team of Anglo-Burmese commandos sailed there from India on Royal Indian Navy motor launches, the rough equivalent of American PT boats. The plan was to insert the raiders, who would paddle ashore in rubber boats, make camp in the jungle, then obstruct Japanese military traffic along an important supply line.[10] But they encountered rough seas—waves as high as fifteen feet churning among offshore rocks, and the team's scout balked. Eifler jumped in the water and took the lead, manhandling four boatloads of men and about one thousand pounds of equipment onto the beach in the dark. He stayed to help with the unloading, forming a human chain with the commandos to pass food, guns, and ammunition from water's edge to the shelter of the jungle. He then shook hands with each man and started back out to sea with the boats. The traces of the landing needed to be gone by dawn.

Two of the boats slipped from Eifler's grasp in the rough water. Still among the rocks that lay between the beach and the open sea, he jumped into the water to corral them. He regained control of the boats but fell against the rocks. The waves battered his head until he was dazed and bleeding. Still, he somehow pulled his load out to sea and found the motor launches by the sound of their anchor chains. (Having given up hope, the crews were weighing anchor, preparing to sail home without him.) It was an amazing feat of strength, endurance, and determination, one with which Stilwell would reward him a Legion of Merit even though the mission was a dismal failure.[11] Within a few days, all of the commandos were either killed in action or captured, tortured, and beheaded by the enemy, apparently betrayed by a local fisherman who found a small but telltale piece of gear on the beach.

This was a turning point for Eifler. Though he reported to Donovan that his wound was "nothing serious and nothing that time will not cure," he did not bounce back.[12] Headaches and blackouts plagued him; he was short-tempered with trusted subordinates; he dismissed the advice of the Detachment's redoubtable doctor, Lt. Cmdr. James C. Luce, USN, who had been wounded at Pearl Harbor on December 7 and was now practicing medicine in the jungle, even leading troops in combat. Eifler treated himself instead with bourbon and morphine, which combined to worsen his condition. In July, this was clear to an unwelcome visitor from Washington on a fact-finding mission.[13]

Even under the best of circumstances, Eifler and Duncan C. Lee were unlikely to find common ground. Though the distant relative of Robert E. Lee wore the uniform of a captain in the US Army, there was nothing military about the youthful-looking lawyer with a "cellophane" commission (so-called because you could see through it) who Donovan had imported from Wall Street to help manage his office. He was an intellectual, a former Rhodes Scholar, who made his way through life with ideas and words, not with the kind of deeds that Eifler admired. Lee appears to have been a guest of the Detachment for the better part of a month, more than enough time to observe Eifler at his worst and draw conclusions about his fitness for command. One account has Eifler angrily rebuffing Lee's attempts to question him, shouting that a mere captain had no right to do so.[14] Another has Lee focusing on the quality of the Detachment's intelligence reports about the Japanese in Burma, and finding them wanting. Eifler had come to Burma to run special operations. Not surprisingly, his men lacked the ability to do anything more than rudimentary intelligence work.[15]

Perhaps on account of Lee's visit, Eifler yielded to the pressure to check into the US Army's Twentieth General Hospital at Ledo, India, in August 1943. Even then, he insisted on bringing his radio and radio operator with him so that he could remain in command. There were some decisions he would not delegate, like the authority to grant "permission to assassinate a certain native official who was being uncooperative."[16] Traitors, real or perceived, were beheaded; recalcitrant villages were bombed.

Eifler's brutal tactics soon earned him a written rebuke from Orde Wingate, the British pioneer of irregular warfare who also operated in Burma:[17]

> I remember your mentioning the execution of a traitor. . . . There is a danger in bumping off without proof, apart from the fact that it is, of course, a crime. From the military point of view it tends to create distrust. . . . I would rather risk loss of men through treachery, and enjoy cooperation, than isolate myself . . . by action designed to frighten [other] possible traitors.

By the fall of 1943, it was clear that the hospital stay had done little good. Eifler's condition did not improve. Even so he had no intention of abandoning his mission, especially after being promoted to full colonel.[18] When the message arrived, his men staged an impromptu ceremony to pin the eagles on their commander. For the occasion, he managed a smile through the less-than-military beard that he was growing.

Duncan Lee's report about Detachment 101's early failure to produce good intelligence could not have been timelier. When he returned to work in October, Donovan's office suite was under siege from within. The director had fought off the attacks from his enemies outside OSS only to find that the loyal opposition inside OSS had prepared a manifesto: the agency needed to change; it was trying to do too much and accomplishing too little. Senior officers headed by the intense, thoughtful deputy director, Brig. Gen. John L. Magruder, wanted to limit the founder's day-to-day "muddling" and get on with the war.[19] In a note to Donovan, the career officer was brutally blunt:[20]

> I shall not waste your time belaboring the point that our present intelligence organization is indefensible. The set-up has been incredibly wasteful in manpower and, except for a few spotty accomplishments, has been a national failure. It has created frustration in Washington and chaos in the field.

The suggested remedy was for a chief of operations (perhaps Magruder himself, who was an orderly artilleryman with some prior experience as an intelligence officer) to run OSS while Donovan became something like the chairman of the board. The agency would "confine its energies to intelligence, support of resistance groups, and high echelon subversion," shedding field operations "such as sabotage, guerrilla warfare, rumors, clandestine radio and so on."[21] This meant that OSS would devote more energy to the kind of work that Eddy and Murphy had done in North Africa than to what Eifler was doing in Asia; it would become less of a paramilitary force and more of an intelligence agency, collecting and analyzing information.

Jim Rogers was not part of the cabal but he had been thinking along the same lines: Donovan had been neglecting intelligence, preferring to build up a special operations "private army."[22] It turned out that Stanley Lovell, the head of R&D, had also lost confidence in Donovan; he agreed that OSS lacked direction, its leader was "just a politician, society mad."[23] By October 18, Donovan had angrily rejected the Magruder plan, ridiculing its proposals.[24] He followed up with a blistering memorandum, arguing that OSS was not like a large company or government organization that needed set procedures; it had been created to wage a new kind of warfare. It must be flexible, able to adapt to the unexpected.[25] Unconventional warfare was the wave of the future, even if no one knew exactly what shape it would take.

The director might have added that he needed the freedom to run the agency out of his hip pocket while traveling the globe, injecting himself into OSS operations overseas and, when possible, sightseeing on the battlefield. In July, he had gone ashore in Sicily with the second wave of the Allied invasion of the island, simply to see what he could see, fire off a clip of ammunition at the enemy, and visit with his old friend George Patton.[26] He was gone for about a month. In September he returned to Italy, leaving what Rogers called "a trail of disorder in his wake," but arriving in good time to join the Allied forces landing at Salerno and once again exposing himself needlessly.[27] Now, on November 8, after dictating his response to Magruder, Donovan suddenly returned to his life on the road.[28]

The director said he would be gone for another month; it would actually be three months before he returned to Washington. His travel earned him the nickname "Seabiscuit," bestowed on him by his staff for racing around the globe like the Depression-era champion thoroughbred.[29] Rogers was now close to the end of his tether, and quietly made plans to quit. He credited Donovan with teaching Washington "the elements of modern warfare," something no one else had done.[30] Rogers had never known "a more imaginative, magnetic man."[31] But the range of his magnetic field was short, and its effects quickly dissipated. Rogers expected to be gone before Donovan returned.

Even if he knew of Rogers's frustration, Donovan had no time to focus on it. His itinerary in November included a dizzying series of stops. In Cairo he offered briefings and maps to participants—reportedly including the president—before the Sextant Conference among Roosevelt, Chiang Kai-shek, and Churchill. From Cairo he caught a flight to Delhi, where he arrived in his "carelessly worn, slightly rumpled khaki uniform" with only one decoration: the blue ribbon of the Medal of Honor.[32] There he lobbied the elusive but charming Lord Louis Mountbatten, now supreme Allied commander in Southeast Asia, to make more use of OSS paramilitary detachments. In China he pressed a wary Stilwell to do the same, and in an unseemly show of temper, browbeat the Nationalist Chinese chief of intelligence, Dai Li, for defending his turf. Donovan then flew back to India, this time to Assam, to spend time with Eifler.[33] He arrived at the Allied air base at Chabua on December 7, 1943, two years to the day after the attack on Pearl Harbor.

A private pilot with limited recent experience, Eifler flew up to Chabua to meet Donovan in a de Havilland Gipsy Moth, a one-engine British biplane made of canvas and plywood in the 1920s that he had bought for the Detachment. He loaded the director into the two-seater to fly him back to Nazira. A testy exchange that evening left Eifler with little doubt that Donovan was following up on Lee's report, probing his fitness for command. On a kind of dare, Eifler offered to show Donovan just how much the Detachment was accomplishing. Despite his recent head injury, he would fly the director of OSS behind Japanese lines to a guerrilla base hidden in "an incoherent jumble" of steep mountains and jungle valleys.[34] Donovan accepted, cavalierly reassuring those who told him this was not a good idea. War was about risks; if the plane went down, he would ride it down; if he was still alive after the crash, he would bite into his suicide pill before the Japanese could capture him. The next morning, when Donovan and Eifler showed up at the airstrip, film director John Ford was on hand to photograph the two standing together just before takeoff, looking to one bystander like "a couple of well-fed pilots out of [the 1927 World War I movie] *Wings*."[35]

Though operating close to its weight limits, the Gipsy Moth performed well on the outbound trip of about 275 miles, including

150 miles over enemy territory, which Eifler flew mostly at tree level to avoid any Japanese fighters that might be in the area.[36] Donovan was enchanted by what he saw at the camp, codenamed "Knothead" after its commander, Vincent L. Curl, a tough, long-serving soldier who had followed Eifler out to Asia and was now an officer.[37] Sporting a luxuriant beard and a sarong, cheerfully ignoring a painful knee injury, Curl had been working behind the lines since April 1943.[38] He introduced Donovan to the roughly three hundred Kachin tribesmen he commanded. Not unlike the Sherpas who shared their Himalayan ancestry, they were out of a would-be warlord's dream. Direct, mostly trustworthy, and friendly to Americans, they were at home in the jungle, quick to learn the use of modern weapons as well as the basics of guerrilla warfare. They hated the Japanese, hitting hard before melting back into the bush. Even their souls were on the right track; a Catholic missionary, Father James Stuart, had lived among them for years and was happy to translate for the visitors.

The afternoon takeoff from the short airstrip at Knothead nearly marred the perfect day. Surrounded by jungle, the runway was too short for the small, overloaded plane. By a mixture of luck and skill, Eifler managed to fly through a gap in the trees and avert disaster. As the plane struggled for altitude only a few feet above the surging brown water of the river that ran by Knothead, Donovan's only reaction was an even-tempered comment about the benefits of more powerful engines. This was not a reproach from a man who enjoyed living on the edge.

After they had landed safely at Nazira, Eifler's executive officer, the West Point graduate Lt. Col. John G. Coughlin, took the director to task for his recklessness. When he had heard that Donovan and Eifler had gone to Knothead in the Gipsy Moth, he was sure that he would never see either of them again. The risk was not in any way worth the gain, especially considering what Donovan knew about Allied intelligence. The older man shrugged the lecture off: going to Knothead was something he had had to do.[39]

The fulfilling adventure did not change Donovan's mind about Eifler. The colonel had done groundbreaking work, but his behavior was now too erratic. The creator would have to abandon his creation.[40] Donovan allowed Eifler to save face, telling Stilwell that he was going back to the United States for refresher training.[41] The reality was that another officer, then lieutenant colonel William R. Peers, would take command on December 17, 1943.

Eifler made it to Washington by the end of the year and did what he could to support the Detachment from OSS headquarters while awaiting a new assignment.[42] But first he went in search of Duncan Lee, who was not hard to find in the cluster of offices around Room 109, Donovan's office suite. The big man, who had almost one hundred pounds on Lee, grabbed the bespectacled lawyer by his lapels and pushed him against the wall. Holding him a few inches off the ground, Eifler threatened to kill the "little son of a bitch" if he ever interfered in Eifler's "activities" again.[43]

Eifler would always believe that it was Lee who was responsible for Donovan's decision to remove him from the command of 101, thereby ending the adventure of a lifetime. It was of course Eifler himself who was most responsible for his own fate, for better and for worse. Without much direction or support from anyone, he had created the Detachment ex nihilo largely by the force of his personality and his incredible energy. Impulsive, irrepressible, and irresponsible, he took big risks and made big mistakes. By his own admission, he may have committed or tolerated war crimes.[44] At the same time, he was productive and cost-effective. The guerrilla army he created could do only so much to win the war but it proved that it could perform its core missions at small cost, having rescued downed Allied pilots by the score, as well as providing bits of useful information and harassing the enemy.[45]

Detachment 101 would be fine-tuned by the more orderly Peers, a career army man who would lead it to a string of successes in the last two years of the war. One veteran explained the difference between

Eifler and Peers: if ordered to destroy a building, both would accomplish the mission, Eifler by commandeering a bulldozer and leveling the place *now*, Peers by finding the structure's weak points and using that knowledge to bring it down, then neatly stacking the bricks.[46]

Peers would grow the Detachment to a strength of almost one thousand OSS members; they in turn would train and guide some ten thousand indigenous fighters. The Detachment continued to operate in Burma without much support from Washington; headquarters kept its focus on what was for OSS the main event, the war in Europe. But the CBI commanders, Stilwell and his replacement, Gen. Daniel I. Sultan, came to appreciate OSS for its support of the offensive to retake Burma in the spring of 1944.[47] Its guerrillas secured the flanks and the rear of advancing troops; they guided others through the jungle they knew so well; they occasionally fought as small conventional units. To anyone who cared to look their way, Detachment 101 demonstrated what irregular forces could accomplish: just so much on their own but a good deal more in concert with regular forces.[48]

14

THE OSS, THE NKVD, AND THE FBI

Christmas in Moscow

The next stop on Donovan's world tour was Moscow. He flew from the China-Burma-India Theater to Tehran. In the Iranian capital, two Red Air Force men armed with revolvers, a navigator and a radio operator, boarded his two-engine US Army transport plane for the flight to Moscow. Their somewhat sinister presence, along with the added weight, was not welcome. It would be another low-level flight, this time in the depths of winter. Even though the Soviets had pushed the Germans back from the gates of Moscow to the general vicinity of Smolensk, more than two hundred miles west, the country was still very much at war. But it was another risk that Donovan was willing to take, one that ended safely at a military airport outside Moscow on December 23, 1943. Awaiting Donovan's arrival on the tarmac stood Maj. Gen. John R. Deane, US Army, chief of the US Military Mission to the Soviet Union. Donovan made himself welcome by presenting Deane with a case of Scotch "to brighten" his Christmas.[1]

Donovan was already on good terms with the handsome, clean-cut South Carolinian. As the secretary of the Joint Chiefs of Staff, Deane had gone on record with his support, telling General Marshall in October 1942 that he was "sold on OSS . . . [as] they have rendered and are capable of rendering extremely valuable services."[2] In October 1943, the Pentagon and the State Department sent Deane to Moscow to improve relations with the Red Army. In theory, the United States and the USSR were allies, but in practice there was little coordination or cooperation between them, apart from the massive shipments of armaments from the US. As the western Allies contemplated the invasion of France in 1944, the need for communication channels to the Soviets was plain. That was one of the purposes of the Tehran Conference in November 1943, when Churchill, Roosevelt, and Stalin aired their military concerns in an atmosphere of goodwill. Both Deane and the American ambassador in Moscow, W. Averell Harriman, wanted to build on that foundation. OSS was part of the process.

For his part, Donovan had long wanted to operate in the Soviet Union, which was still a great unknown for American intelligence. Unlike the British, OSS did not have any contacts—let alone agents—in the country. Since January 1943, Donovan had been exploring the options: perhaps sending an OSS officer to serve in the Soviet Union under some form of cover, or enlisting the employees of one of the American companies already working there to do odd jobs.[3] Another possibility was assigning an embassy officer to the OSS account to work jointly with the British SOE representative in Moscow, Col. E. H. Hill. Hill even suggested the Russian-speaking American diplomat Llewelyn E. Thompson for the job. The short-handed State Department frowned on these options but favored the idea of placing an OSS officer in the embassy. The Pentagon concurred. On November 14, General Marshall sent Deane a cable announcing that the president had approved a plan for OSS to open an office in Moscow and offered the Soviets the same opportunity in Washington.[4] A week later, Deane was on his way to Cairo for the summit conference among Roosevelt, Churchill, and Chiang Kai-shek, which proved to be a convenient opportunity for Donovan to pitch Deane and Harriman on his forward-leaning proposal that "he himself come to Moscow to explore the possibilities of an OSS Mission."[5]

The Soviets were remarkably well prepared for Donovan's visit. They seemed to have already formed a positive impression of the OSS director.[6] Deane was amazed at the unwonted speed with which the Soviet bureaucracy moved after his arrival. On Christmas Day 1943 (which was not a holiday for a good communist), Ambassador Harriman introduced Donovan to Foreign Minister Vyacheslav Molotov, the "rectangular block of a man" with the nickname of "Mr. Nyet" who had negotiated the cynical Hitler-Stalin Pact that had set the stage for Nazi aggression in 1939 and secretly partitioned Poland between Germany and the Soviet Union.[7] On this day, however, Molotov was in a positive frame of mind, telling his visitors that the Soviet leadership was already "prepared for prompt, serious discussions" about opening a channel between the OSS and the NKVD.[8]

Two days later, at 5 p.m., Donovan, Deane, and their translator found themselves at the entrance to the Lubyanka, the ornate yellow brick headquarters of the NKVD in downtown Moscow, originally built for an insurance company in Tsarist Russia but put to a very different use by the country's current rulers. It was now both an office building and a prison where enemies of the state—including the famous British spy Sidney Reilly—had been interrogated, tortured, and often executed. Guards escorted Donovan and Deane through the eerily quiet corridors to a conference room where they met two senior NKVD officers: Lt. Gen. Pavel M. Fitin, introduced as head of the Soviet External Intelligence Service; and a subordinate, introduced as Maj. Gen. Alexander P. Ossipov, said to be responsible for "conducting subversive activities in enemy countries."[9] Fitin seemed young for his rank, about forty, with blond hair and blue eyes. Ossipov was short and less attractive, with a full head of brown wavy hair and "a sallow complexion."[10] Deane found Ossipov somewhat sinister, picturing him in his mind's eye as "the boon companion of Boris Karloff," the stage name of the actor who played Frankenstein in the

1930s. But the Soviet spoke American English well, and, much like Fitin, went out of his way to be pleasant and forthcoming.

Donovan chose a seat across from Fitin and Ossipov where a light would shine directly in his face and, with heavy-handed humor, told the secret policemen he was ready for the third degree. They did not take the bait, but they did accept Donovan's offer to describe the "aims, scope, and operation of the Office of Strategic Services."[11] Fitin was, he said, "interested in all aspects" of the subject, and "listened with the closest attention." Donovan's main point was that he had come to Moscow to invite cooperation, as much or as little as the Soviet government wanted, anywhere in the world where OSS maintained a representative. Fitin asked for specifics. Donovan was willing to exchange intelligence about the enemy and coordinate plans for operations against them. Fitin agreed that might be useful, then struck a discordant note, asking Donovan if he had any ulterior motives. Did he have any other intentions apart from sharing information? Deane snickered at this bit of Soviet suspiciousness, something that he had become accustomed to, but Donovan did not join in, replying instead "in a cold indignant manner" that he had no such intention.[12] Fitin quickly reverted to the role of gracious host, heartily welcoming Donovan's proposals. Officials on both sides would have to work out the details. In the meantime, Donovan could prepare to send his representative to Moscow. The meeting ended on that cordial note at about 6:30 p.m.

Donovan was confident that he had more than accomplished his mission; this was the breakthrough he had hoped for, one that would redound to his agency's credit.[13] He was ready to resume his travels. Ambassador Harriman offered his personal plane, a converted B-24 bomber that had been christened "Becky" by its crew. With four engines, Becky could fly to Tehran in about six hours, much faster than the two-engine transports, which would have to make intermediate stops. But the Soviets balked, refusing to share the weather reports needed to plan a flight. Donovan decided to break the impasse, and at 11 p.m. one night drove out to the airport with Charles E. Bohlen, a Russian-speaking American diplomat. Neither the sentry nor the officer on duty knew what to make of the very demanding American general who insisted on seeing the weather map, but eventually they relented and gave him what he wanted. The map promised clear weather; they could fly out the next day. On the way back to Moscow, Donovan gloated that he knew how to deal with Slavs, as he had known many Slavic Americans in Buffalo, criticizing "you State Department people" for not being tough enough on the Soviets.[14] Bohlen, who would become an ambassador in his own right, held his tongue both that night and the next morning when, packed and ready, Donovan and his party found that the Soviets had moved Becky to another airport and refused to say when she could fly. The result was that, every morning after Donovan's intervention, Harriman, Bohlen, and Donovan would all get up around 6 a.m., drive to the airport, and wait to see if Becky would be allowed to take off, only to have the tower deny clearance, citing weather.

Though comfortably housed at Spaso House, the 1913 mansion that served as both embassy and residence, Donovan was stranded at the mercy of the Soviets. On the night of January 4–5, Ossipov telephoned Donovan at 2 a.m. with positive news: the "directive echelon" of the Soviet government (possibly the State Committee on Defense run by Stalin) had approved the plans for official contact with OSS. Donovan insisted that Ossipov immediately come to the embassy to talk. Amazingly, he agreed. In the small hours of the morning, Donovan detailed his vision for forcing Bulgaria to pull out of the war. He told Ossipov that the US was displacing Britain in Yugoslavia and Greece, which he cited as part of the reason for the American interest in working more closely with the Soviets.

Donovan's sally into foreign policy did no apparent harm. A few hours later, he once again failed to get permission to fly out, and spent another day, the fifth of January, at Spaso House. That night, Harriman invited Fitin and Ossipov to dinner with Donovan, Deane, and Bohlen for what was probably the first social gathering of American and Soviet spies. After dinner, Harriman invited his guests to watch the recent hit musical *Yankee Doodle Dandy,* about the life of the late entertainer George M. Cohan, the man said to have owned Broadway. The Soviet apparatchiks' impressions of the upbeat story in which Cohan supposedly comes out of retirement to play Roosevelt and tap-dances down the White House steps are not recorded. They stayed until 12:30 a.m. before returning to Soviet reality at the Lubyanka. Only after January 6, once Harriman had appealed twice to Molotov, was Donovan finally able to fly out to Tehran.[15]

Soviet Spies in America

Deane was right. The Soviets had been well-informed about Donovan well before his arrival in Moscow. What neither Deane nor Donovan knew was how; it was thanks to the spies who reported to Fitin and the man who called himself Ossipov. He was actually Gaik Ovakimyan, arguably his directorate's leading expert on North America. As the former New York NKVD *rezident,* or station chief, he was known to the FBI. In 1941, the Bureau had ended his eight-year tour by arresting him for violating the Foreign Agent Registration Act by not registering as an agent of his government. A *New York Times* reporter who attended his bail hearing was struck by how mysterious and important he seemed to be: the supposed engineer with the visitor visa seemed to be telling the Soviet consul general, a senior government official, what to do.[16] After Hitler invaded Russia, the US let Ovakimyan go in a gesture of goodwill, and he returned home before anyone had thoroughly interrogated him. The FBI did not learn until much later that Ovakimyan was one of the founders of Soviet espionage in the United States.

Like almost everyone in his country's ruling class, Ovakimyan had *conspiracya* in his blood. This Russian word connotes far more than its equivalent in English, calling to mind a way of life governed by secrets, deception, and mistrust. The Soviets had relied on a massive conspiratorial apparatus from the first days of their revolution in 1917. It spread like a giant inkblot, both within their borders and abroad. It was their primary means of protecting themselves and advancing their cause. At its heart was a ruthless and efficient spy service, far better than its American or British counterparts at recruiting and running spies.

After the Roosevelt Administration established diplomatic relations with the Soviet Union in 1933, the NKVD and the GRU, the smaller Soviet military intelligence agency, stepped up their operations in the United States. At least in the short term, their goals were not to spread revolution, but to understand and influence American policy and to steal technology. Within a few years they had recruited dozens of spies, many of them from the ranks of the Communist Party of the USA (CPUSA), and were running sophisticated operations against largely oblivious American institutions.[17] Before the war, they infiltrated the US government. Among their stars were Alger Hiss and Harry Dexter White, senior officials who reported privileged information from the Departments of State and Treasury.[18] There was even a spy in the White House, Roosevelt's special assistant Lauchlin Currie. During World War II, other spies reported on the race to develop the atomic bomb, a closely guarded secret that the United States had no intention of sharing with anyone but Britain and Canada. As an intelligence agency, OSS was a prime NKVD target, and by 1943, the young spy agency was riddled with Soviet spies.[19] Prominent among them was Duncan C. Lee, the lawyer who worked in the director's office and had incurred Carl Eifler's wrath for questioning his ability to command Detachment 101.

Lee was first drawn to the left by the effects of the Depression on the United States. In the mid-1930s, the Yale and Oxford graduate had fallen in love with an attractive Scottish radical named Ishbel Gibb and then with her politics. Lee married Gibb and, in 1939, secretly joined the CPUSA. At roughly the same time, he went to work at Donovan's Wall Street law firm. It must have been a jarring experience: on the one hand, Lee openly defended the interests of capitalists and received a handsome salary for doing so, and on the other, he secretly served the vanguard of the proletariat.[20] At Donovan's behest in mid-1942, he moved to Washington to join OSS. From its CPUSA contacts, the NKVD learned of the move and immediately realized Lee's potential for espionage: he had direct access to the director and the messages that crossed his desk. Lee's membership in the CPUSA morphed imperceptibly into spying for the NKVD.

Donovan's stated willingness to use communists or ex-communists for war work applied to *known* communists; it never extended to secret communists who shared official secrets with Moscow on their own. Lee knew this, and was terrified of getting caught. He knew he would lose his job on Wall Street and possibly go to prison.[21] To protect himself, he refused to supply the NKVD with documents—that was too risky. But he did have an excellent memory, and turned into a productive spy by memorizing documents and then dictating them to one of his two handlers, both American women with whom he could meet socially near his home in Georgetown without attracting attention.[22] He also shared his insights on Donovan and OSS. Fitin and Ovakimyan avidly consumed his reports; the NKVD record shows that Moscow regularly pressed for more.

In mid-January 1944, Lee secretly reported to Moscow on Donovan's visit, assuring the spymasters that his intentions for the visit were sincere; the director had returned home brimming with enthusiasm. Lee used the word "enthralled," adding that Donovan now regarded Stalin as "the smartest person heading any government today."[23] While he knew that the NKVD had undercover representatives in the United States, and harbored reservations about the resurgence of Russian nationalism, Donovan reportedly no longer saw any reason to fear "communist domination of the world." Lee likely erred on the side of telling Moscow what he thought it wanted to hear—or someone edited his report after it arrived there. That a capitalist lawyer like Donovan had been "enthralled" by the cold, dark, war-torn capital and its government—especially after visiting the Lubyanka and waiting for days to fly out—is hard to believe.[24] But Donovan's relationships with both Roosevelt and Churchill had been strained by 1944, and he might have had a word or two of praise for Stalin's acumen. Certainly Lee's basic message was right: Donovan was pleased with what he thought he had achieved in Moscow.

Hoover vs. Donovan

By the time Lee reported to Moscow, Donovan's achievement was under concerted attack from the man who had more than once been an adversary, but had yet to become an outright enemy: J. Edgar Hoover.[25] A few months earlier, Hoover and Donovan had exchanged brief, civil notes, Hoover congratulating Donovan on his promotion to brigadier general, and Donovan thanking him a few days later.[26] Now in high dudgeon, Hoover did not address Donovan directly, but went through Harry Hopkins to Roosevelt. The letter to Hopkins showcased Hoover as the country's protector from communist spies:[27]

> I have just learned from a confidential but reliable source that a liaison arrangement has been perfected between the Office of Strategic Services and the Soviet Secret Police . . . whereby officers will be exchanged between those services. . . . I think it is a highly dangerous and most undesirable procedure to establish in the United States a unit of the Russian Secret Service which has . . . for its purpose the penetration into [our] . . . official secrets.

Donovan lost his temper, telling Lee that Hoover was a "fool" for opposing the exchange: it would be relatively easy to keep track of Soviet spies operating out of a building with the hammer and sickle above the front door.[28] For his part, FDR asked Admiral Leahy's opinion. Leahy took Hoover's side. Apart from Leahy, the Joint Chiefs remembered that they had approved the initiative, and did not change their position.[29] Attorney General Biddle, the target of another Hoover screed, lined up with the FBI director, pointing out the potential political cost if Roosevelt's Republican opponents learned of the deal.[30] On March 7, Roosevelt had a working lunch with Donovan in the Oval Office at the White House and what must have been a difficult discussion.[31] Then on March 15, Roosevelt cabled Harriman that the deal was off: "the question presented has been carefully examined here and has been found to be impracticable at this time."[32] Harriman was to "inform the Marshall . . . that for purely domestic political reasons which he will understand it is not appropriate just now"—a reference to the fact that Roosevelt had told Stalin that 1944 was an election year in the United States, which limited the scope for politically risky maneuvers.

Harriman and Deane pushed back energetically, writing that, with Donovan's help, they had "for the first time . . . penetrated one intelligence branch of the Soviet Government."[33] It was a large step in the right direction, one that might lead to more contacts. Harriman felt so strongly that he offered to fly back to Washington to brief Roosevelt and the Joint Chiefs in person. Roosevelt refused to change his mind, cabling Harriman again on March 29 that the timing was wrong.[34]

The official exchange of OSS and NKVD representatives would never take place. Fitin and Ovakimyan did not seem particularly disappointed when Deane told them the news, which was not surprising considering they already had their own sources in the OSS. Deane offered to serve as the intermediary between the two spy services, a proposal that was acceptable to everyone. A fair amount of material passed through his hands, more from the Americans to the Soviets than the other way around, but both sides received things they wanted.[35]

Donovan did not get over the lost opportunity—direct liaison would have been more productive—or his anger with Hoover. In December 1944, almost a year after the talks at Lubyanka, Donovan was still angry. Lee would report to Moscow that he had not stopped criticizing the president and his advisors for doing "a very foolish thing," especially when the Soviets must have information on Japan that the US could use in the Pacific war.[36] The young lawyer working at Donovan's side had watched his relationship with Hoover worsen; he now wrote of the "hatred between them."[37]

The FBI in Wartime

Both Donovan's and Hoover's stances made sense.[38] If the Americans dealt with the Soviets at arm's length, any exchange of information about the enemy would likely bring mutual benefit. Hoover might have opposed the initiative out of spite. But he was not wrong when he claimed that it would give the Soviets another base for spying in America, and that monitoring their activities would strain the Bureau's resources. Though still focused on the Axis, Hoover was beginning to understand that the communist threat in the United States was not just from radical workers and union leaders. Soviet officials like Ovakimyan also needed to be watched. In mid-1943, the Bureau had launched two significant investigations of Soviet spies in the US known as CINRAD and COMRAP.[39] That meant the Bureau was even more hard-pressed to come up with the manpower necessary to accomplish all of its wartime missions.

This was true despite previously unimaginable growth. Early in 1940, the number of special agents was under one thousand, and by 1943 it was close to five thousand, not counting the thousands of employees who backed them up, making the FBI and OSS roughly comparable in size.[40] Even with the added staff, the FBI was stretched thin. Hoover was moving forward on a variety of fronts, most connected with national security.

One of Hoover's flagship programs, started in the spring of 1940, was the Custodial Detention Program (CDP), basically a long list of suspicious characters to be picked up in a crisis. The problem was that the list was overbroad and arguably illegal: in addition to enemy aliens, it listed Americans whose only offense was holding radical views.[41] Attorneys General Jackson and Biddle were uncomfortable with the CDP and sought to rein it in, but only with limited success. Complementing the CDP were wiretaps. In May 1940, Roosevelt had acknowledged a Supreme Court decision that made wiretapping illegal but decided unilaterally that the Court must not have intended for its decision to "apply to grave matters involving the defense of the nation."[42] This enabled the FBI and the Department of Justice to move ahead with hundreds of wiretaps between 1941 and 1945, initially of foreign embassies.[43] An aggressive publicity campaign of articles, books, radio programs, and films warned the public about possible threats that the Bureau could neutralize. Hoover himself was the focus of one tribute, *The FBI in Peace and War*, which included the whopper that one of his greatest attributes was his "utter irrelation [*sic*] to the petty, scheming, wire-pulling, job-grabbing world around him."[44]

In the realm of counterespionage, the Bureau redeemed its reputation after the embarrassing denouement of the Rumrich case in 1938. Hoover pushed his men to learn, on their own and from the British, the ins and outs of this arcane trade that was so different from catching criminals. The first "counterspy" (or double agent) that the Bureau ever ran was a German American named William G. Sebold, who walked into the FBI and claimed that the Abwehr, German military intelligence, had coerced him into spying for Germany.[45] Starting in early 1940, the FBI ran the case for more than a year and a half. A special agent transmitted misleading information—ostensibly reports from Sebold—back to Germany from a shortwave radio on Long Island. In a Manhattan office—again, ostensibly run by

Sebold—technicians secretly filmed meetings between him and bona fide German spies.

Two other sets of enemy agents traveled by submarines to the East Coast in 1942 and 1944 to, respectively, commit acts of sabotage and attempt to steal American atomic secrets. An Abwehr agent named George Dasch quickly ended the first operation by turning himself in to the FBI; the second operation lasted a few weeks but ended when the American-born agent William Colepaugh also turned himself in. In the 1942 case, the attorney general and the director took the time to personally serve as the prosecutors in the secret trial of the saboteurs (most of whom received the death penalty). Hoover carefully controlled any information about the Germans that left the Bureau, especially if destined for the White House or the press. Even though the first breaks in each case came from the spies themselves and were not the result of FBI investigations, Hoover ensured that the Bureau received its share of credit—and then some.[46]

In order to preserve its secrets, the one program that Hoover did not publicize was SIS. Foxworth's adjustments and Berle's support had enabled it to grow and even flourish. By October 1943, SIS reached its peak strength, with 331 special agents in the field in more than twenty countries in Latin America.[47] The agents were a mix of undercover SIS men (now better trained to pose as employees of American corporations), special employees with local knowledge, and legal attachés openly working at American embassies. Their targets turned out to be real, if never as fearsome as advertised.[48]

The US government had created SIS on the assumption that there was a dangerous fifth column in the Western Hemisphere. The assumption was mostly just that; there was little evidence for it. Wartime sweeps uncovered a few actual spies, mostly German, reporting home, sometimes using the same methods as SIS—secret writing and shortwave transmissions. German, Italian, and Japanese immigrant communities in South America did nurture potentially dangerous political movements. SIS addressed these threats by collaborating with British intelligence, local authorities, and other US agencies. The Federal Communications Commission's Radio Intelligence Division, which deployed to a dozen countries in Latin America, searched for suspicious transmitters. Sifting through millions of letters, British censors at overseas transportation and mail hubs, especially the mid-Atlantic island of Bermuda, became ever more adept at detecting the anomalies that signaled a spy was at work.

The process was not always tidy. Heinz Lüning was a twenty-nine-year-old German who had spent some time in the Dominican Republic and spoke Spanish.[49] He joined the Abwehr because he did not want to go to war as a foot soldier. At the training academy in Hamburg, he failed to master many of the basic skills that a stand-alone spy would need, but was still allowed to process for deployment; the service was under pressure to expand its coverage of Latin America. Even though he had never been to Honduras, he traveled on an alias Honduran passport. In September 1941, he arrived in Havana, whose seaport was one of the crossroads of the Caribbean. Lüning soon picked up bits of gossip about Allied shipping in waterfront bars and reported back to Germany through the international mail using secret inks.

On a hunch in October 1941, British censors in Bermuda pulled one of Lüning's letters for examination and discovered the secret writing it contained. They unobtrusively copied and resealed the letter before sending it on to its destination. The British shared their discoveries with the US Department of State, which in turn passed the information on to the FBI. After a letter in August 1942 contained Lüning's actual return address (a dangerous lapse on his part), the SIS tipped off the Cuban police, who arrested the hapless spy. He confessed immediately. A search of his lodgings yielded various bits of spy gear, including a half-assembled radio that he clearly did not know how to work.

On October 31, Hoover publicly credited the Cuban police with the arrest.[50] A few days later, the Cuban police chief decided that he would benefit more from a well-publicized execution than from waiting to see where Lüning's information might lead, and after a quick trial, Lüning faced a firing squad on the tenth of November. The FBI spent the next two and a half years running every conceivable lead to ground—it turned out that the German had traveled in the United States—but found almost nothing suspicious.[51] Growing more after his death than during his life, Lüning's FBI file eventually reached a record-breaking four thousand pages.[52]

The Lüning case was, in the words of the FBI official history, "the most important German investigation handled by SIS Agents in Cuba," and the only one in Latin America that ended in an execution.[53] The FBI had not pressed for Lüning's execution; he was more valuable alive. But his spy gear and confession made the case gratifying. It was concrete evidence. The other work performed by SIS in Cuba and elsewhere was more common fare for a counterintelligence service: sifting through a mass of inconclusive background information, official leads, and unofficial allegations.

True to its roots, SIS naturally gravitated to investigating espionage and countering subversion.[54] Whether the young service would also report on other issues—a question supposedly resolved in 1940—never went away. On October 25, 1943, Assistant Secretary of State Berle—still well-disposed to Hoover and the FBI—wrote the director a polite letter suggesting that SIS leave everything but its counterintelligence work to the Foreign Service.[55] American diplomats did not handle spies, but they were the traditional experts on politics and economics. This commonsense proposal struck a raw nerve, since Berle seemed to be implying that the SIS lacked the requisite qualifications. Within two days, Hoover wrote a few hundred words in reply, telling Berle that he certainly did not want the FBI to duplicate anyone else's work.[56] Especially now that the Axis threat in the Western Hemisphere was receding, he would simply reassign many of his agents. He immediately ordered the recall of 136 special agents, who left their posts so suddenly that their departure alarmed the Foreign Service.[57] Hoover's assistant Tamm felt a need to call a meeting with Berle and explain the facts of life to him, hinting at the costs of criticizing the Bureau and its sensitive director.[58]

Berle gave in to the bureaucratic blackmail. He reportedly even ordered the October 25 letter to be stricken from the file. In writing, on the twenty-seventh of December, the Department of State officially requested that SIS report on political and economic matters as well as subversion and espionage.[59]

Though once again free to do what it wanted, the Bureau continued to focus on its traditional areas of expertise. The FBI shared the results with other agencies and especially with the White House. Hoover signed reports without any evaluation or context in the belief that it was his Bureau's job to report the facts and let the reader draw his own conclusions.[60] This was the fatal flaw that FBI reporting shared with other reporting streams that flowed to 1600 Pennsylvania Avenue; the president and his staff needed guides to point the way through the turbulent waters, something that only Alfred McCormack seemed to have learned. The situation worsened as the war went on and the streams surged out of their banks. In the three and a half war years after Pearl Harbor, Hoover sent some 1,600 reports to the White House, most having to do with South America.[61]

FBI/SIS reporting was not unlike OSS reporting. With the exception of R&A products like *The War This Week*, Donovan forwarded a similar mix of standalone and half-processed material to Roosevelt.[62] The reams of paper went to a gatekeeper, like "Pa" Watson or Grace Tully, who was asked to brief the president.[63] But the best chance for getting FDR to focus on a report was to find a different path, ideally during a personal meeting or through other White House insiders.[64] While FBI and OSS both claimed to be in the foreign intelligence business, neither routinely provided finished intelligence to the first customer or even captured his attention for long.

By the end of 1943, an overall comparison of OSS and FBI/SIS would have revealed more differences than similarities. OSS was a wild, once-in-a-lifetime assemblage of men and women who signed up to fight Hitler; SIS agents fit the mold for American law enforcement that Hoover had cast years before. SIS was fighting a shadow war, mostly in the Western Hemisphere; OSS was in shooting wars in Asia and Africa. The Bureau was playing a largely defensive game, countering spies and subversion on the right as well as the left. In Latin America, it operated like a backup domestic security service, one freed from the writ of habeas corpus. OSS was playing an offensive game, proving successful on the margins of the battlefield. SIS men quietly took their share of risks; Foxworth, for example, died in the crash of an army C-54 in Surinam in 1943 while en route to North Africa to investigate a case.[65] OSS service was at once mysterious and glamorous—perhaps as glamorous as being a G-man had been before the war—with world-class filmmakers like John Ford recording the OSS story. Donovan was developing a publicity machine that rivaled Hoover's. Hoover had one vision; Donovan had many visions. FBI/SIS was as overcontrolled by its director as OSS was jostled by Donovan's "mustang plunges."[66] By early 1944, each was starting to contemplate its role in the postwar era. Their differences foretold a struggle for the future of American intelligence.

15

BREAKING CODES, FORGING LINKS

Across the Atlantic

The years of 1941 and 1942 were ones of tremendous growth for the various American intelligence services. They welcomed new members and developed machines and capabilities at a rate that the United States had never seen before. This was as true for the FBI, army, and navy as it was for COI and its successor OSS, the newcomers to the field. Despite a few steps toward collaboration with one another, each worked largely on its own, enjoying at least a modicum of success and sometimes a good deal more: the FBI against Nazi subversion in the Western Hemisphere, the American Navy against the Japanese Navy in the Pacific, and the army in Washington against the Japanese Foreign Office. COI's original mission of coordinating information for the executive branch had largely fallen by the wayside. Its successor, OSS, now working for the Joint Chiefs of Staff, was officially supposed to meet the military's needs for intelligence and special operations; its unofficial mission was to fill whatever niches Donovan could find on his breathless world tours.

Though not particularly keen to work with one another, American spies proved eager to work with their British counterparts. This was plain to see after the summer of 1940. Desperate for help, the British reached out, and more often than not, Americans welcomed their overtures. "Little Bill" Stephenson had started out by courting J. Edgar Hoover, but soon switched the object of his affections to the more receptive "Big Bill"—Donovan. The switch angered Hoover and contributed to a decline in Stephenson's influence.[1] Hoover proceeded to cultivate a relationship with MI5, sending two special agents to London in December 1940 to learn about counterintelligence from the self-declared masters of the art. The senior of the two, Mississippian Hugh Clegg, looked "tough as a gangster" to Guy Liddell, the gentlemanly MI5 director for counterespionage. But Liddell quickly decided that he was actually "a very good fellow," there to learn as much as he could.[2]

The FBI-MI5 relationship waxed and waned, fraying a few months later after a contretemps over the star double agent Duško Popov. The Germans believed that Popov was working for them when in fact he was working for the British, who fed misleading information to the enemy through him. In the summer of 1941, the Germans directed Popov to travel to New York, giving him a list of requirements to service. The British asked the FBI to support the operation. To British eyes, the FBI came close to sabotaging the case. Hoover and his special agents were outraged that Popov was renting a penthouse on Park Avenue that no federal employee would ever be able to afford—even though the lavish lifestyle was part of his cover. They tried to control him, and only succeeded in making him angry. But MI5 could not afford to alienate the Americans and had to swallow back its bile.[3] Collaboration would thus continue between the FBI and MI5, deepening over the course of the war.

Even before Pearl Harbor, the US had started to develop relationships with British codebreakers. One of the happy consequences of breaking Japan's Purple code in 1940 and replicating the Purple machine was that it provided the US with something to trade. Though not yet the G-2, General Strong had been an early proponent of sharing cryptanalytic breakthroughs with the British. While in London in August and September 1940, he met with the British Chiefs of Staff and reviewed the bidding: how the United States had broken a major Japanese code two years earlier, how they had shared some of the resulting information with the Chief of the Imperial General Staff, and how the British ambassador in Washington had proposed a full and free sharing of information. The time had come, he believed, "to put into effect a scheme for regular exchange of codes and information."[4]

Secretaries Knox and Stimson and then G-2 General Miles did not object. Nor did the more conservative uniformed navy officers like Safford—at least not strenuously enough to derail Strong's initiative.[5] The president deferred to the experts. The most prominent expert, Friedman, supported the exchange. In December 1940, he made plans to travel to Britain to liaise with his counterparts. As an experienced cryptanalyst, Friedman would know how to conduct himself. It was roughly a week before the scheduled departure that he literally collapsed at the office from overwork and was hospitalized. One of Friedman's brilliant hires from the early 1930s, Abraham Sinkov, was chosen to take his place.[6] In his capacity as a major in the army reserves, Sinkov would travel with Leo Rosen, the young codebreaker who had helped Rowlett solder the connections on the Purple analog. Two junior navy codebreakers, Lt. Robert H. Weeks and Lt. (j.g.) Prescott H. Currier, would also make the trip to England.

The twenty-eight-year-old New Englander Currier was a one-man intelligence machine. He had taught himself Japanese and could work as an intercept operator, cryptanalyst, translator, and report writer—a combination so remarkable that ONI had wanted to place him under surveillance on the grounds that no one man could have so many skills (a dramatic reflection of the differences in the mentalities of conventional navy intelligence officers and navy codebreakers).[7]

The day after Christmas in 1940, Currier received a note from Joe Wenger at Main Navy calling him back to Washington from the hearth he was enjoying with his new bride. After he made the sometimes difficult winter trip back to Washington from New England, he was told he would be going to England, but "almost nothing else" about the trip.[8] He and Weeks received no briefings beyond "watch your step;" they were not told what to do nor what their duties would be.[9] Nor did they sit down with the army team before leaving Washington to produce a joint plan—but there was nothing unusual about that. Currier knew who the army codebreakers were and bore them no ill will, but he also remembered the days when "nobody cooperated with the Army under pain of death."[10]

What Currier did know was that the team would be bearing gifts for the British, packed in six wooden crates that weighed in at some 720 pounds. Inside were two Purple analog machines and two of the machines that had been constructed to break its predecessor, Red, along with reams of written material.[11] On the day of departure in January 1941, the army and navy teams traveled separately from Washington to a dock in Annapolis, Maryland, where men and equipment went into two open-motor launches for a rough ride in driving rain out to the new Royal Navy battleship, *King George V,* that had just brought the new British ambassador, Lord Halifax, across the Atlantic.[12]

The trip to Britain was not particularly fast—taking almost two weeks—and included tangential swings from the route, first to the south to help convoy a load of Argentine beef and then far to the north. Upon arrival at Scapa Flow, the Americans transferred to the light cruiser HMS *Neptune,* whose crew made the crates fast on the deck. By chance they attracted the attention of a passing Nazi pilot who proceeded to strafe the ship at lunchtime, making a noise like a chain being dragged across the deck that made Currier's mouth go bone dry. "I was so scared," he remembered later, "that I couldn't swallow my soup."[13] Instead he thought of the crates, the wonderful machines being shattered, and of the safety briefing he had received: even in a life jacket, no one could survive in the cold water of the North Sea for more than three or four minutes. But the life jackets stayed dry, and survive they all did, including the precious gear: an hour and a half later, he ventured up on deck and found that the crates were intact, save for a few pockmarks.

A motorcade took the teams from the port of Sheerness some fifty miles northwest of London to Bletchley Park, the fifty-eight-acre estate that was home to British codebreaking. Sometime after 10 p.m., the motorcade pulled up to the headquarters of the Government Code and Cypher School (GC&CS) that had once been "the self-glorifying residence of a local . . . tycoon."[14] (In 1938, Adm. Hugh Sinclair, Menzies's predecessor as head of MI6, had purchased the property, reportedly with his own money.[15]) Mostly Currier was struck by the darkness—with streets and buildings blacked out as part of the defenses against German bombers—and would remember passing through a series of blackout curtains to get to the brightly lighted rooms. There, Director Alexander G. Denniston, a short, athletic sixty-year-old impeccably dressed in a civilian suit, was waiting along with the head of the Military Section, John Tiltman, who had donned a formal Scottish uniform complete with Tartan trousers for the occasion.[16] The highly experienced codebreakers welcomed their guests with sherry, unaware that it was not the drink of choice for the Americans who had celebrated their triumphs with Coca-Cola. But it did not matter. Generally well-disposed to their visitors, Denniston and Tiltman were not rank conscious or social snobs.[17] Neither seemed to mind that these Americans, said to be on a high-profile mission, turned out to be so junior.

The next day the British sent two cars, one for the navy team and one for the army team, and started the Americans out on separate itineraries even though they were all staying at the same comfortable mansion, Shenley Park, a few miles from Bletchley.[18] In daylight, they could see that the nineteenth-century brick headquarters at Bletchley was a strange mix of styles—Tudor cottage, Victorian gothic, Dutch baroque—and looked more like a short row of town houses, each designed by a different architect with his own sense of whimsy, than one building.

The other buildings sprinkled around the estate had a drab sameness, either Nissen huts—the British equivalent of the American Quonset hut, a half circle of corrugated iron planted in the ground—or rectangular wooden buildings with concrete floors and plain wooden desks and chairs, but no bathrooms. The huts were numbered, each dedicated to a specific function, such as deciphering German Army codes or translating messages into English. Upward of one thousand Britons, both civilian and military, worked at Bletchley under these Spartan conditions in 1942. For the academics and professional codebreakers, every day brought new challenges, but for the rank and file, the work was boring and repetitive. The Scottish aristocrat Lady Jean Fforde, who volunteered to do her bit for the war effort, spent the year in Hut 8, where for hours on end she searched for three-letter combinations in messages, and when done, pushed pieces of paper through a hatch to the next room.[19] The near-total silence and unrelenting pressure she experienced were standard.

The British were uniformly friendly as they absorbed Rosen's demonstration of the Red and Purple machines. Tiltman viewed the gift of hardware and information as significant, "the first gesture which puts everything on the right lines," not unlike an unusually thoughtful gift that a houseguest presents on arrival that sets the tone for the visit.[20] The gesture had a once-in-a-lifetime feel since it was so rare for governments, even friendly governments, to share codebreaking secrets. The process was off to a decent start, but it would still be difficult.

At first the British held back, only too aware of a crucial difference between the two nationalities. Britain was at war, her very survival at stake. She had few advantages over the Germans in early 1941. One was that Bletchley had started breaking into German traffic in early 1940 and was reading more and more (but hardly all) messages encrypted by the Enigma, the principal wartime German code machine.[21] Known as the Ultra secret, this breakthrough would remain an advantage only as long as the Germans did not suspect that it had occurred. While the British desperately wanted American cooperation and support, in codebreaking as much as in munitions, they wondered whether they could trust the Americans, who were not yet at war and would have far less interest in protecting the core British secret—about which they initially knew little or nothing.[22] After a few days of taking the Americans' measure, the staff at Bletchley decided to seek permission from "the highest [authority] in England to be perfectly frank" with Sinkov, Rosen, Weeks, and Currier—to take what amounted to an even greater leap of faith than the Americans had taken by giving Bletchley the Red and Purple analogs.[23] Denniston allowed Tiltman to broach the matter with Menzies, who in turn approached the chiefs of staff, who shed lingering reservations before informing Churchill, who minuted his agreement.[24]

Denniston took the extraordinary step of sharing the British dilemma with his American visitors. Codebreaking, he explained, was the "lifeblood to our [war] efforts[,] while for them . . . this work could not be more than a new and interesting problem." He asked for—and received—assurances that they would share what they learned only with their immediate superiors, what Sinkov remembered as "a condition imposed before we were given . . . information about the Enigma."[25] On March 3, 1941, Lieutenant Weeks handwrote his commitment "for the preservation of the secrecy of the work," a remarkable document that eventually found its way into the National Archive at Kew outside London.[26] Now the four Americans could receive general briefings on how the British were attacking the Enigma.

The machine itself looked something like a sturdy typewriter set in a varnished wooden box, and used an electromechanical system of rotors and circuits to transform plaintext into code and vice versa. Every day, German stations on the same net needed to align the three rotors in the same way in order to communicate. After setting the rotors, a code clerk would then type in the plaintext, and the machine would transform it into code. A second clerk would copy the results, and the message would be ready to send over the airwaves. After ensuring that his machine was on the same setting, the recipient would reverse the procedure to break out the plaintext. An enemy codebreaker needed the hardware—an Enigma machine (or its equivalent)—and the daily setting, which he could obtain by capturing a codebook—"pinching" in wartime slang, hopefully without tipping the owner off—or by using "bombes," electromechanical machines that could quickly run through thousands of possible settings.[27] (The Polish codebreakers who had invented the first bombes before the war had bestowed that name on them for reasons that are still unknown—perhaps because they exploded codes or because they sounded like time bombs when you turned them on.)

For the Americans, the orientation to Enigma and the bombes was like an intensive introductory college course. The British even handed out three pencils and a pad of paper to each of the visitors and, at least some of the time, let them take notes.[28] What they saw and heard was different from their work on Japanese systems. Currier and Weeks did not understand everything. Currier thought that it would make more sense to Sinkov; Sinkov hoped it was making sense to Rosen, who had a flair for machinery.[29] Years later he would not remember actually seeing a bombe. But he would remember that he had learned enough to grasp the problem and its solution, and probably wrote down enough to jog his memory.[30]

Currier admired how much the British appeared to be doing with limited resources, starting with a great investment of intellectual capital—it was no accident that Bletchley was halfway between Oxford and Cambridge. Both universities had contributed the brilliant academics who were solving the Enigma problem.[31] But there was still a whiff of the preindustrial at Bletchley. The British codebreakers were craftsmen when it came to designing and building machines. The process was totally unlike an assembly line in America. Currier was taken aback that his British counterparts seldom used typewriters.[32] Instead they relied on paper and pencil, and would do so throughout the war. They used sheets of paper fourteen inches long, which were bigger and more unwieldy than the US standard, 8½ × 11. Outlying stations recorded intercepts by hand, then sent "large cardboard boxes with thousands and thousands and thousands of pieces of paper" to Bletchley and London for processing, also by hand, a method that was already unthinkable in the United States.[33]

The Americans went home after roughly ten weeks, using their embassy in London to ship crates bulging with notes and documents as well as some direction-finding hardware that the Royal Navy was willing to share. But GC&CS did not part with any of its machinery, let alone any deciphered German traffic. Since the US Army and Navy codebreakers did not meet—at least not officially—to digest the results of the trip, their parent commands reacted differently. Friedman was generally satisfied while the navy was not, arguing that the Americans had been shortchanged.[34] They had traded crown jewels for paper. This led to an American request, in writing, for the British invention used to break Enigma codes. GC&CS was "aghast at the letter."[35] First, they could not spare a single one of their machines. Second, they tried whenever possible to avoid putting anything on paper about Enigma.

The well-meaning Denniston decided to make the trip across the Atlantic in the summer of 1941 to explain himself to the Americans and to examine, firsthand, the state of their codebreaking.[36] If at all possible, he wanted to build on the foundation of the February visit. The sixty-year-old traveled courtesy of the US Army Air Transport Command, probably on a C-54, and set himself a backbreaking schedule. His handwritten notes on the elegant stationery of the Hay-Adams Hotel (located on Lafayette Square across from the White House) attest to lengthy meetings with army and navy officials. They mention Safford and John Redman, and hint at his distress over the lack of cooperation between the army and the navy.[37] On August 15 at Main Navy, Denniston sat down with Joe Wenger of Op-20-G.[38] The day after, he went next door to the Munitions Building to meet for

German Army troops march into Paris on June 14, 1940, an event that changed the world and would bring the United States into the fray.

BUNDESARCHIV, GERMANY

Franklin Roosevelt and Winston Churchill met at sea off Newfoundland for the Atlantic Charter Conference in August 1941. By then Churchill and his government had already spent a year urging the Americans to start matching British intelligence capabilities.

NAVAL HISTORY AND HERITAGE COMMAND

The prosperous lawyer and world traveler William J. Donovan, pictured here in the late 1920s

LIBRARY OF CONGRESS

President Roosevelt presides over the swearing in of Navy Secretary Frank Knox in July 1940.

NAVAL HISTORY AND HERITAGE COMMAND

The Japanese Sneak Attack on Pearl Harbor, a drawing by navy combat artist Griffith Coale that suggests the extent of the destruction as well as American outrage

US NAVY ART CENTER

A stylishly dressed FBI director J. Edgar Hoover posing at his desk around 1940

FBI

Henry L. Stimson, a man of many talents and vast experience. He opposed codebreaking in 1929 as secretary of state but supported it in 1940 as secretary of war.

LIBRARY OF CONGRESS

Herbert O. Yardley, the founder of semi-official American codebreaking

NATIONAL CRYPTOLOGIC MUSEUM

The US Army's William F. Friedman is widely regarded as the founder of modern American cryptology.
NATIONAL CRYPTOLOGIC MUSEUM

Frank B. Rowlett, a leading member of Friedman's team
NATIONAL CRYPTOLOGIC MUSEUM

Laurance Safford, the officer who founded the profession of codebreaking in the navy
US NAVY

Genevieve Grotjan, the unassuming civilian who made the break into a major Japanese code in 1940
NATIONAL CRYPTOLOGIC MUSEUM

Both army and navy codebreakers started out in nearly identical "temporary" buildings on the National Mall.

NATIONAL ARCHIVES

Joseph Rochefort, the largely self-educated officer whose hard work enabled victory at Midway

NATIONAL CRYPTOLOGIC MUSEUM

Ensign George H. Gay, Jr., occupied a ringside seat at Midway—in the water after being shot down amid the enemy fleet. He would be the first to confirm that three Japanese carriers had sunk.

US NAVY

Admiral King (*left*) was the demanding commander in chief of the US Navy, while the somewhat more easygoing Admiral Nimitz commanded effectively in the Pacific.

NAVAL HISTORY AND HERITAGE COMMAND

Alfred T. McCormack, the Wall Street lawyer who added much-needed value to army codebreaking. He dedicated this portrait to William Friedman.

NATIONAL CRYPTOLOGIC MUSEUM

Major General George V. Strong, the Army G-2 who supported McCormack but had little patience for Donovan

US ARMY

The multitalented British commander Rodger Winn, RNVR (*left*), taught Commander Kenneth Knowles, USN, how to turn information into intelligence and use it to kill U-boats. By late 1943, the US Navy had learned its lessons well, equaling and even outstripping the Royal Navy in lethality at sea.

US NAVY

The Germans famously used the well-designed, sturdy Enigma (*left*) to encode messages. The Allies crafted ingenious devices to defeat German and Japanese codes. Early on, the results looked and sometimes were homemade, like the Purple analog machine (*right*), originally cobbled together by American codebreakers after hours. Amazingly, it worked.

NATIONAL CRYPTOLOGIC MUSEUM

The tension is almost palpable in this photo from the deck of a coast guard cutter defending an Atlantic convoy from U-boat attack.

US NAVAL INSTITUTE

Attacked by waves of carrier aircraft in the mid-Atlantic on June 12, 1943, U-118 had little chance against the US Navy's hunter-killer system driven by intelligence.

US NAVY

The manor house at Bletchley Park that served as British codebreaking central in World War II

EVENING STANDARD, GETTY IMAGES

Inside one of the huts at Bletchley, a workplace without frills

BLETCHLEY PARK TRUST,
GETTY IMAGES

By 1945, American codebreakers worked in far more modern and well-equipped industrial spaces.

NATIONAL ARCHIVES

Eifler (*left*) and Donovan, about to risk their lives on a flight over enemy territory in Burma

OSS

From Burma, Donovan made his way to wartime Moscow, a grim and uninviting destination in 1943.

CREATIVE COMMONS, NOVOSTI ARCHIVE

Wartime London, sometimes grim but usually inviting to Americans from 1940 on

NATIONAL ARCHIVES

A dapper young Allen W. Dulles, the lawyer who found his calling in wartime intelligence

LIBRARY OF CONGRESS

Dulles associate Gero von Schulze-Gaevernitz (*in dark jacket*) looking relaxed in Bolzano, Italy, on May 12, 1945, in the company of Wehrmacht general von Vietinghoff and to his left SS general Wolff (*in light tunic*). Zimmer and Dollmann stand in the background.

NATIONAL ARCHIVES

Harry S. Truman looking confident in the company of two more experienced wartime leaders, at Potsdam in 1945. Admiral Leahy looks over Truman's right shoulder.

BUNDESARCHIV, GERMANY

General Eisenhower insisted on congratulating army codebreakers after the war. Rowlett and Friedman (*in civilian clothes*) stand at right, Friedman showing signs of the stress he experienced for years.

NSA PHOTO

Farewell to OSS: Donovan (*left*) and Magruder (*clapping*) at the ceremony in September 1945

US ARMY

three hours with Friedman, Sinkov, Rowlett, Kullback, and Rosen. The meeting established a plan for a four-day orientation to the army's work that would make Denniston marvel at how mechanized the SIS was, thanks to the IBM machines that Friedman had gone to great lengths to obtain.[39]

Denniston's return home was, if anything, less comfortable than his flight over: he was wedged in a bomb rack for the fifteen-hour-plus trip from New York to Newfoundland to Scotland.[40] He did not take the rest that he had earned, but returned to work, putting his impressions on paper and preparing for another grueling trip. Denniston wondered whether the Americans weren't *too* mechanized; he was a staunch believer in the human element.[41] He praised the Americans for their work on Magic, while noting that it was their only real area of strength. He found both the navy and army codebreakers competent but was more positive about the army than the navy. The army had benefited from Friedman and his international reputation, and seemed to prize academic achievement.[42]

Friedman was even becoming a personal friend, one who cared enough to send his British colleague to a doctor in New York who could treat his bladder stones, first diagnosed in February but lingering painfully through the spring and into the summer because the treatments cost more than the senior British civil servant could easily afford.[43] The navy, on the other hand, seemed to have a hard edge. One problem was "a female Knox"—a reference to the codebreaker Agnes Meyer Driscoll—who told Denniston that she could beat the British at their own game.[44] Denniston had his doubts, but was willing to placate her, and let her try her hand on copies of all the main German naval messages intercepted by the British in 1941. Above all, he wanted to nourish the hope that "we are just at the beginning of a liaison which may have far-reaching results when America . . . makes up her mind to join in."[45]

16

ADMIRAL DÖNITZ'S UNINTENDED CONTRIBUTION TO ALLIED VICTORY

The US Navy and the British

Denniston's patience eventually paid off. The first working partnership was—somewhat ironically—between the British and the particular service that fostered an aversion to working with anyone else, even other Americans, let alone foreigners.

Starting with Admiral King, many American Navy officers seemed ambivalent, if not hostile, toward their counterparts in the Royal Navy, considered the senior service in Britain.[1] (There was even a high-end brand of British cigarettes, Senior Service, its elegant logo designed to evoke the navy and appeal to discriminating tastes.) Then came the Battle of the Atlantic with its own dynamic, far stronger than anyone's personal attitudes. Britain's lines of communication across the Atlantic, a matter of life or death for the island kingdom, were all too vulnerable to the German Navy. After the war, Churchill would famously declare that the only thing that ever really frightened him was "the U-boat peril."[2] For him, the gravest losses came during the twelve months from July 1940 to July 1941. By the spring of 1941, the Germans were sinking hundreds of thousands of tons of British, Allied, and neutral shipping; the total for June was a staggering 318,740 tons.[3]

The picture brightened somewhat in the middle of the year. In early July 1941, President Roosevelt ordered the US Marine Corps to send a provisional brigade to Iceland, and the US Navy began to escort American ships bound for Reykjavik.[4] In September, the administration expanded the navy's remit, allowing it to escort transatlantic convoys between Newfoundland and Iceland. At roughly the same time, Bletchley Park was beginning to read the German Navy's Atlantic U-boat messages. From Bletchley the information traveled by wire to the Admiralty in London, which was both an administrative and operational headquarters. Inside a new but bombproof attachment—the functional reddish-brown cinderblock Citadel nicknamed "Lenin's Tomb" was an eyesore among the ornate imperial office buildings in Whitehall—the Royal Navy had established an Operational Intelligence Center (OIC).

In the OIC's Submarine Tracking Room, the staff centralized information from a variety of sources, plotted the locations of U-boats, and forecast their courses of action. This enabled the navy to route convoys away from danger. When the system worked, German submariners would find only empty ocean instead of lines of ships stretching from horizon to horizon. It was a clear-cut example of how intelligence could make a difference, codebreakers enabling quiet victories. The result was not as dramatic as Midway—enemy aircraft carriers were not bursting into flame and sinking—but it was still crucial. Thanks in large part to British codebreakers, more friendly ships and their crews were surviving to deliver their precious cargoes.[5]

Even so, tensions between the allied navies persisted. In November 1941, while Denniston was promoting Anglo-American cooperation, the British embassy in Washington was reporting "grave unrest and dissatisfaction."[6] Rear Adm. Noyes, then director of US naval communications (which made him the senior officer in

Op-20-G's chain of command), had complained that the British were still shortchanging the Americans. Washington had fulfilled its obligations under the "virtual" transatlantic agreement but London had not. The material that Denniston had sent over for Driscoll—the raw German Navy messages that London "had been unable [to] cope with"—constituted "a very poor exchange" for the Purple machines and the priceless information about the Japanese systems. Noyes wanted to withhold further cooperation unless he received "full reciprocal information on European work." A somewhat exasperated Denniston replied on the first of December, detailing his understanding of the unwritten agreement and arguing that the British had been as forthcoming as they could be.[7]

When Germany declared war on the United States on December 11, 1941, the picture darkened again. It did not take the German Navy long to discover many lucrative and poorly protected targets along the Atlantic and Gulf coasts of the United States. Near large cities, the prey was even backlit at night. A U-boat could sit in darkness a mile or two off Miami Beach, perhaps even on the surface, and wait for merchantmen to present themselves, one by one, as if they were walking across a stage between the boat and the beach. In January, the Germans sank more than two hundred thousand tons of shipping in American waters; in March that total rose to over five hundred thousand tons, at almost no cost to the attacker.[8] The US Navy was unprepared for this threat. Poorly organized and desperately short of escorts, it fell back on appeals to yachtsmen to keep an eye out for U-boats. There was no mechanism for quickly disseminating intelligence or marshaling far-flung commands, which were decentralized after the navy's fashion. Not surprisingly, the US did not sink any of the marauding U-boats until April 1942. The Germans called it "the happy time."[9]

The British relentlessly badgered the Americans to respond to the disaster unfolding off their shores. The loss of shipping affected everyone, not just the Americans. The failure to form convoys seemed almost criminally negligent. First Sea Lord Adm. Sir Dudley Pound prodded

Admiral King and complained to Churchill. Starting in February 1942, the prime minister expressed his concern and then his exasperation in pointed notes to Harry Hopkins and FDR himself.[10] In what was for him strong language, Roosevelt was moved to comment that King seemed to be taking an unconscionably long time "to get things going"; King replied, as evenly as possible, that the navy was doing the best it could with the limited resources at its disposal.[11] He was not against coastal convoys, only against dispatching them without adequate protection.[12]

Unprompted, the British decided to send twenty-four anti-submarine trawlers and corvettes, complete with crews, to the United States, a reversal of the usual flow of materiel across the Atlantic that reflected the magnitude of their distress at the losses in American waters.[13] At roughly the same time, a delegation headed by John Tiltman of GC&CS traveled to Washington. He came bearing gifts: "everything I could possibly bring in the way of solutions."[14] The British delegation held marathon talks with Op-20-G, educating the Americans on the ways that Bletchley Park and the Admiralty were fighting U-boats on the high seas.[15]

By April 1942, the US Navy had started to craft a response. Daytime coastal convoys were morphing into an interlocking convoy system.[16] King then established an office under his control with the intention of making a better use of intelligence. COMINCH Operations Division, Operational Information Section (F-35) would focus on the undersea threat.[17] It was no accident that King did not call on ONI, which, in the words of one wartime staff member, "was too static to play an important role in as dynamic an activity as the U-boat hunt."[18] But COMINCH still lacked the means to do what the Admiralty was doing so well: using intelligence to influence outcomes by issuing timely warnings. It took another British officer to jolly the Americans on.

Rodger Winn was one of those exceptional men who surfaced in World War II and shaped operations in a way that was unthinkable in peacetime. In 1939, the thirty-seven-year-old trial lawyer with a sharp tongue volunteered his talents to the navy, the service he had long favored, thinking he could interrogate German prisoners. By chance—one story has it that a personnel officer confused his name with that of an experienced submariner—Winn wound up in the Admiralty's Submarine Tracking Room as a civilian employee, one manifestly unfit for almost any kind of military service.[19] No uniform could ever fit right. The polio survivor had a crooked spine and walked with a limp, his powerful shoulders testimony to the role that his arms had to play to keep him upright. His head seemed a trifle large for his body. But his infirmities did not hinder his work; on the contrary, they might have motivated him to work harder and achieve more. A Royal Navy report found that his "acid dialectic . . . courage and . . . formidable personality" combined with his irregular status to phenomenal effect.[20] Not unlike Rochefort, he discerned patterns that eluded others. He developed an uncanny knack for thinking like the enemy and anticipating his moves.[21] That enabled the DNI, Admiral Godfrey, to overcome the Royal Navy regulars' resistance to giving Winn a commission; in early 1941 the lawyer became a Temporary Commander in the Royal Naval Volunteer Reserve (Special Branch) and formally took charge of the Tracking Room.[22]

In the spring of 1942, Vice Adm. Richard L. Ghormley, USN, toured the Submarine Tracking Room and saw Winn at work. He was so impressed that he wrote to King, setting the stage for Winn to visit North America in late April.[23] Once in Washington, the Royal Navy liaison directed him to Admiral Wilkinson of ONI, who "explained . . . with some bitterness" that King and his staff were "very unresponsive towards ONI."[24] After hearing this remarkable confession of impotence, Winn worked his way through a series of offices at Main Navy before finding himself in front of Vice Adm. Richard S. Edwards, deputy chief of staff to King.

Edwards listened to the lecture by the British commander who was not even a regular officer—he could tell Winn's status by the wavy stripes on his sleeve—then told him that he was not buying Winn's "boloney." The US Navy would learn its own lessons, even if it meant sacrificing a few ships. Winn countered that some of those ships were British;[25] the Royal Navy was not "prepared to sacrifice men and ships to your bloody incompetence and obstinacy."[26] Edwards was taken aback; American commanders did not talk to American admirals in that way.[27] After a fraught moment of silence, Edwards laughed and invited Winn to lunch. It was, Winn reported, "an alcoholic lunch," possibly served in the admiral's suite or the senior officers' mess at Main Navy, one that enabled them to "mellow."[28] In the words of the British report, "Winn [now] felt that he had won his . . . battle with . . . U.S. Naval complacency." Like Ghormley, Edwards wound up recommending Winn to King, and arranging an appointment for them to meet in Main Navy 3048, COMINCH's spacious but not particularly grand wartime office. Winn found King surprisingly friendly and receptive (but assessed him, somewhat unfairly, as a bully who lacked substance). After hearing Winn out, COMINCH endorsed the plan to create an American submarine tracking room to mirror the one in London.[29]

The next step in the process was to find a US Navy officer to do the same work as Winn—perhaps another iconoclast, someone like Joe Rochefort or one of his assistants. King's staff remembered Kenneth A. Knowles, a Naval Academy graduate who had been forced into early retirement by failing eyesight in the late 1930s while only a lieutenant; the prewar navy still put a premium on the ability to actually see the enemy from the bridge of a warship.[30] The "tall, lean, severe-looking and intense but genial" Knowles stayed in touch with his shipmates, editing *Our Navy,* a semiofficial magazine with a news roundup and the occasional eye-catching cover, not unlike *Look* magazine, then serving as an instructor in the Naval Reserve Office Training Corps at the University of Texas.[31] Knowles was not an intellectual as much as an energetic and intelligent officer who got things done. When the war broke out, he wanted a command at sea, ideally a destroyer, but the navy continued to consider him unfit for sea duty. That made him welcome a somewhat mysterious phone call from an old boss, asking if he would be willing to serve in Washington. Anything that brought him closer to the war than Austin would suit him, he replied.[32]

The former gunnery officer did not know much about intelligence but was willing to learn. Since he was embarking on "a novel experience," not just for himself but also for the navy, it made sense for him to travel to England to learn from Winn.[33] In the summer of 1942, while Robert Ely and Joseph Eachus, two lieutenants from Op-20-G, studied procedures at Bletchley, Knowles put in ten-hour days at the Citadel, watching as the experts plotted U-boat positions and tried to divine German intentions. An important part of the experience was the personal relationship that Knowles was forming with Winn. The British officer was not only knowledgeable but also "very witty" and fun to work with.[34] Above all, Winn wanted Knowles to succeed; the American would remember him as "a marvelous supporter of [his] activities."

In August 1942, after his return to Washington, Knowles organized a tracking room at Main Navy for COMINCH that would be known as F-21. Its remit included all the major fronts in the Battle of the Atlantic; its specialty would become tracking U-boats.[35] In September 1942, the navy authorized Op-20-G to enter into negotiations with the National Cash Register Company of Dayton, Ohio, for the manufacture of four-rotor bombes that would enable the United States to develop its own capability to decipher U-boat traffic.[36] In October 1942, US Navy Communications and GC&CS signed an agreement for cooperation in the secret war against the German Navy.[37] Unlike the outcome of the Sinkov mission in 1941, this was a formal commitment to work together. The navy and GC&CS would share technology and help each other break into difficult German naval systems, vastly increasing overall codebreaking output.[38] Ultimately, Knowles would have the benefit of British and American codebreaking, as well as the acumen of Rodger Winn: he would have a private line to Winn and a teleprinter to the Op-20-G annex on Nebraska Avenue a few miles away where the bombes were performing their magic.

Compounding the Allies' advantage were continuing improvements in high-frequency radio direction finding. Some thirty to thirty-five stations from Iceland to South Africa to Brazil were set up to take fixes on transmissions, however brief, from U-boats. Each fix produced an azimuth, or line, from the transmitting U-boat to a station within range; two or more fixes yielded lines that crossed, showing the transmitter's approximate location. A program known as "Tina" even analyzed the operator's "fist," the way he tapped out a message, making it possible to identify a particular operator and his boat. When all of the pieces fell into place, Knowles could establish the location of a particular boat within a fifty-mile radius.[39]

In May 1943, Admiral King created an unusual organization that he called the Tenth Fleet. It would translate the submarine tracking room's products into action. Working out of Main Navy, this fleet did not have any ships, only a powerful, exquisitely informed staff. Knowles functioned as its intelligence officer. King was the nominal fleet commander, but its chief of staff was also a flag officer, able to make "recommendations" that were tantamount to orders. These recommendations were death sentences for many U-boats as the US Navy perfected its anti-submarine tactics in the summer of 1943.[40]

More than one encounter involved a "milch cow"—German slang for a large U-boat outfitted as a supply ship for attack U-boats. A decrypt, originally from Winn in London, would yield the approximate point and time in a remote corner of the Atlantic for the rendezvous between a milch cow and another boat. The Tenth Fleet would "recommend" an attack. A navy task force built around a "jeep carrier"—a small aircraft carrier—would steam toward the enemy. A direction-finding "fix" or a blip on a radar screen might bolster the original decrypt. Then the carrier would launch its aircraft to scour mile after mile of empty ocean in the hopes of spotting the enemy. A Grumman TBF "Avenger"—the ungainly torpedo bomber that was one of the US

Navy's workhorses—might catch a glimpse of two U-boats on the surface and speed to the attack. The Germans might dive or fight back with antiaircraft guns. Soon, more Avengers along with F4F Wildcat fighters would fly out from the carrier. Determined to get the job done, the young navy pilots would press the attack home, repeatedly strafing the enemy and dropping depth charges or acoustic torpedoes. When they sank a supply sub there would be a cascading effect, as the Germans struggled to compensate for the failed rendezvous by sending messages to reschedule meetups with boats that were now low on fuel. This in turn presented new opportunities to the Americans. More than once in July and August, much the same scenario played out until the navy had sunk a total of twelve U-boats.[41]

The string of kills prompted Winn to message Knowles privately that the result was "too true to be good."[42] The clever variation on the usual aphorism (which was also the title of a 1932 work by Irish playwright George Bernard Shaw) was Winn's way of suggesting that the Americans had become too aggressive for British comfort and risked the Ultra secret that the Allies were breaking German codes. The preferred British tactic for exploiting a decrypt, rerouting a convoy, saved ships without fireworks. The Americans were not against rerouting convoys, but they wanted to eliminate the threat and shorten the war at sea.[43] And so, without waiting for British concurrence, the Americans went to work. The results were dramatic. Before the Tenth Fleet was created in May 1943, American forces had sunk only thirty-six U-boats. In the remaining seven months of the year, that number leaped to 101.[44]

Winn was right; the Germans could not fail to notice and wonder what had changed. As early as November 1941, Admiral Dönitz, *Befehlshaber der Unterseeboote* (or BdU, commander of U-boats), had ordered his staff to consider whether the British had broken into his communications system.[45] Now, in 1943, as attacks on his boats seemed to be coming out of nowhere with such deadly effect, he was asking the same question again. The answers were always the same: the probability that anyone could break into a code generated by Enigma was tiny.[46] It was even smaller when the enemy was British. Or so the Germans thought after breaking into *their* codes. From the autumn of 1940 through the summer of 1943, German Navy codebreakers in the Beobachtungsdienst (or B-Dienst) were reading some useful British and other Allied messages.[47] Not only were the Germans unimpressed by the quality of the codes, but they could also find nothing that suggested that the enemy was reading their messages.[48]

Without good reporting from other sources, such as refugees or spies, the Germans explored other hypotheses. They imagined Allied technological breakthroughs—perhaps something like infrared rays—or treachery, even a traitor at BdU headquarters. As time went on, their misplaced reactions became ever more far-fetched: hunting for spies; painting U-boats to deflect rays; enhancing communications protocols; limiting access to secret spaces; and generally tightening central control—many things apart from the two that might have helped: replacing the Enigma or, better yet, drastically cutting transmissions from U-boats.[49]

Unable to keep up with Allied innovations, the Germans fell further and further behind in the cycle of move and countermove. Relying on their own specialists, they did not enjoy the benefits of the evolving Anglo-American system, which was a mix of collaboration and competition. The teamwork of Winn and Knowles was almost unthinkable in the German Navy: a brilliant civilian and a medically retired officer from two different countries, cooperating across thousands of miles, developing information that got results. Knowles saw it as a near-perfect fusion of intelligence and operations. When he looked back twenty years later, he wrote that intelligence had been the first and governing step in most operations. He was unable to recall "a single operation that was laid on without full review and use of all intelligence factors."[50] The victory over U-boats in 1943 might not have been as pure a triumph of intelligence as rerouting convoys in 1941—now there were other factors at play like the jeep carrier task forces and lethal new munitions—but it was a key part of the equation.

The US Army and the British

The army codebreakers did not know every detail of every transaction between the navy and the British, but, especially after the Sinkov mission and the return visits by the British, who typically visited both services, they were hardly in the dark, either. That made them want what the other service had: their own written agreement with GC&CS and something like the stream of intelligence that flowed from the Admiralty to Op-20-G. Carter Clarke and Alfred McCormack, the army team that produced the Magic Summaries, had long suspected that London had more to offer them.

In December 1942, Clarke put the issue to the thirty-four-year-old Telford Taylor, the lawyer who had recently come to the War Department by way of Harvard and the Federal Communications Commission, where he had been the general counsel at a very young age. Taylor had drawn duty at the Pentagon for Christmas, which he spent reading old files about Pearl Harbor.[51] Early the next morning, as he was getting ready to shave at a bathroom sink, in came Clarke. The Roosevelt-hating, long-serving regular was deciding that this particular elite New Dealer was not turning out so badly after all, and asked if he was "one of these guys that joined up only on the belief that he'd stay in the Pentagon building."[52] No, he was willing to travel, the personable Taylor replied. Clarke went on in his colorful way:

> You know these bloody English. We don't get anything they're getting out of the German traffic. . . . That Naval guy they have here . . . is . . . [an obscenity]. They've got all this stuff and we ought to have it. . . . We've decided . . . to send a liaison team over there [to get what we need] and you're "it," if you want to do it.

Clarke's inquiry was not as casual as it sounded but part of a series of transactions. In December and January, British representatives were in Washington on a variety of errands. Then colonel John Tiltman, the friendly head of the Military Section at Bletchley, traveled across the Atlantic again, this time with the great mathematician and cryptanalyst Alan Turing.[53] Tiltman was there to attend a routine conference on how GC&CS and the US Army were handling diplomatic traffic, which was still the army's specialty.[54] Turing wanted to consult with the navy on Enigma. As one of the British experts on voice scramblers, he also wanted to see the prototype of a new US scrambler.[55] This seemingly routine request caused a crisis in relations between the two allies.

The army, in the persons of Colonel Clarke and General Strong, balked, faulting the British for not going through the proper channels. Like the navy, they also demanded reciprocity: the British seemed to believe it was more blessed to receive than to give. Tiltman worked the issue for Turing. After finding his way down a maze of hallways that were still under construction in the Pentagon, he endured seemingly endless meetings with Strong. In his slow, deliberate manner, Strong talked around the issue, at one point musing aloud that Tiltman must think that he, Strong, had "horns and cloven hoofs."[56] Tiltman was not sure whether to say "Yes Sir!" or "No Sir!" but opined that relations between the army and GC&CS would improve if Strong were to station a representative in Britain.[57]

The Turing matter sputtered along for another exasperating month, and even became a make-or-break touchstone. Somewhat disingenuously, GC&CS continued to insist that it was not holding anything significant back from the army *and* that Turing must see the scrambler before returning home. The senior British military representative in the United States, Field Marshall Sir John G. Dill, worked the issue at his level, virtually bombarding army chief of staff Marshall with notes, following up with phone calls and face-to-face chats. Marshall was civil, but stood up for his officers."[58] Dill eventually delivered a note that was an ultimatum, even if it was politely phrased: produce the scrambler or face the consequences. Marshall opted for a compromise. In the end, Turing received official permission to see the device which, it turned out, he had already seen, but only unofficially.[59] Still, the bitterness lingered for weeks as the British and the army struggled to define their relationship going forward.

In February 1943, Friedman reported to his chain of command that Arlington Hall would have operational bombes within two months.[60] To make the system work, he needed input from the British, including know-how and raw intercepts (which were hard for the Americans to capture from US sites). He knew the British were not enthusiastic. They continued to fret about losing control of the process: it was dangerous and unnecessary to duplicate Bletchley's capabilities. The Americans, whom they saw as less disciplined, might inadvertently compromise Ultra. This was not unlike the army's and navy's reasoning for keeping OSS out of codebreaking. Admiral Redman, who did not trust the army, trusted OSS even less and cautioned Tiltman never to discuss codebreaking with Donovan if he wanted to stay on good terms with the US Navy and the US Army.[61]

In February, Royal Navy Captain Hastings, the GC&CS representative in Washington who by now had made himself unpopular, conveyed another thinly veiled threat: if the army persisted, it could jeopardize all transatlantic cooperation on signals intelligence.[62] As if that were not enough, he added disparaging comments about army codebreaking and questioned the value of Purple.[63] In the end, the moderates once again prevailed. Telford Taylor, one of the peacemakers, demonstrated both his grasp of the subject matter and his ability as a negotiator, advocating steps that accommodated British concerns. According to Taylor, the Americans should soft-pedal their intent to create their own system knowing that, no matter what happened in the short-term, they would become independent in the long-term.[64]

It was at this stage of the process that Taylor went to Britain with Friedman and McCormack. The trip started on a Friday afternoon in April 1943 at Washington's National Airport.[65] Although the Army C-54 configured for passengers was luxurious compared to a bomber, it was not a comfortable experience for Friedman. At 16,000 feet, he had to use an oxygen mask and was terribly cold; when the pilot descended to 2,500 feet, the warmer air was rough, and Friedman was buffeted by waves of nausea. But they arrived safely in Britain only twenty-five hours after they left Washington.

The small, compatible team, nominally headed by McCormack, divided its time between London and Bletchley Park.[66] Early on they called on "C"—then brigadier general Menzies—at Broadway, the MI6 headquarters. "C" displayed little enthusiasm for bringing another group of Americans into the fold, but his visitors found him "very dapper and pleasant"; "whether he liked it or not," Taylor would remember, "he was thoroughly reconciled to the necessity of it."[67] Four days later, the Americans called at 7–9 Berkeley Street, a seven-story apartment building in upscale Mayfair where GC&CS handled diplomatic traffic from a variety of countries. Here, the welcome was whole-hearted; Friedman's friend Denniston was in charge. He had been demoted after leading GC&CS (which handled both military and diplomatic traffic) to its phenomenal early victories over Enigma.[68] The ostensible reason was that he was not adapting fast enough to mechanization. His fate resonated with Friedman, who had been demoted after being hospitalized at Walter Reed in the wake of his team's victory over Purple. Both men had been deeply hurt but soldiered on for the greater good.

Berkeley Street became a base for the visitors who were at home in the world of diplomatic codes.[69] For his part, Denniston was happy to share his secrets and discuss how best to structure Anglo-American cooperation. He told McCormack, "I have arranged that you should see every section and every detail."[70] The Americans were impressed by how well and how much the British were doing with so little. Taylor noted the "two elderly civil servants . . . [who were] practically palsied" but doing the work of twenty young men at Arlington Hall.[71] McCormack saw a "well[-]established operation" suited for the long haul in both war and peace.[72] He noted the careful staff work of filing, cross-referencing information, and ensuring that it reached the customers who needed it.[73] His cables home were so effusive that the War Department, apparently fearing that he was getting too friendly with the British, directed him to tone down his discussions with Denniston. In reply he surmised in a cable that Arlington Hall was preparing the "Washington Monument for [the] appropriate part of my anatomy."[74]

When the Americans visited Bletchley in early May, the reception was somewhat more formal, but still welcoming. More of a presence than his predecessor Denniston, whose nickname was "the little man," the new director Edward Travis bore the nickname "Jumbo." He was a burly, energetic man, described as "five foot seven inches in height and substantial with it."[75] On the first day, Travis treated his visitors to lunch in a private dining room in the main building, the former rich man's home that struck Friedman as "a terrible[-]looking structure."[76] Lunch started with a choice of gin and bitters or scotch, and lasted for some two hours (after which McCormack embarrassed Friedman by falling asleep).

Having dedicated himself to the mechanization of Bletchley, Travis was happy to offer a quick introductory tour of the huts. At his leisure later in the month, Friedman was able to immerse himself in the day-to-day work of the Park, where the focus was on German military traffic enciphered by Enigma machines. With his expertise, he was better able than any American to appreciate that the British had succeeded by mastering the many Enigma subsystems that changed constantly, the process that was so different from breaking Japanese diplomatic codes like Red or Purple.[77]

Friedman's overall impressions were similar to Taylor's and McCormack's at Berkeley Street: conditions were poor by American standards, but the work force, now more than four thousand strong, was cheerfully performing miracles. The cold and damp work areas, "compared to ours, [were] . . . veritable rabbit warrens."[78] The furniture was primitive, the decor nonexistent. Neither the spaces nor the employees were particularly clean after more than three years of war. Even the principals looked "seedy," their clothes "frayed, dirty, unpressed." But the output was "a sheer wonder of organizing achievement," far exceeding the Americans' expectations. Friedman was impressed by the per capita volume of production. The Americans may have had an edge in "machines, mechanization, fine offices, etc." but "these *amateurs* [emphases in original] . . . have largely surpassed us in detail, attention to minutia, digging out every bit of intel possible and *applying* high-class thinking, originality & brains to the task." This was high praise indeed from a man who was a true professional in what was still a very small field.

The British, who had long been frustrated by having to deal separately with the US Navy and the US Army, did not mind playing the Americans off one another. Travis offered to let Friedman see the Park's naval Enigma machinery—against the express wishes of the US Navy—and asked him to say nothing to Op-20-G. Then, in early May, Travis set out for Washington. Before Friedman could see the naval Enigma, Travis sent word back to Bletchley withdrawing his offer, apparently after talking to Op-20-G himself. Friedman was outraged at the "severe reflection [on] my own status & trustworthiness," and resolved to have the matter out with Wenger when he returned home.[79]

But first there was also good news from Washington. In mid-May, Travis negotiated an agreement formalizing relations between the GC&CS and the army that included a commitment to share intelligence from Axis military traffic, as well the technology used to process it and the security protocols that the British wanted.[80] Under the agreement, army liaison officers, starting with Taylor, would be assigned to GC&CS. They would be allowed to read all decrypts and summaries and select material for secure transmission to US

Army commands and staffs. On May 26, Menzies appeared "rather suddenly" for a formal lunch at Bletchley.[81] He sat at the head of the table, with Friedman to his right. Friedman sensed that "something special [was] brewing." With characteristic circumspection, "C" did not specifically mention the agreement, but was "particularly nice" to his American guest.

The mission that Clarke had sketched out to Taylor in December—overcoming the obstacles to getting what the army needed from the British—had been accomplished. The outcome was that and more, another buttress for the growing transatlantic codebreaking partnership. For the Anglophile Friedman, the experience was both professionally and personally fulfilling. His British counterparts had fulfilled his requests and shown him great respect, more than he sometimes received at home. His work schedule was exhausting, but when he found the time, he reveled in the local culture. Friedman recorded his keen observations of life in wartime Britain in a diary: taking in the "great destruction" from the Blitz on a walk around St. Paul's Cathedral in London; being accosted by good-looking streetwalkers who spoke softly, almost tenderly, to him as he passed in blacked-out Piccadilly; listening to soapbox orators at Speaker's Corner in Hyde Park exercise their right of free speech; watching not particularly military female auxiliaries march British style, backs straight and arms swinging high, to the music of a military band; having a "delightful conversation" with two Shakespeare devotees on Denniston's staff; marveling at centuries-old books piled high in the offices of the Oxford University Press; walking through the medieval courtyards of Corpus Christi College, Cambridge, little changed since the college was founded in 1352. He liked Cambridge best, drinking in its essence "in great gulps."[82] Here Friedman found "the solidity that is England," ancient buildings "devoted to learning and democratic institutions and the dignity of man," words as good as any politician's declaration of what the Allies were fighting for.

17

INTELLIGENCE AND THE MAIN EVENT

Life with British Codebreakers before D-day

On the morning of June 11, 1943, Friedman, McCormack, and Taylor marked the end of their mission by once again calling on "C" at his office on Broadway. Travis and Tiltman had come down from Bletchley for the friendly sendoff; the Americans reciprocated by taking Travis and Tiltman to a nearby US Army mess where the food was always plentiful for a lunch that lasted well into the afternoon. That evening, Friedman and McCormack caught a sleeper train from London's Euston Station to Prestwick, Scotland, where they would board another US Army Air Transport Command C-54 the next day. Before the long flight home, they fortified themselves with three double scotches each.[1]

Taylor was now on his own, left in Britain to represent McCormack and Clarke. He started in Denniston's Berkeley Street offices, immersing himself in diplomatic traffic, looking for material that would help McCormack prepare his top-secret digest for Washington decision-makers. It would have been a daunting task for anyone, but especially for someone with only a few months of experience who was not a trained intelligence officer or cryptanalyst. No one told

Taylor specifically what to look for, but, since he had spent time at Arlington Hall reading Japanese diplomatic traffic before coming to England, he knew almost instinctively what the US needed when he saw it. Under the terms of the Anglo-American agreement, he would then ask the British to transmit his selections to Washington on their (supposedly more secure) communications circuits.[2]

After about six weeks, Taylor moved to Bletchley Park. He found the work there more challenging. Not only was military traffic new to him, but there was so much of it; the quantities were increasing as the British broke into new systems and the demand for information exploded. Taylor had to sort through a mass of information to find the bits that could be useful to the US Army; it was almost like learning a new language and having to use it at the same time. Thankfully, his sharp intellect and pleasant manner combined to good effect; he meshed well with his British counterparts, who, with rare exceptions, helped him succeed. This was true even though few of them, even if they came from Oxford or Cambridge, had met any Americans before the war. One Cambridge historian, Asa Briggs, found that he could talk to Taylor "as easily about history as about cryptography," and that the American had come to appreciate the unusual community at Bletchley.[3]

As the summer wore on, Taylor was reinforced by other, hand-picked officers. Some of the British were standoffish at first, resenting the small wave of well-dressed and well-fed invaders who could escape to London once or twice a month for forty-eight hours, stay at the finest hotels at "ludicrously cheap rates," and buy cigarettes or alcohol for next to nothing at an American post exchange.[4] But by 1943, the brilliant British amateurs were a little bored with their own company. Even if they were in uniform for the duration, the mathematicians and German scholars were not very military, and did not stand on ceremony or rank. They believed in what they were doing and were willing to welcome others who shared their goals. The British went on to incorporate the Americans in their social and professional lives, inviting them to join in killer chess games, excellent amateur operas and variety shows, and even cricket matches.

The British Attend the Birth of American Counterintelligence

Less than thirty miles away, a group of Americans was doing similar work, but only indirectly for the US Army or the US Navy. The detachment from OSS had appeared in April 1943 in St. Albans, a small, quintessentially English city that sits a little more than halfway on the road from Bletchley to London.[5] While Bletchley was defined by a rail junction, St. Albans had been dominated for hundreds of years by an ancient church that grew into a striking cathedral. A small, pleasing river flowed through the city. Since it was an unlikely target for the Luftwaffe, St. Albans had become the wartime home of MI6's archives—the asset that every intelligence service safeguards—along with the counterintelligence officers who needed the files to do their work. Headed by the hardworking but not particularly imaginative Felix Cowgill, a former policeman who had come up through the colonial administration of India, Section V's job was to defend Britain from enemy spies.[6] Ideally, it unmasked them before they entered the country; if they were headed for the island kingdom, it cooperated with the internal security service, MI5.

Section V sifted through mountains of leads in order to identify suspicious activities, then compiled card files that could be checked and cross-checked. Leads might come from intercepting the mails—like the British operation that Donovan had witnessed at Bermuda—or from a station abroad that had come across useful tidbits of information.

The case of Ángel Alcázar de Velasco occurred early in the war, attracted high-level attention, and set precedents. The short, energetic thirty-one-year-old Spaniard had already led a colorful life. He had trained to be a bullfighter before reinventing himself as a founding member of the far-right Falange—he was so far to the right that Franco's men had thrown him in jail for two years. But he was not without charm. The British press attaché in Madrid portrayed him as a likable fanatic who could down a fair amount of whiskey and spin tall tales, occasionally firing his pistol into the ceiling for emphasis.[7] In January 1941, when the Spanish government selected him to serve as its press attaché in London, MI6 in Madrid messaged MI5 in London that Alcázar was a "very dangerous man" who had declared that he was "going to work for Germany in England," presumably as a spy for German intelligence, the Abwehr.[8] Other British officials chimed in that he seemed to be "a tricky little customer" and "a complete snake," but they did not oppose his appointment, partly on the grounds that he would be easy to monitor.[9]

After he arrived in England, the British tapped Alcázar's phone, read his mail, and searched his luggage, all to little avail, then paid a massive bribe to obtain the contents of his personal safe, which yielded little more than clippings from British and American newspapers.[10] Taking another tack, they launched a double agent against him. Posing as a Welsh nationalist who wanted to work against Britain, Gwilym Williams met with Alcázar during the summer. Even after a few months it was still not clear what Alcázar was up to. Was he running other cases in addition to meeting the double agent? John C. Masterman, the somewhat otherworldly Oxford don who did counterintelligence work for MI5 during the war, remembered that "we were [at the time] still obsessed by the idea that there might be a large body of German spies above and beyond those whom we controlled" or knew about.[11]

In late 1941, Bletchley broke the Abwehr's codes, and it became clear that Alcázar was more of a snake than anyone had imagined: most of his sub-sources did not exist. He had fabricated them. The same went for the information he was selling to the Germans.[12] Ultra had just introduced itself as a very powerful instrument of counterintelligence, far more powerful than traditional instruments like telephone taps and mail covers. Thanks to Ultra, the British did not have to make inferences from incomplete evidence; instead they could read Alcázar's reports to the Germans, and learn what his handlers thought of him.

With this kind of information, the British saw they could build a double-cross system. To control enemy operations, they could catch spies when they landed in Britain and then double them back against their handlers.[13] MI5 started by offering the spies a choice between execution—their usual fate when caught in wartime—or collaboration. Some German spies refused to collaborate and were hanged. The rest told the British what they knew about the Abwehr. The British could use that information to catch other spies. They could also deduce German plans and intentions from the spies' instructions.* If they went one step further, they could influence those plans and intentions by crafting false information for the double agents to report back to the Abwehr.

This strange scenario would have made sense to a handful of special agents in the FBI, especially those who were learning at the feet of MI5. But few in COI/OSS had any feel for such work early in the war. There was no counterintelligence branch in the young agency; Donovan did not want one. He was no more interested in organized counterintelligence than he was in the strategic planning that Jim Rogers was urging on him. Such constraints on operations were not suitable for a man of action, especially in wartime.[14] But the British wanted to share secrets with the Americans about various matters, especially double-cross operations, and they needed to be sure that the colonials could protect secrets. That was why "C"—Menzies—asked Stephenson, his man in New York, to tell him which American agency could liaise with London about double agents. Having moved away from the FBI to court OSS, Stephenson stretched the

---

* If, for example, a spy was told to gather information about a navy base, that might mean that the Axis was planning to attack it.

truth when he replied that OSS was branching out into the field of counterintelligence.[15]

To make reality conform with his story, Stephenson embarked on a campaign to create an OSS office like Section V.[16] He coached Donovan on substance and style; they had to pitch the plan in a way that would satisfy MI6 in London without provoking G-2 at home. By September 1942, Donovan was explaining counterintelligence operations to General Strong as if he were an old hand at it: British information, leads developed by American agents overseas, and "the penetration of the enemy's system of secret communication . . . [were coming together to uncover] the identity of his agents, their activities, and plans."[17]

At roughly the same time, Operation Torch, the Allied invasion of French North Africa, was underscoring the need for American counterintelligence. The US Army still had a lot to learn about keeping secrets and protecting itself from foreign intrigue. In late December 1942, an OSS emissary named George K. Bowden traveled to St. Albans to explore.[18] The British resisted when Bowden suggested that they share information by sending it across the ocean to OSS. But they were willing to let OSS officers read counterintelligence files at St. Albans (which would allow the British to control the paper trail). On his way home in January 1943, Bowden called on "C," who confirmed the offer, even adding that he would erect a hut for the Americans next to Glenalmond, the redbrick manor house with twenty bedrooms that served as Section V's headquarters in St. Albans.[19]

The agreement between Bowden and Menzies occurred before OSS had any counterintelligence officers. It was only in early 1943 that OSS designated anyone for the new specialty. One of the first was Norman Holmes Pearson, a reed-thin, thirty-three-year-old English instructor from Yale who had never fully recovered from childhood tuberculosis. A limp disqualified him from military duty. Though more than willing, the new hire barely knew anything about OSS, let alone counterintelligence; Pearson had not even taken the agency's basic training course when he received his new assignment.[20] In early March, Donovan signed an order creating a small counterintelligence division within the Secret Intelligence Branch, a half measure that was better than nothing. Only at the end of the month did General Strong, who had blown hot and cold about OSS counterintelligence, withdraw his objections to letting OSS proceed.[21] Finally, in mid-April, Pearson led a team of three men and four women across the Atlantic to St. Albans.[22] They arrived ignorant but eager to learn how to set up a counterintelligence program.

Some of the British looked down on the beginners. This was especially true of the chief of the Iberian subsection, H. A. R. "Kim" Philby, who found them "a notably bewildered group."[23] The Americans were more polite about Philby but must have found him odd by any standard; he often dressed in his father's World War I khaki tunic and stuttered so badly that his tongue seemed at war with his brain.[24] (This might have had something to do with the fact that, although Philby was a British counterintelligence officer on the surface, his true allegiance was to Moscow. Recruited by the NKVD in 1934, the Cambridge graduate had already been secretly working for Soviet intelligence for years.) Unlike Philby, his congenial chief Cowgill was determined to help the Americans succeed, granting them almost unfettered access and inviting them to familiarize themselves with the work of Section V.[25]

A wartime member of MI6, the writer Malcolm Muggeridge described the process as "going from room to room and from desk to desk, and listening to particular officers explaining where they were at; whether directing and supervising the operations of agents in the field, devising and planting deception material, or cooperating with other counter-Intelligence agencies connected with the various Allied governments-in-exile."[26] After this orientation, Muggeridge was told to immerse himself in the files, "the prize exhibit here being the yield of cracked enemy ciphers"—the Abwehr decrypts that unmasked German operations. The crux of the OSS contingent's work would be similar: sorting through incoming traffic to find the messages that could be useful to OSS counterintelligence, then preparing them for transmission through British channels to Washington, where clerks would index them and create a new set of files.[27] One of the Americans at Bletchley captured the difference between the two countries and their systems: the run-down, poorly furnished MI6 "Index Room" that held hundreds of thousands of cards, many of them dogeared but cross-referenced and up-to-date, contrasted with the rows of cabinets at the Pentagon that were new and gleaming but empty.[28]

Pearson and his colleagues stayed in St. Albans until July 1943, when they moved with Section V to 14 Ryder Street, a dilapidated, six-story "grotesque Victorian layer-cake building" in downtown London.[29] By then the counterintelligence division in SI had expanded into a branch. It would be known as X-2 (not unlike the British double-cross system, known as XX).[30] The branch remained heavily dependent on the British for tradecraft and raw material. Newly minted X-2 officers came to London to perform what amounted to internships. They would walk through the open offices on the ground floor to the creaky elevator that took them to the OSS offices on the two top floors, where they encountered Pearson. Slightly hunched over, the chain smoker enthusiastically welcomed newcomers. He clearly relished the "clandestine hokey-pokey," which, they thought made him seem like "the epitome of a counterintelligence chief."[31]

The OSS Headquarters in London

Though housed with Section V, Pearson and his colleagues were officially part of the growing OSS base in London under David Bruce, the wealthy, cosmopolitan fifty-five-year-old who had joined Donovan as the first chief of the espionage branch and been tutored by Bill Stephenson's deputy, Dick Ellis.[32] British society embraced Bruce in a way that it embraced few Americans. Not only had he shared wartime risks with Londoners—in the course of his service as a Red Cross officer during the Blitz—but Bruce also seemed to be a member of the old imperial ruling class. When he visited England in the spring of 1942, it felt almost like a homecoming: he found it "very natural to be back at the Ritz, with its . . . indestructible old servants, and moths flying out of every stuffed chair."[33] From that base he interspersed official calls—on "C," Felix Cowgill, and Desmond Morton, the prime minister's special assistant for intelligence—with lavish meals in the city's most exclusive private clubs and restaurants. On a typical evening, he dined in a "glaringly decorated private room" at the Dorchester with Lord Louis Mountbatten, the Royal who was then chief of combined operations. When Donovan came to town in mid-June 1942, Bruce joined him for meetings with "C" at MI6 and Sir Charles Hambro at the Special Operations Executive (SOE).[34] In February 1943, when Bruce returned to London to head the base, he received a warm welcome.

At OSS London, Bruce presided over the exponential growth that reflected the sweep of the war. The Anglo-American campaign in North Africa ended in victory in mid-May 1943. The Trident Conference in Washington followed, its outcome a decision by the combined British and American chiefs of staff to invade France in the spring of 1944. July 1943 saw the Anglo-American landings in Sicily and in the east the epic Battle of Kursk between the Wehrmacht and the Red Army, whose soldiers continued to do most of the fighting and dying in the war. In September 1943, British, Canadian, and American armies started the long, bitter slog up the Italian boot. Nevertheless, the three allies reaffirmed their decision to invade France, and Operation Overlord remained their main effort. The operation was shaping up to be a massive undertaking that must succeed; if it failed, it might be months or even years before the Western Allies could try again. For OSS, there could be nothing more important than preparing to "contribute substantially to the assault on Germany from the west."[35] That was what drove the growth of the London base from a small representative office, manned by a handful of liaison officers in the fall of 1941, to a microcosm of its parent in Washington. By the summer of 1944, OSS London would peak at about three thousand.[36]

The relationship between British and American intelligence was evolving. In 1941, the British had intrigued, plotted, and lobbied; they wanted the US government to create an agency in their own image that they could use to influence American policy. COI/OSS was not exactly what the British wanted, but it was good enough. After the United States entered the war, the two countries officially became allies; the British relaxed somewhat, manipulated less, and accepted OSS as a junior partner.

OSS did not know that it was a partner to two agencies with major flaws. Part of MI6, the GC&CS functioned at a high level, but it was taking the service considerable time and effort to build back from the 1939 debacle at Venlo in Holland that had devastated British spy networks on the continent.[37] Although it benefited from the assistance of the European governments in exile in London, MI6 did not have any spies in the upper echelons of the Nazi hierarchy[38]—unlike the Soviets with their well-placed spies in London, Washington, and even Tokyo, as well as the Manhattan Project.[39] Chartered in 1940, MI6's rival, the Special Operations Executive (SOE), was still experiencing growing pains in 1943. Churchill's charge to set Europe ablaze left a great deal of room for interpretation. SOE considered various approaches. Encouraging citizens to rise up against the occupiers accomplished little. Random acts of sabotage tended to be exercises in futility, and usually suicidal. Targeted killings—like the operation in May 1942 to assassinate SS chieftain Reinhard Heydrich in Czechoslovakia—were problematic. Trained, outfitted, and transported by SOE, the Czechoslovak commandos were able to kill Heydrich before they themselves were hunted down and killed, leaving behind a never-to-be-forgotten legacy of bravery and sacrifice.[40] But the operation also resulted in widespread reprisals against innocent victims. The Nazis razed the village of Lidice, killing the men and imprisoning the women and children, then systematically eradicated the Resistance in Czechoslovakia. In a string of SOE operations in Holland from 1941 to 1943, carelessness cost the lives of fifty British agents.[41]

Donovan would not have known about the fiasco in Holland but he would have known about Heydrich, whose assassination made the news around the world, as did the Nazi retaliation. Still, the operation did not keep Donovan from reaching out time and again to SOE, the organization that was more like OSS than MI6 and seemed to know what to do. Douglas Dodds-Parker, a senior SOE officer, describes one of the American's early visits to the headquarters at 64 Baker Street, the plain modern office building that was a short walk from the mythical 221B, where Sherlock Holmes supposedly worked: "We could show him the groundwork of a Europe-wide organization . . . targets; research and development; codes and signals; air and sea transport. . . . All was clearly set out on the maps in the Operations Room . . . carefully curtained so that only one could be visible at a time."[42] Duly impressed, Donovan asked SOE director Hambro to come up with an operation that included a role for OSS that would give the young agency "a good send off with his masters"—that is, recognition and credibility in Washington.[43] Donovan offered to outfit the operation and, apparently thinking of the Heydrich operation, even supply "suicide squads." Nothing came of this impulse. But it was no accident that, over time, OSS followed SOE's lead.[44]

The partnership had valleys as well as peaks. After gaining confidence in North Africa, the junior partner became more assertive, demanding something like parity. The British pushed back. In January 1943, Hambro was "horror struck" at the suggestion that OSS officers could be anything but liaison officers.[45] Later in the year, Donovan upset the well-respected new director of SOE, Maj. Gen. Colin M. Gubbins, by appearing to signal that he did not plan to abide by the "legalistic" written agreements about the division of responsibilities between OSS and SOE.[46] Yugoslavia, the site of a complicated guerrilla war against the Nazis, was a particular bone of contention.[47] Donovan remembered the country from his tour of the Mediterranean in 1941, and felt confident making what amounted to his own policy decisions two years later. More than once he broke the unwritten rule of civility between allies, even when disagreeing; in the fall of 1943 he stunned the British with a violent outburst against British brigadier Fitzroy Maclean, a man as colorful and forceful as himself.[48] The disputes were bitter and serious enough to attract high-level attention; both Churchill and Roosevelt became involved. By the end of the year, the problems between OSS and SOE were more contained than solved and were overshadowed by Overlord.

Preparing for Overlord

The coming Allied invasion of Europe demanded everyone's full attention. Each of OSS London's fourteen branches supported the preparations to the best of its ability: the academics in Research & Analysis prepared background papers for the army and studied the effects of strategic bombing; the propagandists in Morale Operations devised ways to undermine the enemy's will to fight; the gadgeteers in Research & Development outfitted spies about to deploy to the continent; the counterspies in X-2 planned operations to protect Allied forces from German spies after D-day.[49] First among equals were, not surprisingly, Donovan's warriors in Special Operations (SO), followed at a good distance by the spies in Secret Intelligence (SI).

In early 1943, SO embarked on the journey it would take in tandem with SOE. OSS officers came to London to learn how SOE operated and configure SO London for combined operations.[50] In March they were invited to observe "Spartan," a field exercise in England that included a test of a novel concept "for dropping behind enemy lines, in cooperation with an Allied invasion of the Continent, small parties of officers and men to raise and arm the civilian population to carry out guerrilla activities against the enemy's lines of communication."[51] Spartan would spawn a flagship of British-American-French cooperation in special operations.[52] Under the randomly assigned codename "Jedburgh," SOE and SO/London would jointly recruit, train, and equip three-man teams, each made up of one European officer, one British or American officer, and an enlisted radio operator, for insertion into France, Belgium, and Holland.

The next, bureaucratically intricate steps were to market the concept to the various American and British chains of command. In April, Bruce cabled Donovan asking him to buy into the initiative.[53] In the fall, SO and SOE stepped up liaison with the Gaullist Free French, especially their intelligence Bureau Central de Renseignements et d'Action (BCRA), while OSS officers scoured army bases in the United States for soldiers who spoke foreign languages and were willing to volunteer for dangerous work overseas. OSS subjected the volunteers to pseudoscientific screening and preliminary training before allowing anyone to cross the Atlantic. Lt. Roger Hall was "tested morning, noon, afternoon and most of the night. . . . We filled in blanks, picked numbers, chose pictures, pulled levers, pushed buttons, and wrote page after page."[54] Even the irreverent Hall had to admit that the process seemed to work; apart from a few "goofs, misfits, and glory-seekers," once they were screened and put to work, the commandos would fly high.[55]

In January 1944, an SOE/SO headquarters intended to coordinate "the activities of resistance groups in Northwest Europe with the actions of Allied invasion forces," took its place under the Supreme Headquarters Allied Expeditionary Force (SHAEF), which was commanded by General Eisenhower.[56] In February, the Jedburghs began three months of intense training at SOE facilities. Teams were formed and sabotage targets—primarily roads, railways, and telephone lines—selected. The principal goal was to impede the movement of German forces, especially those that might attack the beachhead in the days after the initial landing.[57] But to keep from signaling that the invasion was imminent, SHAEF decided to delay the first official Jedburgh missions until the night before D-day. This dramatically limited their potential; it made little sense to expect a man to parachute into France on one night, meet his counterparts in the Resistance for the first time, and, the next day or even the day after, be able to cut off German reinforcements on their way to Normandy.[58]

The Jedburghs were not designed to spy on the enemy. That responsibility fell largely to MI6 and the Secret Intelligence (SI) branch of the OSS London. Together they planned and executed Operation Sussex, their largest joint initiative of the war.[59] An officer named Francis Pickens Miller oversaw the project for the London base. A World War I veteran, former Rhodes Scholar, and distinguished internationalist, he was in many ways an ideal choice. Miller played a key role in negotiating the ground rules with MI6 in the fall of 1943: the two agencies would coordinate with each other and the Free French, but would operate on their own in the field.[60] The work called for native French speakers who would not attract attention while gathering "strategic and tactical intelligence" on the enemy.[61] After a monthslong process of recruitment and training, in April, two-man teams started to parachute into France and transmit reports back to England.[62] By the end of May, thirteen teams had been dropped into France.[63] The most valuable report came from a team designated OSSEX/6 that alerted SHAEF to the presence of the German Panzer Lehr Division, an elite formation that had not previously been identified in France.[64]

What Sussex contributed was useful. But given the scale of the invasion, it was a small contribution. Most of the intelligence for Overlord came from other sources—up to 80 percent from aerial photo reconnaissance.[65] Next in very rough order of precedence were reports from the French Resistance, channeled through de Gaulle's BCRA, followed by signals intelligence and captured German documents. Bletchley Park and Arlington Hall far outperformed Sussex when it came to fixing the German order of battle—the identification of enemy units in France—by seeing who was communicating with whom and looking at the itineraries of senior officers inspecting the front, as well as reading the text of their messages.[66]

One of those officers was the tireless Baron Ōshima, the samurai who admired the Nazis, still the Japanese ambassador in Berlin. He toured the fortifications on the French coast in the last week of October 1943 and within two weeks started to send lengthy reports back to Tokyo. American and British codebreakers proceeded to decrypt his transmissions using Purple machines. Already attuned to the value of Ōshima's reporting, the peacetime historian and wartime codebreaker Henry F. Graff quickly grasped their significance and was "too electrified to sleep," working through the night and the following day.[67] His superiors alerted General Marshall's and Admiral Leahy's offices that this remarkable guide—which in American hands took the form of a thirty-two-page pamphlet—would soon be on their desks. In the spring of 1944, Ōshima followed up with firsthand reports of his conversations with Hitler about the impending invasion, and one of his subordinates conducted his own inspection and reported the results to Tokyo.[68] It was little wonder that Generals Marshall and Eisenhower prized this enemy general so highly.[69]

On the Beach with Donovan

In mid-May, Donovan came to Bruce's plain five-story brick headquarters at 70 Grosvenor Street, not far from Grosvenor Square, nicknamed Eisenhowerplatz after the general established his office at Number 20. Miller proudly showed his war room off to the director, describing the various Sussex missions, shown on large maps of France on the wall. Donovan had somewhat of an allergic reaction to the details. He directed Bruce to call a staff meeting as soon as possible. When the base's leaders gathered, Donovan stunned Miller by declaring "here in London you have been doing too much planning."[70] Plans, he declaimed, needed to be thrown "out the window" because they were "no good on the day of battle." Confusing the careful preparations that Miller's work required with rigid adherence to a conventional operations order, Donovan claimed that the Italian campaign had been bogged down by too much planning. If he had been in command, the US Army would have been in Rome in three days after the landing at Anzio in January 1944. Donovan's outburst added confusion to an already hectic time—what exactly did the director mean by telling the staff to throw its plans out the window?—and cost OSS the services of a productive officer. Miller concluded that he could "no longer work for such a man" and needed a transfer out of OSS. Bruce found a job for him at Eisenhower's headquarters.[71]

As May ended and June began, a small window, framed by the alignment of moonlight, tides, and weather, opened for the invasion of Normandy. The massive Allied armada—over 6,500 warships, cargo ships, and landing craft—prepared to sail. They would ferry the soldiers of the five US, British, and Canadian divisions who were to take the precarious but vital first steps in the do-or-die campaign to wrest the continent from Hitler's grasp. Donovan was as determined to make this landing as he had been the year before in Italy. Following SHAEF's strict prohibition against allowing officers who knew the Ultra secret and the D-day plan from risking capture, Generals Eisenhower and Marshall both ordered him to stay away.[72] The energetic new secretary of the navy, James Forrestal, even sent a cable to London mandating that Donovan was not to board any US Navy ship in the invasion fleet.[73]

Bruce and Donovan had to devise a roundabout route to Normandy. On May 30, they came up with what Bruce called "a cover story" to disguise their actual destination from their superiors, traveling first to Plymouth, England, then to Belfast. From Northern Ireland, they boarded the American cruiser *Tuscaloosa,* which was flying the flag of an admiral whom Donovan had befriended years earlier.[74] The director of OSS and the London base chief spent the next few days waiting on events, out of touch with their headquarters. Instead, they studied the plans for Overlord and socialized with British leaders in Northern Ireland as well as the ship's officers. Bruce wrote in his diary that, while the crew was working, he and Donovan had "the free run of the ship ... going from the CIC [Combat Information Center] to the chart room ... [It was] like stopping during the afternoon for gossip in a series of men's clubs—except there ... [was] nothing to drink!"[75] Once they were off the coast of France, the two men watched the tableau of battle unfold: Allied ships stretched from horizon to horizon, firing at targets on the beach and disgorging small boats that headed for the surfline. Bruce saw Donovan prepare himself for combat, going through a familiar ritual: putting on rubber-soled shoes, buttoning his trousers at the ankles, donning an olive drab GI cap.[76] On his uniform he wore the two stars of his current rank and one ribbon, the Medal of Honor.

On D+1, the Germans were waiting to see if the invasion of Normandy had been a feint. Was the main event, a landing by an even greater force, yet to come in the Pas-de-Calais, where the English Channel was so narrow that a German sentry standing on the beach in France could see the sparkling chalk of the White Cliffs of Dover twenty miles away? Almost certainly influenced by an elaborate Anglo-American deception, Hitler was holding some of his most powerful divisions in reserve against that possibility—thereby giving Eisenhower much-needed breathing space.[77] In the meantime, the Allies were consolidating the actual beachheads some 175 miles to the southwest, where things were quieter than they had been on D-day. The fighting had died down considerably, as had the wind and the chop that had plagued the first waves of invaders. Bruce and Donovan were now able to finagle a ride to shore. The two OSS leaders worked their way from the *Tuscaloosa* to a launch, to a destroyer escort, to another launch, and finally onto a DUKW, a thirty-foot hybrid amphibious vehicle with wheels and propellers that was on its way to "Utah," one of the American beaches.

At about 4 p.m., the DUKW was moving across the sand with the two men sitting on its hood. The sixty-one-year-old Donovan still had that sixth sense that comes from having served on the front lines. When he heard the "drone of airplane motors," he knew instantly what was about to happen.[78] Though portly, he was still remarkably agile; he rolled off the DUKW to seek cover on the ground as four German ME-109 fighters swooshed low, a few feet over his head, firing bullets that spattered the hood he had been sitting on. Though fifteen years younger and considerably fitter, Bruce was new to combat and slower off the mark. He fell on top of the general, accidentally cutting Donovan's chin with the rim of his helmet. Bright red blood started to pour down the general's front. Thankfully, it was a superficial wound. After the planes flew on to seek other prey, Donovan stood up, wiped the blood away, and "grinned happily," appearing to relish the prospect of more adventures to come. "Now it will be like this all the time," he told Bruce.

The two men started walking inland toward the enemy, likely in the direction of Sainte-Mère-Église, a town made famous in the early morning hours of D-day when American paratroopers landed around the church that was its namesake. On no particular mission except area familiarization, Donovan and Bruce were forming general impressions. They kept an eye out for anyone who happened to be in the area to see what they happened to know, ideally an OSS officer or agent. But they would settle for the three civilians they could see working a vegetable patch at the far end of a field. Addressing the closest American officer who seemed to be in charge, Donovan explained, implausibly, that the three were his agents. They were expecting him; he and Bruce were going forward to meet them. This cannot have made much sense to the young artillery officer, but he let them proceed, warning them that this was still dangerous ground; the front lines remained fluid for the time being.[79]

As Donovan and Bruce advanced inland, enemy machine-gun bullets were soon whizzing overhead again, this time from German gunners on the ground. The two OSS men again took cover, trying to make themselves as small as possible. The general turned to the colonel and said that they could not allow themselves to be captured because they knew too much. Did Bruce have any pills—meaning lethal capsules of potassium cyanide intended for suicide? No, Bruce replied, he had left them at his hotel in London. Never mind, Donovan said, he had enough pills for both of them, and proceeded to go through his pockets. Still hugging the earth, he pulled out hotel keys, photos of his grandchildren, newspaper clippings, travel orders—everything but the pills. He, too, must have left his pills at the hotel. Thankfully, the need for pills was soon overtaken by events: friendly artillery and aircraft silenced the enemy machine guns. Donovan and Bruce survived to encounter Lt. Gen. Omar N. Bradley, commanding the US First Army, who had little time for them. Donovan asked Bradley to permit Bruce to "spend some time with the First Army."[80] Bradley agreed: he would "be very glad" to see the colonel at his headquarters, but for now the duo needed to "go back to wherever [they] came from."[81] Perhaps heeding Bradley, they would reach *Tuscaloosa* in time for a late dinner. Two days later, they sailed back to Plymouth, having spent the better part of ten days on their *Boys' Life* adventure.

The next day, June 10, Bruce and his branch chiefs reported to Claridge's, the luxurious hotel in downtown London that Donovan favored. Donovan treated them all to lunch. If they expected to be congratulated on their more than respectable contribution to the initial success of the landings, they were wrong.[82] Instead, Donovan took up where he had left off in May, expressing his "continued concern" about their plans.[83] He highlighted a specific opportunity he thought they had missed, due either to their own "lack of vision" or the army's reluctance to use OSS assets. Donovan referenced what he had seen during his brief visit to France: how the lines were fluid and French civilians were free to move about the battlefield. The original plan had been to place OSS teams at senior headquarters, at the army and army group levels. But why not deploy additional detachments to lower levels and try to recruit more Frenchmen to gather information on the Germans as they went about their everyday business?

The chiefs objected. The OSS's mission was strategic, while Donovan's proposal was a tactical initiative. It ran counter to their mutual understanding with the British; the US army was primed to resist new OSS initiatives. This was a considerable understatement. Many regular officers did not know anything about the OSS. Others bitterly resisted moves to augment their staffs with OSS irregulars. One G-2 officer rebuffed the idea, declaring he did not "want a man from OSS, not a dwarf, not a pygmy, not a Goddamned soul."[84] The director waved the objections away, insisting that his will be done.

Two days later, Donovan was on his way back to Washington, where he regaled Roosevelt and Stimson with his tales from the front. They were duly impressed.[85] Unlike Churchill, who was hard on senior officers who knew vital secrets and risked capture, no one in Washington called Donovan to account. That is, not until 1976, seventeen years after his death, when Kermit Roosevelt, TR's grandson who oversaw the official OSS War Report, made it a matter of record:[86]

> In retrospect—and this should have been evident at the time—it is clearly outrageous that individuals with knowledge of Ultra . . . should have been allowed to expose themselves to possible capture. Donovan himself did this three times.

18

A DREAM COME TRUE

The French Summer

D-day brought out the best and the worst in Donovan: he was by turns short-sighted, devious, reckless, daring, and uncommonly perceptive. As the Allies forced a turning point in the war, he had disrupted the operations of OSS London, wasted more than a week of his and David Bruce's time, and disobeyed explicit orders given to safeguard vital secrets. To circumvent regulations, he had cultivated relationships with senior leaders; he had undermined planning and elevated opportunism. Yet his insights about the potential of French civilians and the need to extend support to the division level were on target.

What Donovan had in mind was more tactical than strategic. It was the difference between dispatching an agent to find out if a German division like Panzer Lehr was now somewhere in France and reconnoitering the other side of the next set of hills, something that the army might normally be expected to do for itself.[1] Donovan's plan called for OSS London to dispatch more small detachments to the field in order to mount quick, informal operations to identify, recruit, and task "line-crossers" who had business on both sides of the front lines, then meet and debrief them about German dispositions when they returned. It was risky but not reckless, especially if run by OSS officers who spoke French and worked with the French Resistance.[2]

Bruce shed any reservations that he might have had about the initiative after another trip to France in the second half of June, when he was able to consult with army staff officers and members of the organized French Resistance. One, a Lieutenant Mercader, owned a bicycle business and had gone through the lines nine times since D-day.[3] The army could use the kind of information that he was willing to collect; it needed OSS men as intermediaries since it lacked French speakers and was not used to working with spies. Bruce concluded grandly in a report to Washington that no matter what OSS had done in the past by way of providing strategic or long-range intelligence, from now on "the organization's reputation and prestige will be considerably affected by the success . . . of its field units in achieving the tactical desires of army commanders."[4] In other words, OSS would thrive if it gave its military customers what they wanted.

Bruce added line-crossing to a long list of missions in the summer of 1944. Near the top of the list were the liaison teams that were already serving with the headquarters of two army groups and two armies in France and included representatives from various branches of OSS London, especially SO, SI, and X-2. The liaison teams were conduits for requests for information (from the army to OSS) and intelligence reports (from OSS to the army). The X-2 detachments had another mission: finding and neutralizing German stay-behind agents from the lists that MI6 had helped prepare.[5] Bruce also needed to stay abreast of developments in the Sussex and Jedburgh programs that were repaying the initial investments in planning and manpower. The Jedburghs began jumping into France on the night of June 5–6, and, within two months, had peaked at roughly forty-five teams in-country, while the number of Sussex agents reached fifty.[6]

Bruce decided to split his time between London and northern France. In the field, he traveled mostly by jeep with a small entourage including a driver and one or two others, each armed for self-defense.

Bruce himself carried an M-1 carbine. Though always behind the front lines, they were often in newly liberated territory that was not yet safe. They traveled from post to post, organizing, liaising, and encouraging. Not unlike Donovan, Bruce enjoyed getting away from the "paper work and details in such places as Washington . . . [and] London."[7] A good command post, like that of the First Infantry Division, where the staff was courteous, calm, and efficient, lifted his spirits. So, too, did fine dining. Work permitting, he sought out exceptional restaurants and fastidiously, almost lovingly, recorded the menus in his diary alongside observations about the war. On July 25, 1944, for example, he dined at the Inn of Three Hundred Men near the seaside town of Quettehou. He ate soup, grilled plaice, filet mignon, salad, and cheese, followed by "stewed cherries with the thickest of creams"—all washed down with "a noteworthy bottle of 1937 [white] Meursault, a bottle of red Meursault, and a fine champagne of the highest merit."[8]

Three weeks later, on August 15, the Allies launched Operation Dragoon and landed some ninety-four thousand troops on the Mediterranean coast of France. Dragoon was a smaller version of Overlord intended to put pressure on the Germans from the south. OSS supported the operation out of bases in North Africa just as it had supported Overlord from England, albeit on a smaller scale and with a greater emphasis on SI rather than SO operations. A flamboyant twenty-nine-year-old New York lawyer named Henry B. Hyde assembled and ran a large, productive network of spies south of the Loire codenamed "Penny Farthing."[9] Between 50 and 80 percent of Allied intelligence on the German order of battle before Dragoon reportedly came from OSS operations like Penny Farthing.[10] After the war, Col. William W. Quinn, G-2 of the invading Seventh Army, had fulsome praise for Hyde's "28 agent chains" reporting on "German dispositions, fortifications, aircraft, logistics, order of battle, command posts"—everything that the Seventh needed to know in order to "clobber" the enemy.[11]

Thanks in part to this intelligence, the second D-day was relatively tranquil. Donovan was among those who quietly waded ashore—in his case, near the resort town of Saint-Tropez in the French Riviera. Since his trip to Washington in June, he had traveled around the European Theater. Popping up here and there, he demanded a few minutes of a commander's time; reorganized OSS operations in Italy; dispensed cryptic guidance to his officers; arranged an audience with Pope Pius XII, who was one of the war's greatest fence-straddlers in the struggle between the Axis and the Allies; and even held exploratory peace talks with the German ambassador to the Vatican, Ernst von Weizsäcker, who received the hopeful codename "Jackpot II."[12]

Pressed between Overlord and Dragoon, unable to move very far without attracting Allied fighter-bombers, and harassed by the Resistance, the Germans retreated. One route ran to the north up the Rhone Valley from the Mediterranean, another to the northeast from Normandy. In late August, Eisenhower's armies approached Paris, still held by the Germans and still a vital symbol to both sides. The liberation of the City of Light was another event that Donovan did not want to miss. Bruce was told to place himself at the general's disposition, but Donovan did not appear. Instead, Bruce encountered Ernest Hemingway, the great American writer in France as a war correspondent. Bruce was enthralled, finding that Hemingway (who was actually one year his junior) looked "patriarchal, with his grey beard, red face, imposing physique, much like God, as painted by Michelangelo."[13] The two proceeded to Rambouillet, a small city where Allied troops, war correspondents, and operatives of many stripes were staging for the final push. Bruce deputized Hemingway, allowing him and a band of young French resisters to run the kind of tactical intelligence operations that Donovan had visualized. They based themselves at a small hotel, the Grand Veneur, whose wine cellar had, conveniently for them, survived the occupation.

Bruce and Hemingway were sitting on a volcano. Surrounded by weapons from hand grenades to machine guns, Hemingway drank heavily while he worked. (Bruce seemed to confine himself to drinking with his meals.) One afternoon, the two Americans lent a .32 automatic pistol to a French officer who said he needed it to "kill a civilian traitor"—as if that were just another routine task.[14] All the while, Bruce found it "maddening to be only thirty miles from Paris . . . and to know that our Army is being forced to wait—and for what reason?"[15]

The volcano gave off smoke but did not erupt, and the unusual pick-up team turned out to be productive. Its members improvised, seizing opportunities that presented themselves: interrogating prisoners, conducting reconnaissance patrols, and sending out line-crossers, likely on foot or by bicycle, the most common ways to get around France in 1944. They debriefed local citizens who had just come from Paris or who lived "just within" nearby German lines.[16] Bruce and Hemingway were able to provide Gen. Philippe Leclerc, the commander of the Second French Armored Division who would lead the way into Paris, with detailed information about the routes showing obstacles and German dispositions. Bruce concluded "that [by] relying on the intelligence furnished by us and others . . . the French were bypassing those points where . . . determined opposition could be expected."[17]

Once they were within Paris city limits on August 25, Hemingway and Bruce plunged into a sea of chaos. The citizenry seemed to have risen up on its own. No one knew what came first: the Germans loading up their cars and trucks, or the uprising. But now the barricades were everywhere: mattress frames, abandoned cars, downed trees, even the city's famous cast-iron street urinals, *les pissoirs*. Ecstatic crowds flooded the streets, offering Hemingway and Bruce many kinds of alcohol that they felt they could not refuse. Parisians, male and female, wanted to kiss them. Irregular soldiers traded shots with the few Germans who were still fighting, while many of their comrades fled with as much "soldier loot—chickens, washbasins, mirrors, anything you could imagine"—as they could carry.[18] An OSS officer watched in amazement as housewives poured out of apartment houses to attack the fleeing Germans with shovels and anything else they could lay their hands on. For the first time in four years, the Germans showed little interest in retaliating.

The great monuments that Hitler had toured four years earlier were, by and large, intact—thanks to Dietrich von Choltitz, the German military commander who could have ordered their destruction after the Allies appeared on the horizon and the Parisians had risen up against his troops, but did not.[19] ("Paris is like a beautiful woman," Choltitz is said to have explained, "when she slaps you, you don't slap her back.") This meant that, instead of wending their way through piles of rubble, Bruce and Hemingway could climb the Arc de Triomphe on this near-perfect day, a few white clouds dotting the bright blue sky, and look out over the city in her breathtaking splendor. They could visit the Traveler's Club, which they found nearly deserted, then liberate the Ritz Hotel where they ordered martinis all around—not very good, Bruce thought—and dinner—much better, even superb, he granted—for the hodgepodge of spies, irregulars, and writers whom they had attracted. The diners dated and signed menus for each other. The menu that the military historian S. L. A. Marshall took home with him bore words that were neither entirely true nor entirely false: "We think we took Paris."

Bruce was as upbeat as Marshall. Reflecting on the support that OSS elements had brought to bear since D-day, he wrote in his diary on August 29 that "whatever other part OSS may . . . play during this war, its participation in the French Resistance movement represents a proud achievement."[20] Especially considering how reluctant some army officers had been to allow OSS onto their turf, the chorus of praise from the commands in France was resounding. In the weeks that followed, Bruce heard the refrain again and again, from Brig. Gen. Edwin L. Sibert at the Twelfth Army Group, from five officers at Patton's Third Army, from the Seventh Army, and from each of the latter army's three divisions.[21] The combat intelligence teams that small OSS detachments had organized for each division had accomplished their missions.[22] Separately, the SHAEF G-2 praised SI for "the gratifyingly high" level of accuracy in its reports, as well as the "obviously excellent training which the [Sussex] intelligence teams had been given."[23] From the Mediterranean perspective, a senior staff officer would write about his "satisfaction and admiration—not infrequently mixed with wonderment—for the . . . almost unbelievable achievements" of the secret soldiers of OSS.[24] Eisenhower himself would write to Gubbins and Bruce to praise SOE/SO operations in France in the summer of 1944, singling out "the disruption of enemy rail communications . . . and road moves."[25] The letter contained two important qualifiers: that Resistance forces had been harnessed "to the main military effort" and that "no final assessment of the operational value of resistance operations has yet been completed." The Supreme Allied Commander, a lifelong regular officer, appreciated what special operations could do under the right conditions.[26]

In September, the Allied center of gravity shifted from London—drained of men and excitement after D-day—to Paris, largely untouched by bombs, awakening from the occupation, still stylish and ready to absorb the energy of this Allied Expeditionary Force. The city was once again no more than a few hours' drive from the front. OSS London established an outpost at a posh address, 79 Champs-Élysées, with balconies that offered sweeping views in one direction of the Arc de Triomphe and the Place de la Concorde in the other. The City of Light was where everyone now wanted to be; "every last WAC secretary and . . . lieutenant dredged up the most overpowering reason why orders had to be cut for them to go to Paris."[27] They quickly filled three nearby hotels that OSS had requisitioned, and the base found that it had to halve the original plan to bring seven hundred staffers over from London.[28] After setting up an interrogation center at the Petit Palais, a delightful downtown art museum, X-2 moved in with SHAEF at the Trianon Palace in Versailles, another excellent address previously favored by the Royal Air Force (in 1939 and 1940) and the Luftwaffe (from 1940 to 1944). OSS turned the slightly less posh Hotel Cecil into a reception center where Jedburgh, Sussex, and other field agents could congregate and relax after their missions.

The newly blooded guerrillas trickled into the French capital. They included unusual, determined men like the slight, young officer with bad eyesight, William Colby, who could easily have avoided military service. The leader of Jedburgh Team "Bruce," he would downplay his hair-raising escapades on the battlefield in central France in August, describing his mission casually as "to harass the Germans as much as possible, ambushes on the road, blowing up bridges, that sort of thing."[29] An adventurous soul from Baltimore apologized for her delay in getting to Paris and checking in, explaining that she had had to elude the Gestapo.[30] One of a handful of American women who operated behind enemy lines during the war, she was Virginia Hall.[31]

In 1941 and 1942, Hall had played a key role in running an SOE network in occupied France before escaping on foot over the Pyrenees, no easy task with a prosthesis in place of the lower half of her left leg, lost in a hunting accident in the 1930s. After SOE refused to allow her to return to France because she was known to the Gestapo, she transferred to the OSS, which accepted her as a civilian employee at a low grade, roughly equivalent to that of a second lieutenant. OSS allowed her to deploy to Central France to work with the Resistance as well as Jedburgh Team "Jeremy." From March to September 1944, she served as a radio operator and an organizer behind enemy lines, beating the odds by shunning contact with agents who talked too much and moving frequently, a nerve-racking routine. Her cover required her to accept typically female roles—cooking for farmers over an open fire, milking cows and goats, delivering milk and cheese—but the after-hours routine she described in a report was that of a seasoned operative: "looking for fields for receptions, . . . bicycling up and down mountains, . . . visiting various people, doing my WT [wireless transmission] work and then spending the nights out waiting, for the most part in vain, for deliveries" of supplies from England.[32] In her just-the-facts manner, she added that "nobody [in her circle, including herself] . . . deserves any decorations."

The German Winter

The Allies had high hopes after the liberation of Paris. By the end of August, their armies had advanced through much of France toward the frontier with Germany. OSS was reporting that its "action inside France may be regarded as having come to fruition."[33] For over a year, OSS and SOE had thought of little else but the invasion of Normandy and driving the Germans out of France. The dream seemed to be coming true. Could the end of the war be far off? OSS London's William J. Casey was photographed beaming on the balcony of the Paris office with the Arc de Triomphe in the background and an American flag in the foreground. He could almost sense victory in the air.[34]

Setbacks on the battlefield in Belgium and Holland soon tamped down the euphoria. The Wehrmacht was rebounding from its defeats in France and hardening its positions on the Reich's borders. At the end of September, OSS reported soberly that "the progress of the war in the West had carried [our] . . . activities into a transitional phase."[35] The field detachments were still in place but unsure what they would do going forward. OSS London had drawn down, but OSS Paris was not yet fully functional. Bruce was contemplating his next assignment.[36] By mid-October, it was clear to Casey, who had been sitting at Bruce's right hand, that OSS Europe could take on a new mission.[37]

Casey was in some ways an unlikely fit for the spy agency. Originally from Queens, solidly Irish and Catholic, he had gone not to Columbia or Harvard but to Fordham and St. John's Universities. After receiving his law degree, he did not practice in one of the white-shoe firms in Manhattan but on Long Island, on the side, while helping to direct the Research Institute of America (RIA), a wildly successful startup that offered advice to businessmen who were bewildered by the flurry of New Deal regulations. To one of his partners at RIA, Casey was "a coarse young fellow."[38] Most days he looked like he had slept in his clothes, talked like a deliveryman who had never been out of the city, and seemed unable to sit up straight or keep his feet off his desk. His table manners were an abomination. But, the partner added, he also read voraciously and had "a roving, curious mind." He could quickly absorb and process masses of data, then put the results to good use. Far more energetic than most, he could do a prodigious amount of work in a day.

Casey joined the navy because he thought it was his patriotic duty, and wound up at a desk at Main Navy on Constitution Avenue. Though he would never claim to have the soul of a warrior, this was not the war that he had in mind for himself. OSS thankfully was only a few blocks away on Navy Hill. Casey used his legal network to reach Donovan's law partners, who set him up with a colonel willing to talk to the thirty-one-year-old New Yorker in the uniform of a Navy Reserve lieutenant, junior grade. The colonel was looking for officers to bring a little more order to OSS London. The idea was that Casey would spend the summer in Donovan's outer office, observing how the OSS secretariat worked, and then establish a second secretariat in London. A strong advocate for OSS, Bruce was more than a nominal chief. But before Casey arrived in London, the branches—SI, SO, R&A—had been operating on their own to a large extent.[39]

Casey's biographer has the none-too-elegant young officer presenting himself to the patrician Bruce in London in early November 1943.[40] Casey reportedly arrived with a letter of introduction signed by Donovan that explained why he was there. Bruce seemed to resent the implied criticism. Setting the letter aside, he promised to find something useful for Casey to do.

Bruce allowed Casey to work in the command suite for OSS Europe at 70 Grosvenor Street as a special assistant. He built alliances with senior officers, listened, and strategized. More than a month into his new assignment, he complained to Otto C. Doering, one of Donovan's long-time associates on Wall Street as well as Navy Hill, that there was still no secretariat in London and that he had not "found the scope or opportunity to use . . . whatever ability I have."[41] When Donovan came to London, Casey catered to him, happy to go shopping for the books he wanted and bask in the reflected light of the man who was becoming his idol. While other, older officers resented the disruptions that Donovan caused, Casey gushed that "we all glowed in his presence."[42]

By D-day, Casey was officially managing the flow of paper for Bruce's office. After D-day, his writ ran to Paris and beyond, with the grand title of Chief of the OSS Secretariat in the European Theater of War (OSS/ETO). He became Donovan's unofficial eyes and ears, traveling extensively around the theater, mostly on his own, sometimes with Donovan. When they were together, the two high-energy Irish lawyers from New York, one from Wall Street nearing the peak of his career, the other from Queens still working his way up, brainstormed about the best way to operate in Germany.

Donovan still could not stop dreaming of martial glory in the wake of the special operations in France: in early September he signed a memorandum calling for OSS to conduct "aggressive subversion behind enemy lines" in Central Europe.[43] He foresaw "bold raids and nicely carried out attacks from hideouts, . . . patrolling by small groups, [and] . . . destroying industrial installations." What OSS/ETO had in mind was more like the kind of secret intelligence work that the Sussex teams had done. On a flight in September, Casey presented Donovan with a plan that he and Bruce had drafted along with Arthur Goldberg, the prominent lawyer who was head of SI's Labor Desk.[44] A month later, Casey addressed a thoughtful six-page memorandum directly to Donovan, reviewing the many challenges—and possibilities—of operating inside the Reich.[45]

One of the key points that the memorandum conveyed was that operating in Germany would be very different from operating in

France. "Controls . . . are now so tight that the establishment of agents inside Germany is likely to be an extremely slow and uncertain process." Nevertheless, the Allies still needed military and political intelligence. The solution was for the various OSS branches in Europe to work together to an unprecedented extent: the mission demanded "much more comprehensive and detailed staff work than has been the case heretofore." As if to illustrate just how "slow and uncertain" the process had been in the fall of 1944, OSS/ETO would later admit that it had only four agents in Germany at the end of the year.[46]

Then, early on a frigid day in December, the Wehrmacht hurled some thirty divisions out of Germany against the American lines in the Ardennes Forest in Belgium—a near-complete surprise and major intelligence failure that would become known as the Battle of the Bulge. Few believed that the Germans were beaten, but fewer still believed they were capable of a major offensive in the depths of an especially cold winter. Clouds had hampered Allied aerial reconnaissance, the principal source of their intelligence before D-day. Nor were enough other sources available. There were no Jedburgh or Sussex teams and few line-crossers on the German frontier in winter, let alone spies in the German high command.[47] Ultra (and Magic) traffic revealed traces of the impending offensive, but not enough to capture SHAEF's attention.[48] Now mostly on home territory, the Germans were relying more on landlines that the Allies could not intercept. The need for Allied agents inside Germany had never been more urgent.

For Casey, Donovan was one general who reacted decisively in the crisis: he "must have reached for an airplane the moment he heard about Hitler's drive into the Ardennes."[49] The director appeared in Paris just after Christmas 1944 along with Whitney Shepardson, the overall head of SI, and met behind closed doors with Bruce and J. Russell Forgan, Bruce's designated successor. Casey remembered the moment when they finally emerged: "Donovan took me aside and said, 'Bill, you're the new chief of SI for Europe. I'm giving you carte blanche.' 'To do what?' I asked him." Casey imagined that Donovan's blue eyes twinkled as he answered, "Why, to get us into Germany."[50]

Casey swung into action even before the German offensive had ground to a halt. The unassuming Walter Lord, a Princeton graduate who could have avoided serving in the war but was doing his duty at OSS London, remembered how Casey's "contagious, almost innocent eagerness and idealism" captivated others.[51] First, Casey dealt with threshold issues, like wearing a rank that did not match his new responsibilities. Admiral Stark, the former CNO who was still the senior American naval officer in Europe, suggested that Casey shed his uniform and work as a civilian.[52] Another issue was getting the British to agree to let the Americans act on their own. The two allies had carefully negotiated their respective roles before D-day, in part because the British were still reluctant to relinquish control over operations in Europe, and in part because the Americans needed their help. This had now changed. Casey called on "C" at MI6 and Maj. Gen. Gerald W. Templer at SOE. (Wounded in combat, Templer was running the German desk at SOE during his recovery.) Both British officers doubted that OSS would be able to get agents into Germany or, if it did, get their intelligence out. But they had no objection to letting the Americans proceed.[53]

Another threshold issue was understanding the challenge—what was the situation on the ground? What did SHAEF want? After a tour of the front, Casey and J. Russell Forgan, who was now the senior OSS officer in Europe, went to Versailles to meet with Eisenhower's deputy G-2, Brig. Gen. Thomas J. Betts, in his palatial office. Betts gave the OSS officers a general idea of the war plan—to advance eastward on a broad front while the Soviets moved west, penetrating Germany from the other direction. What he wanted was the German order of battle, especially information about the flow of reinforcements through road and rail hubs toward the front.[54] Casey's impression was that Betts would happily take whatever OSS produced—SHAEF was OSS/ETO's most important customer—but only if it did not involve much effort on his part. When asked for support, Betts's answer ran along the lines of, "You have your orders. Now carry them out."

Back in London, Casey thought through the problems that he would have to solve to meet SHAEF's requirements. He would build on the Sussex program, adding his own variations and improvements. For him, as for Francis Miller before D-day, details mattered, and he addressed them diligently: recruiting agents who would fit in without attracting attention; training them for weeks at a facility outside London known as "Area F"; developing cover stories to explain their presence in Germany; counterfeiting German documents to support those covers; procuring authentic European clothes; even inventing more secure ways for agents to communicate by radio from behind enemy lines. By the end of January 1945, Casey was confident that he would be able to deploy newly trained agents from London within three to four months.[55]

They were an unusual group of men and women. Many of them were Polish, Dutch, Belgian, or French nationals who could pass as foreign workers, like the millions the Nazi regime had forced to work in Germany. This meant that their backgrounds would be difficult to check. It also meant that they could not be expected to know much about Germany, let alone speak German, or have access to decision-makers. An overlapping category comprised refugees from European labor movements, including German communists. This gave the budding anti-communist ideologue Casey pause. But if it was what "the old man" Donovan wanted, he would do it.

In a violation of the spirit (and arguably the letter) of the Geneva Convention of 1929, Donovan approved the screening and selection of German prisoners of war who were anti-Nazi. Another category were German-Americans, including Jewish refugees, who were willing to return to the Reich under cover. The risks for all were high. Most would be parachuting deep into the enemy's homeland at night, in winter, by twos or threes. They would mostly be going in blind with no one waiting for them on the ground. They would have to make their own way, develop access, and report by wireless. If caught, they could expect to be tortured and executed. With growing admiration, Casey watched agents prepare for their missions, and he made sure they received whatever they wanted by way of creature comforts before they deployed. OSS even lodged one group in a French chateau.[56]

In the end, Casey—and OSS—exceeded expectations. A handful of agents who were already in the pipeline, especially those who had come up from the Mediterranean, deployed in January or February 1945.[57] The newer agents hit their stride between March 17 and April 25, when twenty-eight teams from London and another fifteen from southeastern France deployed.[58] The detail work paid off; the cover stories, expert forgeries, and old clothes held up to a remarkable extent. The new Joan-Eleanor communications system turned out to be a lifesaver, allowing agents to broadcast up to a circling aircraft rather than across the continent, thereby limiting the risk of detection. There were casualties, but far fewer than expected. Casey estimated the rate at about 5 percent. He would report in July 1945—two months after the German surrender—that all but seven of his agents had been "recovered," his bureaucratic way of reporting that they were once again in friendly hands, presumably alive and as well as could be expected after operating in Nazi Germany.[59]

Casey always believed that the game was worth the candle. SI/ETO had to do what it could to support the troops at the front for as long as the war lasted. But for all the hard work, the take was relatively modest. Years later, Casey would say that "the intelligence produced was of marginal value."[60] This was not espionage on a grand scale—the sort of thing that the Soviets were doing by cultivating spies in the offices where the British and Americans made strategy, as in the cases of Kim Philby and Duncan Lee. Instead, it was sophisticated line-crossing. The operations produced "strategic" information only if the word referred to what was happening on the ground in Germany up to fifty miles from the front lines. By-products included acquiring close-in tactical intelligence, persuading local German commands to surrender, and helping to rescue Allied prisoners. Some agents even penetrated local Gestapo offices and gleaned information about them.[61]

The two Belgian agents of Team Painter who parachuted into Bavaria in early spring started with the address of another young Belgian in Munich; he proved willing to help them register and find work—in the Gestapo motor pool where he himself worked. The team moved into the employee barracks on the premises, hiding their radio and even securing permission to run its antenna on the roof (ostensibly to listen to air raid warnings). A suspicious Gestapo officer eventually confronted them; they countered with a vague promise of lenient treatment after the war and a large role of banknotes. The corrupt officer took the money and returned with rosters that would enable the US Army to find and arrest a slate of some sixty war criminals after occupying Bavaria in May. Casey appeared on-site while military police was still making the arrests. He sought out his agents, who gave him a tour of the motor pool, and told him their story "from beginning to end."[62] Listening with rapt attention, he stayed with them "late into the night."

Last but hardly least, Casey proved that the Americans could operate on their own, without help from the British. Given more time, he would write in his final report, his agents could have done more. Nonetheless, they could take pride in having carried out their work on their own "in a professional and effective manner" and achieving "some worthwhile results."[63] "We had," he concluded, "done enough and seen enough, in southern France, in northern France[,] and finally in Germany to be convinced that . . . [our] vision could be carried from the realm of fancy to that of fact." By any standard, it was a remarkable performance, especially for a young business lawyer who had learned on the job in a few weeks. Years later Casey would call running SI/ETO in 1945 "the greatest experience of his life"; his biographer would label it his finest hour.[64]

19

ALLEN DULLES'S NEARLY PRIVATE WAR

Believing in Yourself

David Bruce liked Bill Casey well enough. But they were hardly on the same plane. During the war, the American aristocrat reserved his enthusiasm for other greats like Hemingway, the writer he thought looked like God as painted by Michelangelo, or a man who, in his eyes, was "one of the chosen ones of the earth," Allen Welsh Dulles.[1] This was a quite a compliment, since Dulles was both colleague and rival. Each was developing a prototype for modern American intelligence. In London and Paris, Bruce worked at the crossroads of the Anglo-American military alliance, figuring out how to meet its needs for intelligence. Meanwhile, in Bern, Switzerland, Dulles sat at Hitler's doorstep, surrounded by enemy territory, left largely to his own devices to do what he thought an American spy should do.

Bruce was right. More for service and accomplishment than wealth or breeding, Dulles and his family had been chosen. Born in 1893 in Watertown, New York, to a liberal Presbyterian minister, Dulles could count two secretaries of state among his relatives, one of them Robert Lansing, who served in Woodrow Wilson's cabinet during World War I. Dulles went on to attend Princeton, teach in India after graduation, and in 1916, join the American Foreign Service. His first two postings occurred at the height of the First World War in Vienna, the capital of Germany's ally Austria-Hungary, and Bern, the capital of neutral Switzerland.

After the war, Dulles remained in the Foreign Service but devoted his nights to studying law at George Washington University in the hope that a law degree would further his career as a diplomat. When it turned out that the degree made little difference to the Department of State, Dulles felt he could not refuse a generous offer from Sullivan & Cromwell that his elder brother had arranged for him. John Foster Dulles was one of the principals at the Wall Street law firm that had been founded in 1879 in support of big business. As staid and formal as it was powerful, Sullivan & Cromwell was not as good a fit for Allen as it was for John Foster. Corporate law never displaced his fascination with foreign affairs. To keep his hand in, he joined the prestigious Council on Foreign Relations as well as the Room, Vincent Astor's private intelligence club. Usually leaving his wife and children at home, he traveled widely, especially in Europe. Sociable and charming, even fun, the Republican internationalist developed a remarkable network of contacts around the world, not least among them a long string of lady friends.[2]

Mutual interests led Donovan to recruit Dulles for COI after Pearl Harbor. It was David Bruce who actually made the phone call and asked Dulles to run an office for SI in New York. Dulles did not tell his wife, ignored his brother's mean-spirited warning that lawyers who left the firm for the duration should not assume that their jobs would be waiting for them after the war, and started working at COI for no pay in February 1942.

Thanks to his connections and experience, Dulles was far less dependent on Donovan than other new hires and able to chart his own course. When Donovan was growing his outposts overseas, Dulles resisted the director's suggestion that he go to London and work for Bruce. He argued successfully that he could do more in Bern, where he had served in the last war. He already had a foundation there: he knew the people and two of the languages spoken in Switzerland, German, and French—although he never shed his American accent. He even enjoyed a good relationship with the US minister to Switzerland, Leland B. Harrison, a Foreign Service blue blood (and an alumnus of Eton, Harvard, and Harvard Law) who welcomed him to the legation in November 1942 and bestowed upon him the ambiguous title of special assistant. Overseas as in Washington, the title covered a range of duties and powers and was certainly more impressive than third secretary, the title Dulles had held in Bern in 1917. As long as he appeared to respect Swiss neutrality, he would have a great deal of leeway.

Dulles seemed to relish the challenge of operating on his own in a country surrounded by hostile territory: Nazi Germany to the northeast, fascist Italy to the south, Vichy France to the northwest. Like Donovan, he started by assembling a strong team. His base was a handful of officers, secretaries, and code clerks, reportedly less than ten, who worked at the legation or one of the outlying consulates.[3] He promptly added American expatriates who happened to live in Switzerland and would fall somewhere between regular employees and principal agents. One of the key members of Dulles's team was a naturalized American born in Germany, Gero von Schulze-Gaevernitz. He was bilingual and bicultural, able to relate both to Germans and to Americans. Dulles valued the good-looking businessman in his forties for his "beguiling personality and . . . great capacity for making friends."[4] An added bonus was that Dulles had known Gaevernitz's father, a liberal professor and politician, from his days in Berlin before the United States entered World War I. The family was even connected to Germans who were opposed to Hitler.

The American expatriate Mary Bancroft was another important addition to the payroll. A member of the family that once owned Dow, Jones, and Company, the fit, attractive thirty-nine-year-old was married to a Swiss financier and lived in Zurich, Switzerland's financial capital. After Pearl Harbor, she let it be known that she wanted to support the American war effort. Gerald Mayer, the Office of War Information representative, got in touch and, in December 1942, introduced her to the newly arrived special assistant.[5]

Bancroft's first impression was that the pipe-smoking forty-nine-year-old with his neatly trimmed mustache, casual bow tie, and "keen blue eyes behind rimless glasses" was no one's assistant.[6] Her interest was piqued. The man had presence; he was energetic, even charismatic. She was a little surprised when he invited her for dinner at his well-appointed apartment in Bern. Its address, Herrengasse 23, would become part of the Dulles lore; the solid, five-story building, built in 1690, sat on a ridge above the Aare River with the mountains of the Bernese Oberland in the distance. A path led from the river through grape arbors and vegetable gardens to the discreet back door of Dulles's ground-floor apartment.[7] This evening the dinner, served by an impeccably turned-out butler, proved to be something between a job interview and a first date.

After a second dinner in Zurich a few days later, Dulles declared that their romance would cover the work, and the work would cover the romance; each would be the apparently harmless explanation for the other. At first she was not sure what he meant; there was no romance yet. But, drifting in a failed marriage, she decided that she did not mind the arrangement he was proposing—she was falling in love with Dulles. To Bancroft, he was so alive and interesting; she admired the way he solved problems, quickly and ingeniously. It would be exhilarating to work with him. Besides, she reasoned, there would be no unmatched expectations; he was almost disarmingly forthright about what he wanted: work and sex.[8]

Dulles employed Bancroft and Gaevernitz as talent scouts. Most mornings he would telephone Bancroft and give her instructions using American slang and double-talk that they imagined no one else could understand. Her first case was Anna Siemsen, a respected

German socialist and former professor at the University of Jena who had been forced to flee from Nazi Germany.[9] Dulles wanted Bancroft to cultivate Siemsen as a potential asset to see what and especially who she knew. If Bancroft built enough trust, Siemsen would be willing to share her network of contacts, some of whom could prove useful to Dulles. Siemsen was happy to welcome Bancroft to her countryside chalet for long chats but apparently did not become a spy.

Gisevius

Both Bancroft and Gaevernitz worked the important case of Hans Bernd Gisevius. He was the German vice consul in Zurich—at least on the surface. He was an unlikely vice consul. Born in 1904, the doctor of laws was already a veteran civil servant, his career having started in the Prussian state government in 1933.[10] The fair-haired North German stood six feet, four inches tall and wore baggy suits that made him seem even larger. Bancroft called him "a giant of a man;" Dulles would eventually take to referring to him affectionately as "Tiny."[11] He spoke with passion and urgency, usually in a loud voice. He was both opinionated and political, a conservative nationalist well to the right of center, but also a believer in the rule of law who was outraged by the Nazis' crimes. Although he had been in Switzerland since 1940, Gisevius did not have any official duties at the consulate general, or even an office to call his own. This was due to the fact that, one level below the surface, he was an Abwehr officer, officially sent abroad to spy on the enemy; on the next level further down, he was a long-time member of the German Resistance to Hitler, in Switzerland to open a link to the Allies.

For two years, Gisevius had a complicated relationship with MI6 in Bern. The British appear to have been willing enough to take his information. MI6 files record the claim that one of their agents obtained enough information from him between August 1940 and

December 1942 for twenty-five reports to London—including the claim that the Germans had broken one of the codes that Dulles was using.[12] But Gisevius was frustrated that they were unwilling to discuss foreign policy with him, let alone to engage with the German Resistance. In his memoir, Gisevius complained that they "stuck to the old-fashioned [idea that] . . . the 'enemy' was . . . solely an object of espionage."[13] Dulles, he continued, turned out to be different—in his eyes, Dulles was "the first intelligence officer who had the courage to extend his activities to the political aspects of the war." Gisevius wanted to be treated like an emissary, not a source. Along the way, he was willing to share information that might help the Allies, but that was not his main purpose.[14] To protect himself, he needed to do both in secret, which made OSS a good interlocutor for him.

Gisevius's relationship with the Americans began in January 1943 after his relationship with the British tapered off.[15] Gaevernitz arranged for Gisevius and Dulles to meet. Before long, the two were sitting by the fire in the study at Herrengasse 23—Dulles's idea of a secret meeting.[16] The German spent even more time with Bancroft. The cover for their meetings was work on a massive manuscript—1,415 double-spaced pages in Gisevius's "very involved and difficult" German prose.[17] The manuscript was equal parts memoir, polemic, and political manifesto. Its central focus was how Hitler had cemented his power over Germany, and how a band of conspirators was secretly working against him. It was tantalizingly incomplete; the final chapters had yet to be written, both literally and figuratively. Bancroft's task was to build trust while she translated the manuscript, eliciting useful information along the way.

The ploy worked. Over the course of 1943, Bancroft and Dulles learned about the German Resistance from Gisevius—the loose conglomeration of politicians, officials, and soldiers—labeled "the Breakers" by OSS Bern.[18] At various turning points from 1938 on, the resisters tried many things, from staging a coup d'état to killing Hitler, but each time they had failed. Their failures were partly due to their bad luck and his good luck—providence seemed to be on his side as he changed an itinerary at the last minute, or as a bomb failed to detonate. But it was also because the regime enjoyed overwhelming popular support, especially in its fight against Soviet Russia. Most active-duty generals just could not bring themselves to support the conspiracy, no matter how many crimes Hitler committed nor how great the looming apocalypse.[19]

One of the hopes that Gisevius conveyed was that the western Allies would find ways to encourage Germans to oppose Hitler. In early 1944, Dulles asked Washington politely for guidance: "any indication with which you could supply me regarding what you would be interested in achieving via the Breakers."[20] But, as Dulles learned, no one in the Roosevelt administration wanted to encourage the resisters.[21] On the contrary, the policy remained unconditional surrender: Germany's only option was to surrender, without any conditions, to all the Allies at once.[22] Complementing the policy was Roosevelt's refusal to distinguish between "good" and "bad" Germans. As he declared more than once, he felt the German people as a whole needed to learn their lesson, not just their Nazi leaders.

Dulles did not give up. He kept the stream of reports from Bern to Washington flowing, in April conveying the resisters' hope that, if they removed Hitler, the "Anglo-Saxon powers" would negotiate with Germany.[23] By mid-summer 1944, Dulles abandoned any pretense of merely reporting the facts; he was passionately arguing the case on the Breakers' behalf. Could the Allies declare that the policy of unconditional surrender applied to the regime, but not to the German people, especially if they overthrew Hitler? By now, Dulles claimed, the situation had evolved since Roosevelt had announced the policy: "with the Russians at the door of East Prussia, with the [Normandy] invasion a success, and Italy practically lost to the Germans . . . encouragement to internal revolt may be given without much risk that the Hitler legend could rise again on the myth of a stab in the back."[24] Donovan forwarded Dulles's plea to Roosevelt. But the president was not about to change his policy—and might not have even read the memorandum. On July 13, he left Washington on a lengthy trip to the Pacific, bound for Pearl Harbor first by train via Chicago, where the Democrats were holding their national convention. He would not return to Washington until August 17.

Dulles might have advocated for the resisters, but he did not hold out false hope to them; he told Gisevius that they could not count on any special treatment from the US or the UK.[25] Gisevius passed the word back to Berlin. Undaunted, the resisters proceeded with their plans to kill Hitler and overthrow the Nazi regime. Gisevius traveled back to Berlin on July 11 to be on hand for the coup attempt, and moved cautiously around the capital, aware that the Gestapo suspected him of treachery. On July 20, 1944, the Breakers' Col. Claus von Stauffenberg flew to a military airfield near a town called Rastenburg in East Prussia. From there he drove to the Wolf's Lair, the bunker complex in a dense forest that served as Hitler's forward headquarters, and, during the dictator's midday briefing, activated a bomb—a captured SOE explosive device—that he placed under a table. He left the room to make a phone call. He waited for the detonation. After he saw the column of smoke rise from the building, he and his aide sped off to the airfield to fly the four-hundred-some miles back to army headquarters in Berlin. There Stauffenberg met his fellow plotters, including Gisevius. He reported what he had seen and energetically turned to issuing orders for the coup d'état that was predicated on Hitler's assassination. Sadly for the plotters, Hitler turned out to have survived the blast. Troops loyal to the dictator seized the headquarters, ending the coup. A few hours later Stauffenberg died in front of a firing squad shouting, "Long live our holy Germany!" Other plotters committed suicide. The unlucky ones were arrested, tortured, tried, and executed by slow hanging. Gisevius managed to survive because he happened to be outside the headquarters building when the tide turned against the plotters. He went underground, staying with friends for months.

Thousands of miles away, Roosevelt's chief of staff Leahy learned of the attack on Hitler not from OSS but by reading the newspaper.[26] This was an amazing reflection on American intelligence in the fifth year of the war. How could this be? Dulles argued that the relationship with Gisevius had been worthwhile because of the stream of intelligence that it had produced.[27] OSS had diligently forwarded his information around Washington, notably to the Departments of State and War.[28] Donovan himself had signed the notes to Roosevelt's secretary.[29] True to form, the president seldom reacted.[30] One problem was still overload. Donovan's office was well on the way to forwarding a staggering total of roughly 7,500 reports to FDR by April 1945.[31]Another problem was still quality. In Washington in-boxes, Dulles's reports typically appeared as raw telegrams with a cover letter. It did not help that he wrote like a diplomat recording his contacts with foreign officials and adding policy recommendations; at other times he seemed like a journalist trying to capture what people were feeling about the war. He focused more on the big picture than on underlying details. While some OSS officers defended Dulles, others in Washington condemned his work. According to SI chief Whitney Shepardson, the War Department "discounted" news from Bern "100%" for being tendentious and unreliable.[32] A few months later Shepardson followed up with a scathing judgment: "your information . . . is now given a lower rating than any other source"—and urged Dulles to use the "greatest care in checking all your sources."[33]

Kolbe

Fritz Kolbe had more than a little in common with Gisevius. Born in 1900, he was only four years older than the doctor of laws. Also a civil servant, Kolbe was a walk-in with an incredible story. In August 1943 he traveled to Bern, where he asked Ernesto Kocherthaler, a trusted German-born friend, to do him an enormous favor: approach the British embassy on his behalf and convey his offer to share official secrets. The receptionist tried to turn Kocherthaler away, but to no avail. He sat and waited until a Colonel Cartwright from MI9—the part of military intelligence that dealt with British prisoners of war—agreed to see him, but only long enough to glance at a mimeographed copy of a cable that Kocherthaler offered as proof of his principal's good faith before politely turning him away. Though it had occurred three years earlier, the Venlo incident on the German-Dutch border still cast a long shadow; for the rest of the war, British officials tended to see every German approach as a possible trap.[34]

Kocherthaler went on to try the Americans, eventually meeting with Gerry Mayer, the same official who had originally met Bancroft. The German-speaking Mayer heard Kocherthaler out, who this time brought not one but sixteen secret cables and offered to leave them for Mayer to peruse. The cables looked authentic enough, stained with the lifeblood of a bureaucracy—routing stamps, underlining, and initials. Mayer knew he should get in touch with Dulles and arranged to host Dulles, Kocherthaler, and Kolbe for a meeting at his apartment.[35]

Dulles thought that the forty-two-year-old Kolbe, with his prominent cheekbones and frank, almond-shaped eyes, had "typically Prussian-Slavic features."[36] Nearly bald, he seemed about five foot seven and was solidly built. It turned out that Kolbe had been a member of the *Wandervögel*, the back-to-nature German youth movement that got its members outdoors for strenuous activities, including long hikes that could seem like forced marches. If Gisevius, the secret policeman with a law degree, was a high-strung racehorse, the less sophisticated but still poised Kolbe was a workhorse who knew his way around the farm. He worked as an aide to Ambassador Karl Ritter at the Foreign Office in Berlin. Ritter was responsible for liaising between the Foreign Office and the military; Kolbe's job was to pick out the incoming messages from overseas embassies and major commands that Ritter needed to see. He had been able to bring the sixteen cables to Bern because he was also a diplomatic courier, charged with loading and escorting diplomatic pouches. Carefully sealed, pouches were immune from police or customs inspections—which made them a near-perfect mechanism for a spy.

Kolbe was willing to bring copies of documents and his handwritten notes to Bern for Dulles. He was also willing to share information that he had absorbed at work—like the layout of Hitler's headquarters at Rastenburg, the identities of German spies abroad, or their successes in breaking Allied codes. Dulles wanted to know why Kolbe was risking his life. Kolbe did not want money. He had for a long time been working out how to fight Nazism and concluded that the only way to purify Germany was for her to lose the war and start anew. That meant first helping the Allies win.

Dulles assigned Kolbe various code names and numbers—the best-known would be "George Wood"—and started to send messages to Washington and London. Bern wanted "Wood's assistance"; the case could have "vast consequences" if it remained "highly secret."[37] It was also overwhelming. OSS Bern could not handle large volumes of traffic—the take from just one meeting was too much for the staff—and it had to ask MI6 for help in processing the material, which would now go to Washington and London. Since by Dulles's own admission the walk-in seemed too good to be true, the case would have to be carefully vetted. For that reason, it would be handled in counterintelligence channels—Section V of MI6 and X-2 of OSS—where British and American officers would puzzle over it for many months.

The Kolbe case was the most significant human penetration of the German government by an American or a British intelligence service in World War II. But, far from generating accolades, it ignited a furious controversy on both sides of the Atlantic. Claude Dansey, the deputy director of MI6—once described as a man so wicked that he should be taken straight to the Tower of London for execution—could barely contain his anger.[38] He had worked in Switzerland, and was already prone to second-guessing Dulles. The Germans had, he reported, compromised at least one of the codes that the American used.[39] Now he added the charge that the Germans had duped—he used the word "stuffed," which had vulgar connotations—the gullible Americans. Dulles's first unlikely defender was Kim Philby, the MI6 officer who was also a Soviet spy. He came up with the idea of sharing some of the Wood material with the former head of GC&CS, Denniston, who was now responsible for decrypting diplomatic traffic.[40] Denniston was excited to find some exact matches between Wood's cables and cables he was pulling out of the ether, which made the material extremely valuable for British codebreakers as a check on their work.[41]

In October, Kolbe reappeared in Bern, bringing a fall cornucopia to Dulles: copies of ninety-six cables and ten pages' worth of other information. It took Dulles until mid-November to process the take.[42] A few days later, on November 23, X-2/London guardedly expressed its judgment in writing: the material was genuine; OSS Bern should continue to run the case and encourage Wood to keep passing official documents.[43] OSS London then suggested that Washington obtain a second opinion from Alfred McCormack, the army's brilliant Wall Street lawyer turned wartime-intelligence analyst, and the rough equivalent of Denniston in London.[44] He would surely be a good judge of the take from Kolbe.[45]

In late December Kolbe once again appeared in Bern, which meant that Dulles spent the holidays with him—often late at night and for hours on end—in his study at Herrengasse 23, taking delivery of another two hundred documents, pages of notes in Kolbe's crabbed handwriting, and tidbits on some thirty more subjects. Among the documents was evidence that a spy had walked in to the German embassy in Ankara with official British documents—a startling development and another good indicator that Wood was acting in good faith.[46] Dulles cabled Washington. He believed in Wood and would stake his reputation on the case.[47] In January, Donovan felt confident enough to excerpt some of Kolbe's information for FDR.[48]

By now McCormack had started to review the Wood material, a process that would continue for months after yet another massive haul from Kolbe in April.[49] The McCormack survey ended with a final report in May 1944 that was not unlike the X-2 report a few months earlier.[50] It was positive, but qualified. The material appeared genuine, and some of it had been useful to military customers, especially the observations of German military and air attachés who had recently toured Japanese territory. However, the report continued, the material was not as useful as it could have been. This was partly due to the fact that it was not timely, given the lag between original date of transmission and processing in Washington, and partly because it contained secondhand information as well as the opinions of diplomats. McCormack might as well have openly compared signals and human intelligence: at its best, the first yielded near real-time intelligence, based on an enemy leader's own words, while the second was more subject to delays and various human foibles.

McCormack then claimed the right to control dissemination because the material could affect cryptanalysis.[51] While this argument was not unreasonable—if the Germans learned that the Allies were reading their cables, they could react by changing their codes and Ultra could go dark—it was another loss of power and prestige for OSS. The older Wall Street lawyer, Donovan, had to yield to the younger Wall Street lawyer, McCormack; codebreaking once again took precedence over espionage. The third Wall Street lawyer, Dulles, would feel the same sting when Donovan directed him to send all Wood material to Washington, and let Washington determine its distribution. Dulles protested angrily—but to no avail.[52]

By the end of 1944, both Kolbe and Gisevius had risked their lives again and again to get information to OSS. Both remained at great risk—Gisevius was still in hiding in Berlin, and Kolbe would continue working at the Foreign Office in the German capital for another few months, dodging Allied bombs as well as the Gestapo. Dulles used people—Bancroft saw how he judged people by their usefulness—but he also did his best to take care of them.[53] Probably at Bern's urging, OSS began contemplating the kind of commitments it could make to Germans "who work loyally for our organization."[54] Possibilities included allowing them to enter the United States after the war, or placing their earnings on deposit in an American bank. On December 1, Donovan posed the question to Roosevelt, writing that OSS needed "authority which only you can give." This was one OSS memorandum that Roosevelt did read and chose to answer. He informed Donovan that he did not believe that the US should offer any guarantees to Germans "who are working for your organization" because the guarantees would be too easy to abuse or misconstrue, and might enable some Nazi criminals to escape prosecution. The president concluded with unusual decisiveness: "I am not prepared to authorize the giving of guarantees."[55] If Dulles wanted to help his Germans, he would be on his own.

Wolff

Around midday on Sunday, February 25, 1945, Dulles took a call from a Swiss military intelligence officer named Max Waibel. He wanted Dulles to meet him that same evening in Lucerne, a picturesque lakeside city a short train ride from Bern. Dulles and Waibel had forged a relationship of professional trust and even of friendship; confident that Waibel would not waste his time, especially on a Sunday, he made the trip with Gaevernitz. Over an excellent dinner of fresh trout, Waibel explained that he had been approached by a man named Baron Luigi Parilli, an Italian industrialist, and Professor Max Husmann, a Swiss schoolmaster, who wanted to convey a message from the Germans. Dulles agreed to let Gaevernitz hear the emissaries out and caught the train back to Bern.[56]

Waibel led Gaevernitz to a nearby hotel. It was the romantic nineteenth-century Schweizerhof, facing the lake, with breathtaking views of the Alps in the distance. What Parilli and Husmann had to say in a Swiss hotel room was like a shaggy dog story that was not funny or sweet; it was rambling and complicated, without much of a punch line. A captain in the SS with the unlikely Italian-German name of Guido Zimmer claimed to be a practicing Catholic and an art lover. He knew that Hitler would lose the war but would never surrender. In the west, the Battle of the Bulge had ended in defeat for the Wehrmacht while the Soviet juggernaut advanced, unstoppable, toward Berlin and Vienna from the east. The Reich's cataclysmic death throes entailed ever mounting destruction. Zimmer wanted to preserve the remaining treasures of Italy.

The SS captain had aired his feelings with another art lover, the scholar and linguist Eugen Dollmann, who had translated for Hitler and Mussolini. Even though he had spent much of his life in Italy, Dollmann was still enough of a German to be a colonel in the SS and work in uniform. He in turn had spoken to the senior-most SS commander in Italy, Karl Wolff, who was said to be like-minded. Zimmer then started looking for ways to contact the Allies. He was friendly with Parilli, who had a relative who had attended Husmann's school. Husmann knew Waibel, who thought of Dulles.

Gaevernitz listened and made no commitments. He observed that it might be worthwhile to talk directly to Dollmann, Wolff, or Albert Kesselring, the field marshal who was the German commander-in-chief in Italy. Dulles and Gaevernitz doubted they would hear any more of this strange matter, and wrote a memorandum for the record that they filed under Peace Feelers.[57] It was not a small category, full of dead ends and wild claims. There was especially good reason to question the bona fides of intermediaries from Italy. A few weeks earlier, OSS had thoroughly embarrassed itself by vouching for reports from a source in the Vatican who claimed that Japan was preparing its minimum demands for a negotiated end to the war. Circulated to a handful of customers in Washington, including the president, the reports turned out to be the work of a skilled fabricator who had written pornographic novels before writing for *L'Osservatore Romano*, the daily newspaper of the Vatican, where he had absorbed enough insider knowledge to master a different kind of fiction.[58]

A few days later, the OSS men were surprised when Waibel called to announce that Dollmann and Zimmer had come to Switzerland and wanted to talk. This time, Dulles decided to send an OSS officer named Paul Blum, an accomplished linguist who made the trip on March 3 to a restaurant in Lugano, another stunning lakeside destination, closer to Italy but still in Switzerland. Speaking French, Dollmann took the lead and probed the Allied position. Would the Americans be willing to negotiate with Himmler? "Not a Chinaman's chance," Blum blurted out in English.[59] What about Wolff? Dollmann did not say that he represented Wolff, only that they were on good terms and he could try to persuade Wolff to come to Switzerland. Perhaps the OSS could hear him out? Blum did not disagree. First, however, OSS wanted a sign that the Germans were serious and operating in good faith, which they could give by releasing two Italians, one a prominent politician, who had fallen into the hands of the SS.

Dulles knew, as he would write in his memoir, that "an intelligence officer in the field is supposed to keep his home office informed of what he is doing."[60] But, he added with a hint of slyness, an officer should be careful not to overdo things lest Headquarters send unwelcome guidance or, worse, try "to take over the conduct of the operation." So, walking the fine line between telling Washington too much and too little, Dulles took until March 5 to send off a carefully worded cable about the meeting in Lugano.[61]

The teleprinter had barely shut down when the two Italians, newly freed from captivity, appeared at the Swiss frontier followed by Wolff himself. Wolff wanted to see Dulles, and through an intermediary sent his résumé with his peacemaking credentials. On top was his business card, with his fearsome title: SS-Obergruppenführer and General of the Waffen SS, Highest SS and police leader and military plenipotentiary of the German Armed Forces in Italy.[62] Attached were pages of references, including Pope Pius XII, the calculating former Vatican diplomat who had never spoken out against Hitler, let alone the Holocaust, and the former deputy Führer, Rudolf Hess, who was delusional and perhaps insane. (In May 1941, after months of careful preparation, he had stolen a Luftwaffe ME 110 fighter-bomber in Bavaria and flown himself to Scotland, where he tried to parachute onto the estate of the Duke of Hamilton who, he very wrongly believed, would be willing to broker peace negotiations between Britain and Germany.) Under Wolff's list of references were letters recording instances of his clemency and his role in protecting cultural monuments. Not surprisingly, Wolff focused on the recent, more favorable past, not on the ten years that he spent as one of Heinrich Himmler's closest aides and as Himmler's personal liaison to Hitler while they were directing the Holocaust. Dulles was not fooled. Though Wolff was for him mostly a name on a chart, what he had heard about the German from the Italian underground was not flattering. He did not, Dulles noted wryly, expect to find in this SS general "a Sunday school teacher."[63]

Dulles would sit with Wolff on March 8 in an apartment that he kept in Zurich "for meetings of the touchiest nature."[64] It was located at the end of a quiet street and looked out on yet another alpine lake. Dulles and Gaevernitz set the stage for the late-night meeting by starting a fire in the fireplace—Dulles's theory being that the crackling flames set his visitors at ease. When Wolff arrived with Professor Husmann, Dulles did not shake hands, just nodded in greeting and offered his guests a scotch.

Wolff was "a handsome man and well aware of it," Dulles would recall. Nordic, well-built (well-fed, some said), graying blond hair, blue eyes, and (especially for a Nazi) good manners. Unlike Hitler, he spoke High German without a regional accent. The forty-four-year-old was, he wrote, "probably [the] most dynamic personality in northern Italy," which meant that he might well have the will and the power to bring about a surrender.[65]

Wolff described what he could and could not do: he commanded the SS in Italy but not the Wehrmacht, the conventional military. Germany had lost the war, and he was prepared to act on his own to surrender the forces under his command. But the result would be far better if he could also persuade Field Marshal Kesselring to surrender the Wehrmacht in Italy. He had a good relationship with Kesselring and, so long as no one betrayed his plans to Hitler, he hoped for a good outcome.

This time Dulles spared few details in the cable he transmitted on March 9 to OSS in Washington, London, and Paris, and Allied Force Headquarters (AFHQ) in Caserta, Italy, the regional friendly command. Despite the cablese, his excitement almost leaped off the page: "I believe this may present [a] very unique opportunity to shorten [the] war and to permit occupation of North Italy and possibly even penetration [of] Austria under most favorable conditions. Also might wreck German plans for establishment of maquis," a reference to the possibility that the Nazis were organizing their own partisans to continue to fight after the regulars surrendered.[66] Dulles emphasized that he had made it clear to Wolff that he could not negotiate: "unconditional surrender was the only possible course."

By meeting Wolff, Dulles was doing work done by others in the past. Holding peace talks, even secret peace talks, was a task for diplomats. The military accepted surrenders. Dulles's primary job had been to gather secret information as well as to support the French and Italian Resistance. But this unusual opportunity had come to him through contacts that he had carefully cultivated, and he was ready to exploit it. The need for secrecy made OSS the right channel. It also made sense in March 1945 to explore an offer like Wolff's. No one knew how or when the war would end; the Allies had been stunned by what the Germans had still been able to do at the Battle of the Bulge only three months earlier. Who knew what other surprises they might yet have in store for their enemies?

The initial reactions to Dulles's initiative were positive. Donovan was reportedly enthusiastic from the start.[67] Not only was the possible surrender an attractive prospect on its own, but it would also be another opportunity to showcase what OSS could contribute and perhaps silence the agency's doubters. The Pentagon was guardedly optimistic. On March 10, AFHQ/Caserta requested and received permission to send two generals, the American Lyman Lemnitzer and the British Terence Airey, to meet with Wolff. Only London wanted to slow the pace; Churchill suggested suspending further action until the Soviets had been notified and invited to join in any further talks in Switzerland.

When notified, the Soviets erupted in paranoia and anger. They accused Britain and America of negotiating with the Germans behind their back, of trying to make a separate peace with the enemy.[68] (It was the kind of behavior the originators of the 1939 Molotov-Ribbentrop agreement understood only too well.) The crisis among the Allies quickly escalated, and the east-west alliance shuddered. It even briefly seemed in danger of coming apart, something no one in Washington or London wanted. Roosevelt asked Marshall and Leahy twice to convey the American position—Dulles was not negotiating with the Germans in Switzerland—while averring that he could "not avoid a feeling of bitter resentment . . . for such vile representations."[69]

In the meantime, the secret surrender had stalled on the Italian side of the border. Field Marshall Kesselring, said by Wolff to be ready to surrender, had been elevated to command the entire Western Front and moved to a headquarters in Germany. His successor in Italy, Col. Gen. Heinrich von Vietinghoff, was reluctant to act, and blew hot and cold. Without the Wehrmacht that Vietinghoff now commanded, Wolff's surrender would be a half measure.

Kesselring's transfer turned out to be only one of the obstacles to proceeding. In late March, Himmler summoned Wolff back to Berlin for a tense round of consultations. In early April, Himmler called to order

Wolff not to leave Italy again, and vowed to periodically check on his whereabouts. The SS chief chided Wolff for moving his family to western Austria and ordered them east back toward Salzburg, where the SS could better care for them.[70] It was an unsubtle threat. Nevertheless, Wolff agreed to incorporate an incredibly brave OSS radio operator into his entourage to open a communications link with Dulles.[71]

On the afternoon of April 12, not long after messaging Ambassador Harriman, who was still the American envoy in Moscow, that he wanted "to consider the Berne misunderstanding a minor incident," Franklin Roosevelt collapsed and died of a cerebral hemorrhage at Warm Springs, Georgia.[72] It was not until the next day that Donovan heard the news. He happened to be in Paris, staying at the Ritz, and spent the morning with Dulles, Casey, and Forgan, Bruce's replacement in London. For more than an hour, Donovan just sat on the edge of his bed, slumped down, mourning the president who, he maintained, had defended him from the wolves at home.[73] He was only somewhat interested in hearing about Casey's operations to penetrate the Reich. But he mustered more than a little enthusiasm for Sunrise, as Dulles had dubbed the surrender negotiations, and wanted to hear the latest. It was, Dulles would write, "just the kind of operation he [Donovan] liked to see us fight through to the end."[74]

Dulles learned that Admiral Leahy had summoned Donovan to the White House to explain what he, Dulles, was up to; Walter Bedell Smith, Eisenhower's chief of staff, had called Forgan "on the carpet in Paris" for the same reason.[75] Donovan admitted that the Soviets had objected and senior officers in the US military had balked. Much of this was news to Dulles, who was not always comfortable around Donovan (or vice versa).[76] Casey watched Dulles fidget in his chair, alternating between outrage and embarrassment as he listened. But, on the whole, Donovan stayed upbeat about Sunrise. He was convinced that OSS could overcome the obstacles; at the end of the day, no one in London or Washington wanted to short-circuit an unconditional surrender that would save Allied lives.[77]

German defenses everywhere continued to crumble in the second half of April. The Red Army was a little more than fifty miles from Berlin, while British and American armies advanced from the west. In Italy, the Allies were ready to penetrate the Po Valley, which meant that the Germans were running out of space at the top of the Italian boot. Yet no one was sure how long they would continue to fight. Dulles still believed that the Germans might be planning a last stand in the Alpine Redoubt. This was one reason why, he argued, Sunrise made sense. It would prevent German troops in Italy from withdrawing into the redoubt. It would also enable the British and Americans to advance to the northeast, securing the province of Trieste on the Italian border with Yugoslavia.[78]

At this point, Wolff sent Dulles a handwritten letter expressing his sympathy for Roosevelt's death, while reaffirming his commitment to end the fighting in Italy.[79] From abroad, Donovan directed Navy Hill to send briefing papers to President Harry S. Truman summarizing the operation and its prospects.[80] But the combined British and American chiefs of staff had had enough, and ordered Sunrise shut down on April 20 (which happened to be Hitler's last birthday), citing as reasons Vietinghoff's reluctance to surrender along with the complications with the Soviets.[81] Dulles's response was a masterpiece of prevarication: he was taking "immediate steps to carry out instructions" but had to move carefully to avoid jeopardizing the safety of the OSS radio operator at Wolff's headquarters and his own relationship with Swiss military intelligence.[82]

On Monday April 23, Dulles learned that Wolff, along with Vietinghoff's representative, was on his way to Switzerland to surrender both the SS and the Wehrmacht in Italy, and so informed AFHQ. Almost immediately, British Field Marshal Sir Harold Alexander, the Supreme Allied Commander in theater, cabled that he would ask the Combined Chiefs to reconsider.[83] Their reply was equivocal: Dulles should not resuscitate Sunrise, but if Wolff talked to the Swiss, and they told Dulles what he had said, that would be acceptable. The back and forth grated on Dulles's nerves, aggravating his gout and causing agonizing pain. (Born with a clubfoot, Dulles had a long history of foot problems, and even in public often wore soft shoes to ease the pain.) On April 24, Gaevernitz sent for a doctor and, out of earshot, arranged for Dulles to receive a dose of morphine. Without that shot, Dulles candidly admitted, he would have been sidelined.[84]

While in Switzerland, Wolff told the Swiss intermediaries about yet another brief but hair-raising trip to Berlin, describing his meetings with Himmler, who was on edge but more rational than the bunker-dweller Hitler. The Swiss in turn told Dulles; he reported the valuable, well-sourced information to Washington.[85] On April 25, Wolff felt he had to return to Northern Italy to direct operations. Once back on Italian soil, he found his way barred by partisans, and retreated to a villa manned by SS border troops, which the partisans proceeded to surround. Dulles and Waibel, his Swiss counterpart, decided to rescue Wolff—his death at the hands of partisans would end the prospects for an early peace.

A strange Swiss-American-Italian team, augmented by two cooperative German border guards, sallied forth and got through to the villa. After persuading Wolff to take off his SS uniform, they bundled him into a car and talked their way through checkpoints back across the Swiss border. Dulles's man Gaevernitz was waiting at the crossing, not intending to speak to Wolff (because that could be construed as negotiating), but only to see that he crossed the border safely. All the same, Wolff walked over to him to say he would never forget what OSS had done for him. One account has Wolff going so far as to embrace Gaevernitz, which would have been another first for an SS general.[86] (Just as likely, they simply shook hands in the manner of Central Europeans.) OSS had just saved the life of a general officer in the SS, and Sunrise was about to occur.

By April 26, Washington had changed course and agreed to a surrender if it occurred at AFHQ in Caserta in the presence of a Soviet officer. Now all too aware that any missteps could hurt

OSS—Dulles could give the impression of being a free agent, which could spark another international incident—Donovan made it clear to Dulles that he was required to follow Washington's instructions to the letter. There was no room for interpretation. Dulles agreed to comply, repeating the instructions back to Donovan in a cable to show that he understood.[87]

Frantic efforts followed to get the SS and Wehrmacht representatives, one empowered by Wolff and the other by Vietinghoff, to Caserta, more than five hundred miles south of the Swiss border. By April 29, the two officers were in place and able to sign the surrender instrument, to take effect on May 2. More complications ensued as Wehrmacht officers and Nazi officials, including Kesselring, balked. They contested its validity, threatening each other with arrest and court-martial. Only after Hitler's suicide on April 30 became known did the path clear. Throughout the war, his general policy had been to forbid German withdrawals and surrenders; no matter what the two sides had agreed to, the actual surrender in Italy—the laying down of arms on May 2—might not have occurred if he had still been alive.

While the Germans were laying down their arms in Italy, Stalin announced that the Red Army had overrun Berlin, now little more than a smoldering ruin. The garrison had capitulated on the night of May 1–2. Five days later, on May 7 at Rheims, France, Col. Gen. Alfred Jodl of the German high command signed the instrument of total, unconditional surrender of all German forces everywhere, including some thirteen million in the army alone, to take effect on May 8.

In Europe, Dulles and Gaevernitz were treated as heroes for what they had done to end the fighting in Italy. On May 4, they made their way to Caserta to receive "a most enthusiastic reception from the Field Marshal" and his staff, who were beside themselves with relief and appreciation.[88] Eisenhower invited Dulles to Rheims for the ceremony on May 7, where he received more congratulations than he thought he deserved.[89] From Washington, OSS deputy director Magruder blessed Dulles for saving lives—including that of his son, who was in Italy and would have had to continue fighting if the negotiations had failed. Lemnitzer, the American general who had met Wolff in Switzerland, outdid Alexander and Eisenhower, writing Donovan to credit him and the OSS "for having in existence the organization, trained personnel, and the means required to take full advantage of the situation."[90] The praise was of course music to the ears of Dulles, who wrote Donovan that Sunrise was "a rather unique example of the type of services which OSS can supply."[91] During wartime, the spy agency could establish "secret contact and secure communications between the leaders on each side of the battle."[92]

In years to come, historians would argue about the significance of Sunrise. Some thought it was not even worth mentioning. Others faulted Dulles for trying to steal a march on the Soviets and helping to start the Cold War.[93] A cross-cutting argument was that, early on, he intended it to be a blueprint for covert action—how to work behind the scenes to change history, and when to sup with the devil for the greater good.[94] The mostly negative evidence only goes so far. Sunrise was primarily about hastening the inevitable through secret negotiations. Dulles himself admitted that such an opportunity might never come again.[95] But Sunrise had lengthened the list of missions that Dulles had pioneered. He was confident that it was practical knowledge that America would need again.

20

WHEN DOING "SWELL WORK" WASN'T ENOUGH

Protecting Magic

From 1943 on, almost everyone in the country was waiting for the war to end, the time when soldiers, sailors, and marines would finally come home and turn to peacetime work. The general hope was that things would not simply return to normal. The 1930s and early 1940s had been fraught with crises, from worldwide Depression to German, Italian, and Japanese aggression, and the threat of communism, usually identified with the Soviet Union. Life after the inevitable victory over the Axis—which almost no one doubted would come, sooner in Europe and somewhat later in the Pacific—had to be better.

While the average citizen had *hopes* for the future, American intelligence agencies wanted to make *plans*. The best planners had the longest memories. Among them were the codebreakers who remembered what had happened after World War I: capacity was cut to the bone, work was entrusted to mavericks like Yardley, and leaders like Stimson were unmovably against the very idea of codebreaking. This time around, they wanted to avoid the worst effects of the inevitable retrenchment.

Both army and navy codebreaking had expanded significantly; centralized operations had led to undeniable successes. Some postwar retrenchment was unavoidable, but the services could try to protect as many of the gains as possible. As early as 1943, General Strong, the G-2 who had clashed with Donovan but protected the interests of army codebreakers, was even thinking in terms of an army-navy merger of some sort.[1] Predictably, the navy had its doubts; Adm. Joseph Redman, then director of communications, did not want to surrender any control. Then commander Wenger, the best strategist in Op-20-G who was soon to be its chief, proved open to evolutionary change.

The result was the creation of a body known as the Army-Navy Communication Intelligence Coordinating Committee (ANCICC) in April 1944. The ANCICC, whose members included Wenger and Carter Clarke, McCormack's partner at the head of the army's Special Branch, started out as little more than a highly exclusive forum for coordinating operations in the Pacific and discussing postwar plans. It would only settle "such controversial matters as can be resolved without reference to higher authority."[2] Still, it was an important step in the direction of establishing joint mechanisms.

In February 1944, illness forced Strong to lay down his duties as G-2 and retire after forty-four years of service.[3] In the months that followed, his replacement, Maj. Gen. Clayton L. Bissell, challenged the system that Alfred McCormack had put in place with Strong's blessing. Known as "a flying general," Bissell had been an ace in World War I but had little experience in the field of intelligence. Demanding that his subordinates stage elaborate daily briefings, he seemed to care more for military procedure than substance.[4] These productions could consume up to one-third of the workday, which infuriated the Wall Street lawyer with civilian work habits, and McCormack avoided them whenever he could. The result was more friction between the temporary colonel and the career general to whom he reported.[5] Another sore point was that McCormack had built up a highly qualified wartime staff to process Magic and Ultra that was not only larger than the less-qualified regular staff, but also made it seem redundant. The "interminable wrangling within the War Department" lasted for most of the spring and summer months of 1944 and led to study after study of the best way to do the work.[6] On June 1, 1944, the same week that Eisenhower invaded Normandy, McCormack and Clarke's Special Branch lost the battle and had to give up its separate identity. Their staff was folded into other G-2 offices.[7]

One by-product of the wrangling was another look at the future of codebreaking after the war. Assistant Secretary of War McCloy—the fellow New York lawyer who had championed McCormack since 1941—declared in August 1944 that "one of the chief pillars of our national security system after the war must be an extensive intercept service."[8] As one of the best sources of intelligence, codebreaking needed to be developed and nurtured so that it would stand "a better chance of perpetuation in peacetime." General Marshall agreed and was prepared to take extraordinary steps on its behalf.

In the summer of 1944, while the Allied armies were fighting their way across France, the presidential campaign was taking shape. FDR was seeking an unprecedented fourth term. His Republican opponent was the young and energetic governor of New York, Thomas E. Dewey. He had a strong record first as the courageous prosecutor who had challenged the Mafia and then as the efficient but humorless administrator of the Empire State. In July, Roosevelt had taken the wind out of Dewey's sails by making listeners laugh about Republican charges that his dog Fala had traveled at taxpayer expense. In September, still searching for a compelling issue, Dewey prepared to charge the Roosevelt administration with failing to prevent the Japanese attack on Pearl Harbor; he claimed that, thanks to Magic, the administration had been forewarned but was too incompetent to take action in time.

The fact that Dewey was wrong—Magic intercepts had shown the approach of war but not the IJN's specific target—would not have mattered. This explosive charge would have generated a great deal of publicity that Tokyo could not help but notice even across the war-torn Pacific. Up to now, the Japanese government had been oblivious to the possibility that the United States was reading its codes despite leaks that had sprung up. The existence of Magic was almost an open secret in Washington. On September 11, a Republican representative named Forest A. Harness had even dropped hints on the House floor about the instructions that the Japanese embassy in Washington had received in 1941—knowledge that came from Magic.[9] When Bissell learned of Dewey's plans, he thought the matter important enough for Marshall to take up with Roosevelt.[10] Instead Marshall decided to act on his own and signed a letter beseeching Dewey not to lay Magic bare. Bissell selected Carter Clarke to deliver the letter, telling him to put on a civilian suit, get on an airplane, and arrange to meet with Dewey and no one else.[11]

Diligently laying out the facts like a good investigator, Clarke reported what happened next in a statement that reads like a police procedural.[12] On September 25 he flew to Tulsa, Oklahoma where Dewey was campaigning. The next morning, he was "up at 0700. Phoned Gen. Bissell at 0730 . . . had breakfast . . . got haircut and shave," then contrived to see Dewey privately without revealing why. Dewey was suspicious, wondering if Roosevelt were behind the initiative. He was clearly reluctant to forgo what he thought was a promising line of attack. Clarke offered what assurances he could, but Dewey did not want to read the letter, fearing that the information it contained would limit his options. However, he agreed to receive "anyone Gen. Marshall cares to send to discuss this cryptographic business or the whole Pearl Harbor mess" at greater length once he was back at his office in Albany.

Clarke flew back to Washington to brief Bissell and Marshall. Marshall decided to send Clarke to Albany, but first he edited the letter to address some of Dewey's concerns. He assured Dewey that the president was not involved. The only other senior officer he had consulted was Admiral King. Not even Secretary Stimson knew what he was doing. Then Marshall laid out his core argument:[13]

> Our main basis of information regarding Hitler's intentions in Europe is obtained from Baron Ōshima's messages from Berlin reporting his interviews with Hitler. . . . These are still in the codes involved in the Pearl Harbor events.

Marshall went on to detail other American victories that had resulted from codebreaking: Coral Sea, Midway, and current operations that were "largely guided by the information we obtain of Japanese deployments." Codebreaking had revealed "their strength in various garrisons, the rations and other stores . . . available to them, and what is of vast importance . . . their fleet movements and the movements of their convoys." He noted that the Roberts Report on the Pearl Harbor attack had been redacted to remove any references to Magic, and implied that if the Japanese learned of even one compromise, they might make widespread changes in their communications systems. He illustrated his argument by criticizing "some of Donovan's people (the OSS) [who] without telling us, instituted a secret search of the Japanese Embassy offices in Portugal" causing the Japanese to change "the entire military attaché . . . code all over the world."

This time Dewey read and discussed the letter with Clarke. He paused at the passage about OSS, asking if the charges were true. Clarke assured him they were. Like Donovan, Dewey had graduated from Columbia Law School and practiced law in Manhattan. Favorably disposed to his fellow New Yorker and his agency, he did not feel toward OSS "like most of the Army and Navy people I have talked to." Many of his own associates were in OSS; they were "damn good, excellent fellows in fact." What was more, they had told him that most army and navy intelligence officers were not specialists but generalists on rotation, which meant that the field was a low priority for the military. (This was the same refrain that McCloy and McCormack liked to repeat.) Dewey then shared that he had known Yardley—the renegade codebreaker had even been a friend; he, Dewey, had been able to persuade Yardley not to publish a second tell-all book about codebreaking.[14]

Dewey was now obviously engaged and understood the bidding. He and Clarke parted on good terms. Clarke did not extract any formal promises from Dewey, but he did not need to. Dewey did not raise the subject on the campaign trail. When Marshall and Dewey later met, by chance at Roosevelt's funeral, Marshall invited Dewey to his office, where the general showed him Magic intercepts and "made as plain as we could to him just what the importance of these matters were." Dewey's attitude, Marshall thought, was "very friendly and gracious."[15]

Marshall's charge against OSS was exaggerated. Over a year earlier, an OSS agent in Lisbon had come across a scrap of information on a low-level code used by the local Japanese naval attaché; Strong had leveraged the incident to complain to Marshall that OSS was endangering Magic, voicing the fear that the Japanese would change their codes because OSS had stolen part of a key. But it turned out that the Japanese did not change their codes and that OSS was not guilty as Strong had charged. OSS tried in vain to set the record straight.[16] Nevertheless, Marshall continued to blame OSS for acting on its own. It was a stark reminder of the supreme importance the Pentagon attached to Magic, and of renewed reservations—even frustration and anger—toward OSS.

Donovan was losing ground that he could ill afford to lose. His original patron, Navy Secretary Knox, had died of a heart attack in April 1944. The other Republican in the cabinet who supported Donovan, Stimson at the War Department, seemed exasperated by the bickering among the intelligence agencies, and wanted it to end. Why, he had pondered in his diary in November 1943, did America need army intelligence, navy intelligence, *and* OSS? It seemed "ridiculous" to him that the US had so many different kinds of intelligence; it would be better somehow to unite them.[17]

Donovan's Last Offensive

Since the fall of 1943, Donovan had been developing his vision for a postwar intelligence agency in between his trips around the world. Mirroring the process of creating OSS and COI in 1941 and 1942, his plan was not fully formed at first. In an early paper, dated September 17, 1943, his thoughts stumbled awkwardly onto the page: "The Need in the United States on a Permanent Basis as an Integral Part of Our Military Establishment of a Long-Range Strategic Intelligence Organization with Attendant 'Subversion' and 'Deception of the Enemy' Functions."[18] Its two principal points were, first, that OSS could morph into a peacetime agency, and second, that the new agency could be the fourth arm of the military, on a par with the army, navy, and air force, which would make it possible to come at the enemy by air, land, or sea—or by secret agents and special operations. He tasked skilled drafters on his Planning Group, Jim Rogers's legacy, with sharpening his arguments. In October 1944, they were at work on a paper entitled "The Basis for a Permanent U.S. Foreign Intelligence Service."[19]

Donovan decided to lobby the White House staff for his ideas. Only the bookish Dr. Isador Lubin, longtime head of the Bureau of Labor Statistics and a New Deal insider, was willing to write a few words on Donovan's behalf, messaging the president that OSS "has been doing some swell work" and that it should continue after the war.[20] Offsetting Lubin's note was one from John Franklin Carter, the former diplomat and sometime journalist who remained one of FDR's favorite unofficial spies. He had heard that Donovan's comprehensive plan for a postwar secret intelligence service was on its way to the president and warned him against endorsing it.[21] According to

Carter, OSS was penetrated by the British. Besides, Donovan's plan would be far more expensive than relying on Carter's private information service. Roosevelt sent Carter's note to Donovan on October 31 for his eyes only, asking opaquely: "Will you be thinking about this in connection with the post-war period?"[22]

Donovan took this as an invitation to finalize his proposal, which he signed on November 18 and sent to the White House.[23] The basic ideas were straightforward: the United States needed a permanent national intelligence service like OSS; the agency had worked under the JCS to meet the military's needs during the war; in peacetime control should revert to the president. The Secretaries of War, Navy, and State could form an advisory board to assist the director, but he alone would have the power to compel them to furnish information and personnel. He would not, however, exercise any police or law enforcement functions. Along with a cover letter, Donovan attached the draft of a directive to establish "a central intelligence service" that the president could sign or refer to Congress to make it law.

The dull prose masks the drama. This was a clear—perhaps the clearest ever—expression of Donovan's aspirations. He was asking the president to espouse a long-term solution that would create a powerful permanent civilian central intelligence agency with three main functions: collecting secret information (like SI), political or paramilitary subversion (like SO), and analysis (like R&A). Neither COI nor OSS had fulfilled its promise—COI because it never quite matured, and OSS because it was subordinate to the Joint Chiefs and therefore circumscribed by them. Donovan realized that the president was his last best hope. His chances of winning over other parts of the government, especially the uniformed services, were poor.

But so, too, were Donovan's chances for strong support from Roosevelt. The chief executive continued to rely heavily on Admiral Leahy to run national security for the White House as well as to chair the Joint Chiefs. Leahy's biographer has styled him as the second most powerful man in the world.[24] At the president's side,

Leahy even stood above Marshall and King, who were themselves towering figures in their own right.[25] One difference between Leahy and the others was the degree to which he shunned the limelight, preferring to quietly exercise his enormous power. He appears in many wartime photographs of the high command, but you have to look for him, usually dressed in his navy blue uniform, looking a little stiff, his jacket completely buttoned, gold braid in place. To relax, he might take the jacket off but not the carefully knotted black tie; a famous photo has him fishing in uniform. There were no affectations to make an impression like MacArthur's corncob pipe and aviator sunglasses. Instead, Leahy usually appears to be focusing on the others in the room. Donovan knew to court him, calling on him occasionally and bringing him small gifts like "a captured German Mann pistol evidently designed for . . . assassins" that Leahy happily accepted.[26] But the pistol did not change the admiral's attitude toward OSS, which ran from bemused to dismissive.[27] In 1942, Leahy had marveled at Donovan's plan to buy friends for the United States in North Africa; in 1943, he had sided with Hoover over Donovan about letting the NKVD open an office in the United States; in early 1944, he had labeled reporting from OSS secret agents in Europe "interesting" but "of doubtful authenticity."[28] Now, in November 1944, it was to Leahy that Roosevelt handed Donovan's memo, directing him to take it up with the "General Board," an old navy term for the staff.[29]

The Joint Chiefs had already been considering how best to shift from wartime to peacetime intelligence. They were not averse to continuing OSS's work in some form, even though it took second place to codebreaking. Now, reacting to Donovan's memorandum, they took the issue up again. Military officers and civilians at the Pentagon debated the best way forward. They did not quickly find consensus but, not surprisingly, the trend in their discussions was to tie postwar intelligence to the Departments of War, Navy, and State, thereby limiting its independence.

The negotiations became more complicated when Hoover weighed in on December 13. His proposal, sent first to Attorney General Biddle, was straightforward. Building on the SIS experience, he wanted to replace OSS offices overseas with FBI offices as Donovan's agency wound down after the war. This would, he argued, not require any major reorganization. FBI men would simply duplicate the system they had established in the Western Hemisphere. Legal attachés would run undercover agents and liaise with local law enforcement and intelligence agencies. They would coordinate their reporting with army and navy intelligence, as well as the Department of State.[30]

A month later, yet another reaction to Donovan's proposal came from the codebreaking community. Adm. Joseph Redman saw Donovan's initiative as a call to action but not in the sense intended by the OSS director. In a memorandum to the navy chief of staff in January 1945, he observed that "the supposedly secret plan of the OSS for coordination of . . . [intelligence] activities is widely known."[31] He implied that it had leaked out from the Pentagon and Main Navy to the general public. The problem was not the leak, but the fact that OSS and "other civilian agencies will [continue to] insist on a reorganization of American intelligence activities." That made it vital for the army and the navy to take "progressive steps" to protect their codebreaking from "the encroachment of other agencies." He proposed "the formal establishment of an Army-Navy Communication Intelligence Board [ANCIB] . . . to ensure that communication intelligence, the most important source of operational intelligence, will be discussed independently of other forms of intelligence." Codebreaking must remain sacrosanct, walled off in its own compartment. To preserve security, not even the Joint Staff, or its Joint Intelligence Committee, would be in that compartment. When their staffs prepared the paperwork to transform the ANCICC into the ANCIB, a board with much more authority, neither King nor Marshall objected.[32]

The new memo war in Washington sputtered along while the Battle of the Bulge was raging in Europe. Only after Eisenhower had checked Hitler's last great offensive did it flare up. On February 9, 1945, the *Chicago Daily Tribune* proved Admiral Redman right about leaks. The paper's front page ran the headlines: "New Deal Plans Super Spy System, Sleuths Would Snoop on U.S. and the World."[33] The lurid misrepresentation of his ideas had Donovan outmaneuvering the FBI, the Secret Service, ONI, and G-2. Follow-up articles gave voice to congressmen who denounced the plan and reprinted the most recent joint staff response to Donovan.[34] In the second story, a photo of Leahy stood alone near the words "Opposes Control by Donovan Agency." It was a correspondent with a long history of embarrassing the Democratic establishment named Walter Trohan who had broken these stories. He was doing exactly what his employer wanted. Col. Robert R. McCormick, the publisher of the *Tribune,* hated Roosevelt and the New Deal with a white-hot passion that had long clouded his judgment.[35]

Donovan was enraged. He immediately launched his own investigation and demanded that the Joint Chiefs do the same, denouncing not only the political attack on the president but also the attempt to sabotage American intelligence.[36] He claimed the leak was tantamount to treason. The investigation by the army's Inspector General was inconclusive; at least forty-eight copies of core documents had passed through 150 to 175 hands, any one of whom could have leaked the information. This left Donovan free to conclude that Hoover had leaked documents that he, in turn, must have obtained from someone on the joint staff. Trohan himself said nothing in 1945. In the 1970s, he made the sensational claim that the leaker was Roosevelt's loyal press secretary, Steve Early.[37] If true, this meant that it was the White House that had wanted to derail Donovan's initiative, which was more or less what happened when Marshall officially deferred the matter.

When the story broke, both the president and Early were thousands of miles away at Yalta in the Crimea, conferring with Stalin and Churchill. It was March before Roosevelt was back in touch with Donovan. The president's schedule shows that they met in the Oval Office for fifteen minutes on March 15, which, Donovan's biographer concluded, they devoted to small talk and not to the Trohan affair.[38] Two weeks later, while Donovan was preparing to return to Europe, he worked with the pliant Isador Lubin to craft a note for the White House economist to send to the president: it was time to reinvigorate the plans for a central intelligence service that had "stalled in one of the subdivisions of the Joint Chiefs of Staff."[39] The OSS director even helped Lubin draft a memorandum for Roosevelt's signature directing Donovan to poll various government agencies and secure a consensus of opinion.[40] Roosevelt was at Warm Springs, Georgia, taking the waters and trying to recoup his strength when he signed the memo on April 5. It was the last time the two men communicated.

A New President

Roosevelt's death on April 12, 1945, all but eliminated Donovan's chances of prevailing in his quest to preserve OSS. The new president, Harry S. Truman, came into office determined to do his commonsense best. He started with little experience in foreign affairs. Though he had been a senator since 1934 and vice president since January, no one had briefed him about such momentous undertakings as the Manhattan Project, let alone Magic. He had to absorb information at a dizzying rate.

The World War I artillery officer seemed to have already formed a bad impression of the more glamorous warrior Donovan—perhaps because the director was a slick Easterner rather than a plain-spoken Midwesterner like Truman, or because he was the wrong sort of Republican, or because Donovan had seemed evasive before then senator Truman's committee, which was charged with investigating wartime waste and abuse.[41]

Within two weeks of becoming president, Truman had been briefed on the discussions about postwar intelligence, probably by Leahy, and certainly by Harold D. Smith, who was still director of the Bureau of the Budget.[42] In 1941 and 1942, his officers had evaluated both COI and OSS, and were not impressed by what they found. Smith looked like an accountant, with his wire-rimmed glasses and dark three-piece suits, but he was actually an expert in public administration, one prized by both Presidents Roosevelt and Truman, and prominent enough to make the cover of *Time* magazine in 1943. Roosevelt valued Smith because he brought a semblance of order to his disorderly administration; Truman valued him because he shared Smith's love of order.[43] Truman would remember their first discussion of the future of American intelligence, how "Smith suggested, and I agreed, that studies be undertaken at once by his specially trained experts in this field."[44] The two men also agreed that it was "imperative" that the experts take the time they needed to do things right. "One thing was certain," Truman would remember saying, "this country wanted no Gestapo under any guise."

Smith reinforced Truman's preference for a measured approach. Neither man was against a permanent civilian intelligence agency. Truman simply refused to be manipulated or rushed by anyone, and knew that, whatever the outcome, he wanted safeguards. America was fighting a war against Hitler's police state; he did not want anything like that at home. Though he placed them in the same general category, the president was thinking more of the FBI than the OSS when he used the word "Gestapo," believing that the Bureau was "tending in that direction . . . dabbling in sex-life scandals and plain blackmail."[45] Truman's attitude toward the FBI would mean that, like Donovan, Hoover would have few chances to realize his own grandiose plans—or even to meet with the president.[46]

Meanwhile, the secret surrender in Italy was playing out and causing as much frustration as satisfaction in Washington, what with vacillating Nazis and angry Soviets who suspected a double cross. It did not improve OSS's standing with the White House. On May 1, Truman shared his thoughts with Attorney General Biddle when they met for thirty minutes at the White House.[47] Truman told Biddle flat-out that he did not like the Donovan plan and ignored Biddle's observation about a possible alternative, SIS—the FBI system of agents and legal attachés. Instead, the president repeated his aversion to Donovan's plan, adding darkly that he "was not convinced as to Donovan's loyalty to President Roosevelt."

A private meeting between Truman and Donovan on May 14, 1945, lasted fifteen minutes and did not go well for Donovan. According to Truman's notes, Donovan told Truman how important OSS was and what it could do for the government.[48] According to Donovan's authorized biographer, Truman lectured Donovan on the need to protect American civil liberties from the OSS. Neither man left happy. Donovan went off to lunch with Lubin, likely to discuss the battle for the future of OSS that Lubin was helping to wage.[49] For the rest of the year, Truman would ignore Donovan's requests for another private meeting.[50]

Donovan would fight to the last round. The months that followed were not unlike the months between November 1944 and April 1945. J. Edgar Hoover intrigued as deviously as he had before, contributing to a sloppy compilation of charges against OSS. Based largely on secondhand information from unnamed sources, this document charged Donovan and his men with epic fraud, waste, and abuse before ending in two recommendations: fire the director and expand the FBI's SIS. Known as the Park Report—after Colonel Richard J. Park Jr., an army intelligence officer who had joined the White House staff in 1944—it allegedly landed on Truman's desk soon after he took office.[51] If Truman read the fifty-four-page report, he did not initial it or comment on it. The gadfly Walter Trohan, however, almost certainly did read it—and with relish.[52]

The Park Report was the perfect basis for another series of Trohan articles between May 16 and 20, 1945. Trohan renewed the old charge that OSS was controlled by the British and riddled with communists, which was true but not in the sense that the writer alleged. (Trohan was thinking of OSS officers like Milton Wolff, a fellow traveler who had openly fought alongside communists in Spain in the 1930s, not of Soviet spies like Duncan Lee, who secretly shared the contents of Donovan's inbox with the NKVD.) Other articles focused on the theaters where OSS had been completely or partly barred: MacArthur's Southwest Pacific and Nimitz's Pacific Ocean Areas. Donovan tried to control the damage as best he could.[53]

Now that the war in Europe had ended, the reminder about the Pacific stung: the Allies were winning battle after battle without OSS. With the notable exceptions of China and Burma, Donovan's agency had little to contribute to the cataclysmic struggle against Japan in the summer of 1945.[54] Instead codebreaking took center stage. Thanks to Magic, decision-makers in Washington saw the evidence that the Japanese were determined to fight on, whatever the cost, even if it ran into millions of dead soldiers and civilians. The Pentagon was even able to count the divisions of defenders behind the beaches that it planned to assault on the home island of Kyushu.[55]

How men like Stimson and Forrestal, or Marshall and King, reacted to any particular decrypt is hard to pinpoint—because they seldom recorded their thoughts on Magic or Ultra secrets and because many factors came together to shape decisions in Washington. Some of their decisions had long histories. The decision to start using the terrible new atomic weapons as soon as they had been assembled and tested, as well as the decision to drop two bombs on Japan, one after another, are good examples. The military had been making plans for months; Truman endorsed and approved those plans; decrypts played a central role in the process.[56] The buildup on Kyushu revealed by decrypts led planners to consider alternatives. A multitude of decrypts affirmed the assumptions underlying the bombing campaign. Even after the first bomb exploded over Hiroshima on August 6, decrypts indicated that the Japanese government was still not prepared to surrender unconditionally.[57] Japanese attitudes changed only after the second bomb exploded over Nagasaki two days later.

Donovan was traveling in China, visiting OSS bases there, when he heard the news about Hiroshima and Nagasaki. Speaking to his China hands, he predicted that America would still need them.[58] The Japanese would now surrender, but OSS had to monitor Imperial forces as they withdrew from occupied territory—and spy on the Chinese and Russian communists who wanted to fill postwar vacuums in Asia and Europe. On the fourteenth, after almost three weeks overseas, Donovan was back in Washington. He arrived just in time for Truman's announcement at 7 p.m. that Japan had accepted the surrender terms.[59]

The next step would be to prepare for the surrender ceremony, which would occur on the deck of Truman's favorite battleship, the *Missouri,* in Tokyo Bay a little more than two weeks later to officially mark the end of hostilities. Still during August, Truman remembered his determination to "demobilize as soon as possible the forces no longer needed" that would apply to wartime agencies.[60] In the second half of the month, Donovan fielded inquiries from the Bureau of the Budget about reducing and liquidating OSS. He cooperated with Director Smith while both renewing his arguments for a smaller but permanent intelligence service like OSS.[61] It did not do any good; the administration's plan did not change. Like other wartime agencies—and the paramilitary Special Operations Executive in Britain—OSS would be liquidated.

In an ironic twist for the man who had so recently condemned leakers as traitors, Donovan now turned to the public. After ordering dozens of OSS files to be declassified, he put officers to work writing press releases and articles for popular magazines. The *New York Times* recorded September 13 as the day OSS "tossed off its cloak . . . to reveal the strength of its dagger."[62] A day later Gen. Jonathan M. Wainwright, left by General MacArthur to surrender his army in the Philippines in 1942, spoke gratefully about the moment when an OSS officer appeared to rescue him from Japanese captivity; OSS proceeded to publicize its laudable record for rescuing thousands of American fliers caught behind enemy lines as well as prisoners of war.[63] An article about Operation Sunrise that appeared in the *Saturday Evening Post* on September 22 followed the Dulles narrative, casting OSS in a positive light.[64] All the while, Donovan continued to send ever more pointed memoranda to the White House. His screed of September 13 bordered on disrespect by holding school on Truman "in the national interest, and in your own interest as the Chief Executive" on the folly of parceling out OSS assets.[65]

For almost three years, Donovan had successfully protected his agency from its enemies—a major achievement since he had proven almost as adept at making enemies as friends. This time nothing seemed to work. Onetime supporters like Stimson and Marshall stayed on the sidelines. Budget director Smith and presidential chief of staff Leahy filled the vacuum. They had made up their minds, and the president stood with them. On September 5 and 13, Smith touched base with Truman, making sure that the chief executive wanted to proceed with liquidating OSS even though Donovan "was storming about our proposal."[66] The president affirmed his desire to act without delay.[67]

This would be an interim solution. The plan was for the administration to reassess its options in a few months. No one was against the core concept. On the contrary, speaking for the Joint Chiefs of Staff, Leahy signed a memorandum from the chiefs of staff to Truman on September 19 that "recent developments in the field of new weapons"—a reference to the dawn of the atomic age—affirmed "the importance of proceeding without further delay to set up a central intelligence system."[68] But while the chiefs liked some of his ideas, they wanted Donovan gone. Leahy drew up a list of qualifications that seemed to rule out the OSS director:[69]

> The Director of the Central Intelligence Agency . . . should be either a specially qualified civilian or an Army or Navy officer of appropriate background and experience. . . . It is considered absolutely essential, particularly in the case of the first director, that he . . . exercise impartial judgment in the many difficult problems of organization and cooperation which must be solved before an effective working organization can be established.

Smith carried the fateful executive order to the White House on the afternoon of Thursday, September 20, for Truman's signature.[70] Its provisions were straightforward: OSS would cease to exist on October 1, only ten days later; the Research & Analysis Branch would go to the Department of State; the other parts of OSS would come under the Secretary of War, which would have the authority to employ or disband them. Smith reminded Truman that Donovan was opposed to the plan but repeated his opinion that "this was the best disposition of the matter."[71] Truman agreed. He was fed up with how hard it was to get agencies to downsize and wanted to act. He looked the order over and signed it along with a nice-enough letter thanking Donovan for his service, and another letter charging his secretary of state, James Byrnes, with developing a plan for postwar American intelligence."[72]

Smith suggested that Truman might want to deliver the news to Donovan himself. This time Truman preferred to pass the buck and told Smith to do it. Smith had no more stomach for confronting Donovan than Truman, and delegated the task to his subordinate, a conscientious civil servant named Donald C. Stone, who trudged up Navy Hill only to find that Donovan was out of town. Saying he would "cooperate in arrangements," an OSS staff officer named Louis Ream accepted the document without fanfare.[73]

21

AN END AND A BEGINNING

The Last Days of OSS

On Saturday, September 22, 1945, three Bureau of the Budget officers walked into Donovan's corner office on the ground floor of the East Building on Navy Hill. They came to talk about the mechanics of liquidating OSS, but had to spend the morning defending themselves from an angry Donovan who blamed their bureau for the executive order that would end his agency's life.

One issue was whether abolishing OSS would literally put its employees out of work. The president's order only applied to the organization itself and to its director—not to anyone else, the budget officers explained. This was cold comfort to the OSS men in Room 109. Another issue was the unseemly rush. Donovan was for once bureaucratically correct, pointing out that it would be hard to do a good job of winding up all of the OSS's business in eight days. Budget officer G. E. Ramsey Jr. admitted that Donovan had a point. After the meeting, he pointed out that State and War had yet to designate the officers who would manage the large OSS contingents—thousands of men and women who would come under their control. If that did not happen immediately, he wrote in the language of bureaucrats everywhere, "it would seem highly desirable to extend the termination date of OSS to at least October 15."[1]

That was not to be. War and State quickly appointed managers to receive the OSS personnel. State's choice was a surprise: Alfred T. McCormack. He likely came to State's attention through the thirty-three-year-old interim assistant secretary of state for administration, Frank McCarthy, who had just come over from War, where he had worked as an assistant to General Marshall and knew the lineup.[2] McCormack, who was by this time thoroughly fed up with General Bissell at the Pentagon, did not take long to accept the offer.[3]

McCormack seemed a good choice to transform the well-respected OSS Research & Analysis Branch into the Interim Research and Intelligence Branch at State. R&A had more wartime potential than actual success.[4] While the branch might have contributed little to the prosecution of the war, it was "an indisputably brilliant episode in the history of ideas, of intellectuals, and of intelligence."[5] Its members had the kind of Ivy League credentials McCormack respected; they had wanted to create products similar to his Magic Summaries but been stymied by Washington dynamics. At War, the choice was Donovan's hardworking, intense deputy, Brig. Gen. John Magruder, the artillery officer who was already familiar with the workforce and its mission; he had even thought through many OSS organizational issues.[6] He would become chief of the interim Strategic Services Unit (SSU). His boss, assistant secretary of war McCloy, conveyed his expectations: some OSS functions had to go by the wayside, "normally shrinking . . . as a result of the end of fighting," but others "must be preserved . . . as potentially of future usefulness to the country."[7] This was, McCloy stressed, what the President wanted: "obviously the whole subject is one for careful and cooperative study."

With the appointments of Magruder and McCormack, the transfers of the 10,390 remaining OSS employees were set to occur on October 1.[8] In the few days that remained, Donovan took ownership of OSS history by preparing to take official records home with him. He ordered his aide Edwin J. Putzell Jr., a former Rhodes Scholar and junior associate at Donovan, Leisure, to join him in copying documents from the director's safe onto 35mm film with a Kodak Recordak Camera, technology that had been introduced in 1928 and was widely used by banks. About the size of a sewing machine, it did the job well enough. Working at breakneck speed, mostly at night, the two men produced an astounding 131 rolls of film, each containing hundreds of frames. Donovan obtained sixty-two more rolls of 16mm films of cables from the OSS Communications Branch, and, as if all these secret files belonged to him rather than the government, shipped the collection off to his law office in New York.[9]

On Friday, September 28, the last workday at OSS, Donovan convened 755 employees for a farewell ceremony at the vast Riverside Stadium, home of three ice and roller skating rinks at the foot of Navy Hill by the Potomac.[10] Looking distinguished in his khaki summer dress uniform, with an honor guard and an array of flags behind him, Donovan delivered a moving speech that celebrated OSS as "an unusual experiment . . . to determine whether a group of Americans constituting a cross section of racial origins, of abilities, temperaments, and talents could meet and risk an encounter with the long-established and well-trained enemy." He then paraphrased the letter of appreciation from Truman, telling his audience that they should feel deeply gratified that the president was planning to base "a coordinated intelligence service upon the techniques and resources that you have initiated and developed." They had opened the way into uncharted territory. After his speech, Donovan handed out fourteen medals to a line of OSS men and shook hands with other soon-to-be former employees. Before presenting a silver tray from the workforce, former deputy director G. Edward Buxton directed a few words of praise to the director, telling him that he had been the man of the hour: his life of "unique experience and training . . . [had] prepared you for your role in World War II."[11]

That role was to upset the status quo and be the agent of change. He had been disruptive and creative at the same time. He formed the first American central intelligence agency out of nothing and sent it into battle. That alone has secured his place in history. In its encounters with the enemy, OSS acquitted itself well enough even if its successes were never as decisive as Donovan would have liked. The OSS record as a spy service was only fair. For various reasons, it did not produce much finished, actionable strategic intelligence. OSS shone brightest not when it was operating alone, but when it supported Allied armies in Asia and, above all, in Europe after D-day.

As the war went on, Donovan became almost a caricature of himself. The energetic maverick showed his impetuous, arbitrary side, contributing to his own downfall and that of his agency; there was only so much of Bill's wildness that Washington could tolerate in September 1945 as the administration searched for solid ground after the avalanche of events, each momentous enough in itself, from Roosevelt's death to defeating Germany to dropping the atom bomb on Hiroshima. But Donovan had set a process in motion that continued after he left office. Even those who did not share his vision had to react to it, and that ultimately would bring about lasting change.[12]

Washington without Donovan

J. Edgar Hoover certainly reacted to Donovan's vision: he rejected it. In September he had done everything he could to take OSS off life support and suffocate its offspring. On September 6, he minuted the new attorney general, Thomas Clark, railing against "General William Donovan's plans for the perpetration [*sic*] of his dynasty . . . under another name in the world-wide intelligence field," meaning Magruder and his unit.[13] Hoover proceeded to make dark allegations against Magruder's family and imply that he would use his new position to "look out for [the] interests" of an unspecified large corporation. No doubt frustrated that he had lost his access to the Oval Office, Hoover urged Clark to argue the case for expanding his overseas intelligence service, SIS, whose writ had originally been limited to Latin America. The day after Truman signed the order to liquidate the OSS, Hoover was at it again, writing Clark in his signature prose that it was "essential to consummate the Government program upon this subject at the earliest possible date."[14] On September 27, he sounded the alarm after hearing that McCormack was "forming a World-wide Intelligence organization" at the State Department.[15] He feared that McCormack was not favorably disposed to the FBI's SIS and once again urged Clark to bring "our views" to the president's attention immediately.

Hoover must have been frustrated when Clark sent the director's memos first to the Bureau of the Budget instead of directly to the White House, and when that bureau's officers subsequently questioned whether SIS was a good choice for running foreign intelligence operations.[16] They distinguished between the international security intelligence and security programs—the FBI's forte for which they gave good marks—and the "economic, political and other basic forms of intelligence," for which, they argued understandably, it was not particularly qualified. The BoB recognized that the FBI was not in exactly the same business as OSS or military and naval intelligence.

Hoover now decided on an end run, sending an FBI special agent named Morton B. Chiles to call on Truman. A Missouri man, the clean-cut Chiles was the son of a long-time neighbor. True to his roots, the president even came out to the reception room to escort Chiles into the Oval Office for a cordial chat on October 2. Chiles explained Hoover's vision—now unabashedly for the FBI to run "World Wide Intelligence"—to Truman, who listened politely and asked a few questions. He even seemed receptive when Chiles suggested that the president meet with Hoover to discuss the matter.[17] But even a fellow Missourian could not sway the president, and that meeting did not happen.

Hoover never gained much traction with Truman. Instead it was the president's dictum that the FBI should operate only in the United States that determined the Bureau's future, removing from consideration the idea of a domestic law enforcement–cum–foreign intelligence superagency.[18] The story of the FBI's SIS would be one of dwindling resources until it was eclipsed two years later. The commonsense idea was that the FBI look inward, while the successor to OSS looked outward. Starting in the fall of 1945, the FBI would do just that, focusing on real and imagined domestic threats, including the well-placed Soviet spies in the US government who had enjoyed a mostly free ride during the war. One of the FBI's first targets would be Donovan's aide Duncan Lee. He was, Hoover informed the White House in November 1945, one of five former OSS employees under investigation.[19]

A dark horse now emerged from the pack in the intelligence sweepstakes. With OSS out of the way, Alfred McCormack tried as hard as Hoover had for a win for his new agency at State. McCormack based his bid on a questionable reading of the letter that the president had written to Secretary of State Byrnes on September 20, directing him "to take the lead in developing a comprehensive and coordinated foreign intelligence program for all Federal agencies concerned with that type of activity."[20] Once again wielding the undefined but considerable powers of a special assistant, McCormack began a monthslong campaign to establish a strong intelligence agency under his control. His views shifted somewhat over time, but the basic idea was to replicate what he had done for the army during the war.[21] He did not see a need for a new, independent central intelligence agency. Establishing a c.i.a. would, he declared opaquely, be as bad as forcing all lawyers in Washington to prepare their cases under "one central organization."[22] All the government had to do was work better and smarter, to put the right people to work in the right way. Under McCormack's plan, existing intelligence agencies would still be free to operate on their own. He thought that competition among them would be healthy. In the legal world, the truth emerged from an adversarial process. Was the work of foreign intelligence any different?

It took some nerve for McCormack to make his proposal—especially considering that he had once mocked State for being too pious to do intelligence work—but it was not outrageous. MI6, for example, came under the British Foreign Office, not unlike the FBI, which came under the Department of Justice. But MI6 controlled both signals and human intelligence. McCormack, on the other hand, had already dismissed the idea of centrality; now he was showing little interest in having the government stand up a robust spy service. This meant that he wanted to sidestep another one of Donovan's core concepts: recruiting spies to steal secrets. No doubt confident that State would continue to receive something like the Magic Summaries, he appeared willing to forgo the benefits of spying and rely instead on a single stream of reporting that, so long as it continued to flow, could supply the government with vast amounts of information. What mattered was what happened to that information. Sounding like himself in 1942, McCormack argued that "the critical process in intelligence is performed at the research desk."[23] He added that he saw no reason to remove codebreaking from the army and the navy. It was a system that had worked during the war, so why shouldn't it work in peacetime?

This was exactly what the two services wanted. While others argued about the future, the codebreakers were moving quietly and inexorably toward their goals. Still collaborating through the ANCIB, the army and navy had agreed unanimously on August 21 that "OSS, Treasury, and FCC . . . be excluded entirely from cryptanalytic activities."[24] They were trending away from any involvement with civilian intelligence agencies. On September 12, the Secretaries of the Navy, War, and State quietly requested the president's permission to continue collaborating with the British.[25] Exaggerating only slightly, citing the value of that collaboration, and adding a reference to "otherwise unobtainable" diplomatic and economic information, they recapped the cryptanalysts' "outstanding contributions to the success of Allied forces in defeating Germany and Japan." They did not, rightly, include a metric, but their claim was undeniable: though impossible to quantify, communications intelligence had made a huge difference in the war.[26] Truman approved their request without any fanfare.

In October 1945, Bissell, Joseph Redman, Clarke, and Wenger gathered in Washington with the British, represented by Travis, who was still the head of GC&CS, to discuss postwar collaboration.[27] That US Army and US Navy codebreakers were still sitting down together on a regular basis was remarkable enough in itself. That their British counterparts were at the same table at the same time, discussing matters in a collegial manner, showed how much the war had changed the landscape. In 1940 and 1941 it was hard for any of the three groups—army, navy, and GC&CS—to see its way clear to sharing precious secrets; now it was now almost impossible for them to imagine operating without each other. The three parties even agreed not to make any more two-party agreements—like the navy's and the army's separate wartime agreements with GC&CS. By November 1, 1945, the Americans and British would agree on the final draft of a US-British Agreement for cooperation in the realm of communications intelligence that would be ratified in the new year and fortify their alliance against postwar challenges.[28]

The decisions to strengthen and extend wartime codebreaking arrangements came almost as a matter of course; the army and the navy were taking care of what they regarded as an absolute necessity.[29] But even they were now willing to acknowledge that the issues raised by Donovan were worth addressing. So, in September 1945, it was back to the conference rooms of official Washington for another series of meetings that would last until January 1946.

A general consensus soon emerged: America needed something different going forward. No one wanted to turn the clock back to 1939. That alone set up a dramatic contrast. Before the war, the very idea of strong foreign intelligence, let alone running spies, had to struggle for breath; after the war it was one of the centerpieces of the agenda. As Leahy had acknowledged, the need was all the more pressing now that the atomic age had begun.[30] From one day in August to the next, the cost of failure had risen astronomically. No other country had the bomb, but that time could come. Washington needed to forestall an atomic Pearl Harbor, the specter that lurked in the background. In the fall of 1945, no one could escape the headlines generated by the Joint Committee on the Investigation of the Pearl Harbor Attack, which was part of the national quest for definitive answers to the events of December 7, 1941.[31]

Moving beyond the starting consensus was insanely difficult. The surrender of Japan had not ushered in an era of good feeling in Washington. If anything, the atmosphere was worse now that Americans did not have to close ranks against the Axis. The old bureaucratic rivalries, overlaid with new anxieties, roared back with a vengeance. It did not help that the leading roles went to such hyper-assertive men. At State, Secretary Byrnes seemed to think that he should be president in place of Truman. At Navy, Secretary Forrestal was himself no shrinking violet, as his permanently broken boxer's nose attested. Even he had to abandon the idea of inviting Hoover, Bissell, and McCormack to settle their differences over drinks and dinner after his intelligence chief warned that the "veiled antagonism" among the three would make for an "uncongenial" evening.[32] Longtime Washington insider Lawrence Houston, a former OSS officer who was a Donovan proxy, found himself in the "toughest bureaucratic fights" he would ever experience. "Tougher than any I'd ever seen before, as tough as anything I saw afterward."[33]

From summer to fall, the arguments swirled back and forth among the various camps: McCormack at State, Magruder at SSU, the army, usually supported by the navy, and the Bureau of the Budget.[34] (To Hoover's chagrin, the FBI found itself largely excluded from the proceedings.[35]) Cabinet secretaries themselves occasionally participated, Byrnes initially backing McCormack, then setting out on extended trips that left his special assistant at the mercy of hostile State careerists who did not want a separate intelligence office, especially one run by an outsider. Even finding suitable candidates to consider for top jobs was tough. Assistant Secretary of War Robert Lovett commented that "the only name he had heard mentioned was that of Allen Dulles who was generally regarded as highly competent. . . . He had organized the best job of the OSS in Switzerland."[36] The favorable mention of Dulles reportedly caused Donovan to spread the word that Dulles was a poor administrator who was not fit to run a large agency.[37]

The basic questions remained intractable: would there be a central intelligence agency; would it be military or civilian; who would appoint its director; would it collect or just collate information? A key issue was whether it would have any power over other stakeholders. More collegial than his former chief, General Magruder held his ground, politely summing up the arguments for an independent central intelligence agency, using words like "safeguard." His c.i.a. would not have any policy-making function; its independence would limit the potential for overreach by any one of the traditional departments.[38] If, say, State added an intelligence service to its existing authorities, it would have too much power. Later, Magruder reminded army intelligence that his SSU had preserved "the essential personnel, techniques, and facilities [needed] . . . for peace-time intelligence procurement," by which he meant espionage and counterespionage.[39] He highlighted the assets of the old SI and X-2 branches of OSS.

By late November it was clear that the various camps would not be able to agree on their own, and Budget director Smith wrote to the president asking him to intervene.[40] By early January 1946, even the usually indefatigable McCormack was talking about "the apparent hopelessness of proceeding without further direction from the president."[41] The Pentagon decided to commit its reserve—the "Missouri gang": officers from Missouri who could appeal to the Missourian in the White House.[42] Between January 8 and 18, they worked through a series of drafts with Budget director Smith. By January 21, there was a letter to the Navy, State, and War secretaries ready for Truman's signature.[43] Along with the president's personal representative, who would be Leahy, the secretaries would form the "National Intelligence Authority" (NIA) that in turn would set up a Central Intelligence Group (CIG), staffed by officers detailed from their departments. The president would pick a director of Central Intelligence who would be responsible to the NIA.

The plan was closer to the old Joint Staff proposals than anything Magruder, Hoover, or McCormack had proposed. It called for something like a pair of special purpose committees rather than a functional agency or board of directors. Beholden to cabinet secretaries, the director of central intelligence would have little control over his own domain, while the NIA itself was set up like a debating society. The result was, in one acid comment, a headless body (the NIA) and a bodyless head (CIG).[44] Director Smith predicted gloomily, this "whole subject of intelligence . . . will not be solved in an orderly fashion and . . . we will go through the usual two, three or more reorganization stages—God bless bureaucracy!"[45]

Smith was right. When he signed and dated the letter on January 22, 1946, Harry Truman marked the end of the history of American intelligence in World War II and the beginning of its postwar history. The infighting would stop for the time being. But, as Smith predicted, there would be no end of commissions and reorganizations for years to come. Taking on the basic structure of OSS, the CIG would morph into the Central Intelligence Agency under the National Security Act of 1947, which would also create the Department of Defense. In 1948 the army and navy codebreakers would finally be consolidated into one agency, the Armed Forces Security Agency. In 1952, it would morph into the National Security Agency, an agency of the Department of Defense not unlike the FBI, which remained a semi-autonomous part of the Department of Justice.

This was the legacy of World War II. America began the war with little useful intelligence of any kind but was able to rely on improvisers like Joe Rochefort. While the FBI kept foreign spies from the door, William J. Donovan developed a prototype that would endure—that of an intelligence agency with branches for espionage, analysis, special operations, and counterintelligence. Codebreakers like Friedman and Rowlett, working in tandem with synthesizers like McCormack, unmasked the enemy's secrets and illuminated the path to victory. Noting their success and potential going forward, the army and the navy proceeded to lay the foundation for an industrial codebreaking enterprise of hitherto unimaginable proportions. By the end of the war, America had many more intelligence resources than on December 7, 1941. But important issues that had cropped up during the war remained unresolved. Would American spies, codebreakers, and FBI special agents learn to work together at least as well as they all now worked with the British? Would the spies incline more toward Donovan's special operations, paramilitary and political, and his bias for action—sometimes for its own sake? Or would they see the merits of his loyal deputies, the Rogerses and Magruders who called for careful planning, espionage, and analysis to satisfy their government's need to know?

EPILOGUE

The guard had already started to change well before Truman signed the letter to the secretaries in January 1946. A few OSS Special Operations officers stayed with SSU, but, like their colleagues in R&A, most returned to private life. The rest were reabsorbed into the mainstream of the army. In the 1950s a handful of OSS veterans emerged as US Army Special Forces officers; the army would honor a tenable but small link between Donovan and the Green Berets.[1]

It was a connection that the war-scarred founder of Detachment 101 in China-Burma-India could do only so much to encourage. In January 1946, Carl Eifler entered Tripler General Hospital in Hawaii to start on a medical odyssey that ended inconclusively six months later at Letterman General Hospital at the Presidio in San Francisco.[2] The doctors could barely control the damage from his wartime head injury. Punctuated by violent outbursts, seizures, and amnesia, his struggle would last for years. Perhaps to atone for wartime guilt—he commented more than once that members of Det 101 had committed war crimes that he did not specify—he eventually obtained degrees in divinity and psychology at Jackson College, a small missionary school in Hawaii. In the 1950s he went on to get a PhD in clinical psychology at the Chicago Institute of Technology, and then worked for the Department of Health in Monterey County, California, where he developed a reputation for compassionate engagement.

William J. Casey was one of the first of the lawyers to return to private life, submitting his resignation at the end of August 1945. While his wartime work had been more fulfilling than anything else he had ever done, he now wanted to make his fortune, and dedicated himself to an imaginative mix of business and legal ventures in New York. He would dabble in politics and return to Washington off and on. In the 1980s, when he failed to get his first choice, which was to be secretary of state, he would settle for director of central intelligence.[3]

Donovan and Dulles followed suit in December 1945. Dulles left government service first, returning to a warm welcome at Sullivan & Cromwell in New York. Donovan arrived home shortly before Christmas. He had spent a few months assisting in the prosecution of Nazi war crimes at Nuremberg before falling out with the lead prosecutor, Supreme Court Justice Robert Jackson. He now planned to take up the law practice he had left in 1941. A few months later, McCormack was also back at his law office in New York. After another furious row with the old guard at State, he resigned in anger in April 1946.[4] He might have wanted to direct American intelligence, but he was always more of a lawyer than a bureaucrat. He was able to re-engage at Cravath and claimed once again to enjoy the work—though he now felt free to admit that his favorite client had always been the French American opera singer Lily Pons, not one of the capitalist behemoths that the firm existed to serve.[5]

McCormack, Donovan, and Dulles all received comparable awards for their wartime service: Donovan and McCormack the Distinguished Service Medal, Dulles the civilian Medal for Merit. In November 1945, Clarke, now a brigadier general, pinned the medal on his former colleague, who was back to wearing expensive dark suits.[6] Truman himself presided over separate ceremonies for Donovan and Dulles in 1946. In January, Donovan seemed unhappy while Truman, looking crisp and decisive, pinned the medal on his uniform—though by the end of the brief ceremony, the president had coaxed a smile out of the old warhorse, and for a brief moment as they shook hands they looked like the best of friends.[7] How Dulles felt on his day in July is not recorded, but his wife, Clover, was reportedly thrilled by the award, and Dulles himself would later say that Truman had always been "awfully nice" to him.[8]

Senior medals typically marked the end of distinguished wartime careers and brought closure—with the exception of Donovan and Dulles. Donovan found it hard to return to his legal career.[9] His partners at Donovan, Leisure suspected that he was ad-libbing rather than preparing his arguments in the courtroom; he appeared more interested in keeping up with OSS veterans and gathering material for the history of American intelligence that he would never finish.[10] Dulles, too, had trouble rekindling his limited interest in the law, and seemed to devote as much time to the Council on Foreign Relations and writing about his wartime adventures.[11] Both men ached to be called back into service, which was what had deepened their rivalry and led Donovan to let it be known that he thought Dulles was not much of an administrator.

The same was, of course, true of Donovan. The two men displayed many of the same merits and flaws during the war. Donovan assembled a remarkable team; he promoted the idea of a strong central intelligence agency with the components that a world power needed. Dulles, too, assembled a strong team and, working largely on his own, orchestrated diverse operations in the foreign field. They were leaders who cultivated impressive networks of useful contacts and motivated their followers. They cared for their subordinates and agents. They had a rare ability to spot and exploit opportunities, narrowing their focus to the case that promised the most rewards in the short-term. Partly on account of wartime pressures, partly because of the way they were, they showed far less interest in careful planning. Donovan carried this attitude to an extreme, denouncing planners as if they were dangerous subversives. Both men occasionally bent or broke the rules. Donovan could be breathtakingly insubordinate and irresponsible; Dulles could make Donovan wonder whether he was following explicit instructions. They occasionally blurred the lines between their personal and professional lives in troubling ways.

Donovan's and Dulles's shortcomings led, justifiably, to questions about each man's suitability to head an intelligence agency. In the postwar world, Dulles had the edge. He was younger by ten years and not polarizing. Most who knew Dulles liked him well enough, while Donovan tended to ignite passion, both for and against his person. Before long he would become an icon for OSS veterans and their offspring. For men like Casey, who put Donovan's picture up in his office, he would always be *the* role model.[12] In the end, though, it was Dulles who would return to Washington to lead the CIA in the 1950s. Donovan had to content himself with an ambassadorship to Thailand that ended in 1954 when he started to exhibit the first signs of dementia. Over the next five years, the disease would slowly rob him of his memory. When he died in 1959, President Eisenhower mourned the loss of America's "last hero."

After his nemesis died, Hoover sent a polite letter to Mrs. Donovan acknowledging his life of service to others.[13] That Donovan had received many more accolades over the years must have grated. While Truman handed out senior awards to Donovan and many others for wartime service, Hoover had had to content himself with the American Legion's Distinguished Service Medal for successfully defending the country from subversion and espionage.[14] Being less appreciated seemed to reinforce his determination to remain at the helm of the FBI. He would stay on until his death in May 1972.

Many of the professional codebreakers also remained on government rolls after the war. In 1946, William Friedman received a well-deserved Medal for Merit for his wartime work but continued to work as a cryptologist, albeit at a slower pace. He often felt out of place in this new world dominated by machines. In 1947, his depression, first documented after the unrelenting work of breaking the Purple code in 1940, returned. He endured a difficult three years, checking himself into the psychiatric ward of the local Veterans Administration hospital and undergoing electroshock therapy. At times he was unable to concentrate and talked about suicide. Friedman had not been wounded on the battlefield but he might as well have been; the years of continuous work and what he called "repression by secrecy restrictions" had taken their toll.[15] In 1955, after two heart attacks, he officially retired. He and his wife, Elizebeth, returned to the subject that had launched their codebreaking careers some forty years earlier. In 1957, they published the last word on Shakespearean ciphers, debunking far-fetched theories that the bard had not written his own plays. Their book won a prize from the Folger Shakespeare Library on Capitol Hill in Washington.[16]

The army codebreakers kept in touch with one another. Friedman was no exception. He occasionally saw colleagues like Sinkov and Kullback, who came to his retirement ceremony to present him with a scroll signed by 1,300 friends and admirers. He stayed on surprisingly friendly terms with Carter Clarke, who left codebreaking and went on to command troops in the occupation of Japan from 1950 to 1953. The two men corresponded about current events and revisited their war. Somewhat guarded, Friedman made an effort to stay upbeat in his letters, writing about the histories that were starting to appear.[17]

Compared to Friedman, Clarke was free with his opinions, hinting at wartime conspiracies, criticizing "faulty memory Marshall" and railing against something he called "the screwball Donovan effect."[18] Clarke would retire from the army in 1954, start a second career as a property manager in Washington, and after ten years retire again, this time to Clearwater, Florida, where he wrote letters to the editor of the local newspapers and answered a few questions for the groundbreaking historian David Kahn, who diligently scoured the country looking for old codebreakers. Unlike OSS veterans but like many codebreakers, Clarke did not believe that their story should be told; it was still too secret.[19] Nevertheless, he wrote Kahn that the Magic Summaries had made a war-winning difference by putting decrypts in context: "To take a raw intercept, unevaluated, uninterpreted and not melded with other intelligence is as futile as reading Holy Writ and trying to determine how the Pharaoh deployed his chariots at the Red Sea."[20]

Clarke was not alone in developing conspiracy theories. For a handful of navy intelligence officers, it was the obsession to which they dedicated many of their retirement years. Theories ranged from mildly to deeply conspiratorial. Edwin Layton, Kimmel's and then Nimitz's intelligence officer at Pearl, was in the first camp. He faulted the Navy Department in Washington for not sharing bits of information that *might* have put the fleet on alert before the Japanese surprise attack.[21] Layton was trying to defend his reputation and that of his commander, Kimmel. The trailblazer Laurance Safford, who almost single-handedly laid the foundation for navy codebreaking, had a different agenda. He was in the second camp, believing that there had been a cover-up. He argued that the Roosevelt administration had intercepted and destroyed messages that foretold the attack because the president wanted war.

Layton's argument was—and is—debatable. Washington could have shared more with the field. But Magic intercepts were about diplomacy; they did not point to any specific military targets. Even if they had, the US did not have the ability to analyze them properly, to do what McCormack's officers first started to do only in 1942. A single message, standing alone, was unlikely to have made a difference. By contrast, Safford's argument had no basis in fact whatsoever; no one intercepted and destroyed incriminating messages. No one could find them, no matter how hard he tried, because they never existed in the first place. Codebreaking prowess did not confer upon Safford the kind of analytic skills that a Cravath lawyer brought to the table. Instead, his arguments placed Safford squarely in the partisan camp that was out to smear Roosevelt, an extension of the campaign tactic that Dewey had considered and wisely discarded in 1944.[22]

Layton and Safford both stayed in touch with Joe Rochefort, the commander of the Dungeon at Pearl, the underground intelligence center that did such remarkable work early in the war. He likely contributed to Layton's work—their retirement homes were only a few hours apart on the California coast. He certainly listened to Safford's arguments, but in the end refused to support them.[23] Rochefort was making his own way through the postwar world, just as he had made his own way through the war. On the January 1947 retirement list, he was one of the first of the old guard to leave active duty. Only forty-five years old, he had never known anything but the navy. As he started to adjust to civilian life, he found that it was not enough to be a retired navy captain. The man who had mastered the Japanese language and learned to think like a Japanese admiral still needed a high school diploma—he had dropped out of school to join the navy in World War I. Without too much trouble, he managed to persuade the principal of Polytechnic High School in Los Angeles that his life experience qualified him for a diploma.[24]

Getting Rochefort the medal he so richly deserved for his wartime service proved far more difficult than getting his diploma. It was a quest pursued not by Rochefort himself but by men who had worked for him. Early on the prime mover was Jasper Holmes, the disabled submariner assigned to the Dungeon who had drafted some of the original paperwork during the war.[25] Now back at the University of Hawaii, where he would become the dean of engineering, he kept the torch burning. In 1957 he approached Admiral Nimitz, who was by then a regent of the University of California. Even with Nimitz's support, the navy bureaucracy rebuffed the initiative in 1958. Rochefort's death in 1976 did not end the quest. Holmes passed the torch to one of his protégés, Donald M. Showers, who had walked down the steps into the Dungeon to report for duty as a newly commissioned ensign in February 1942. Decades later, Showers retired from the navy as a rear admiral, the only member of Rochefort's team to make it to flag rank. His retirement made it possible for him to dedicate three and a half years to the cause. Making good use of material that was going into Layton's book, he assembled a large dossier, circulated memoranda, and buttonholed officials. In November 1985, some forty-three years after the battle that Rochefort did so much to win at Midway, the secretary of the navy approved the award. The navy wanted to simply mail the medal to Rochefort's next of kin, but Showers waited for the right moment to suggest that President Ronald Reagan preside over a ceremony to award the Distinguished Service Medal posthumously. On May 30, 1986, Rochefort's son, Joseph Jr., now retired himself from the army, and daughter Janet accepted the medal at the White House.[26]

The DSM led to other posthumous awards for Rochefort: the Presidential Medal of Freedom later in 1986, and fourteen years later, a place in the National Cryptologic Museum's Hall of Honor, created in 1999 to recognize trailblazers in American codebreaking. Rochefort was now in fine company, with William and Elizebeth Friedman, Yardley, Rowlett, Grotjan, and many others. But even here he was the outlier, still stirring controversy a quarter century after his death. Historian Robert J. Hanyok argued that Rochefort did not belong with the others in the hall.[27] They either made breakthroughs in cryptology or built capacity and institutions that outlasted them. Rochefort performed superbly but he did not do those things. Hanyok is not wrong. Though a uniquely qualified intelligence officer, linguist, and cryptanalyst, Rochefort did not break new ground. It is impossible to deny that, in the same year that Hypo enabled the triumph at Midway, it was already being eclipsed as the navy moved to expand and centralize its intelligence operations.

Perhaps Rochefort belongs in a different hall of honor, one reserved for outliers like him who shone when the work of American intelligence was done by a handful of individuals, before wartime pressures turned it into a quasi-industrial enterprise. They often had little support and faced opposition from those who were comfortable with the status quo. Yet they made outsize contributions to winning the war. The criteria might shift, but many of the same names would be in the hall, especially Friedman and his team. British officers might receive honorable mentions: Denniston, the one-time head of GC&CS who became Friedman's friend and ally, and Winn, the handicapped English lawyer who taught the US Navy how to sink more German submarines.

The most notable addition would be Alfred McCormack, who also regarded Friedman as a kindred spirit. His portrait would go up in the hall for understanding how to produce useful intelligence *and* creating a system to do so in 1942. McCormack's system did not outlast the war. In 1945, his lawyer-analysts went back to their civilian lives and the head of army intelligence did his best to forget the New Yorker. But his wartime work made a difference that is better described than quantified. When McCormack died of cancer at the young age of fifty-five—his death likely hastened by three decades of seventy-hour workweeks—he passed almost unnoticed in Washington. The two paragraphs near the bottom of the editorial page of the *Washington Post* on July 13, 1956, were easy to miss.[28] They proclaimed, "More than any other man, it was [McCormack] . . . who made a smooth, coordinated intelligence operation out of what had been a jerry-built shambles of military information." Whether they knew it or not, "commanders from the man in the White House down to the platoon leader [in the field] stood in his debt."

The anonymous but well-informed writer captured a great deal in a few lines: where the wartime journey had begun, what signals intelligence and McCormack had contributed to victory, and, to paraphrase Churchill, how much so many owed to so few who had quietly done their duty.

ACKNOWLEDGMENTS

This book has had many friends. First of all, let me be clear: it stands on the shoulders of many works by practitioners and scholars who have come before me, too numerous to mention here with a few exceptions. Perhaps my greatest debt is to my good friend David Kahn, *the* historian of codebreaking, who started me out on this path by drawing my attention to the fact that no one had written a good overview of American intelligence in World War II. No one I know has been more generous to fellow writers of history than David; his generosity ran from making introductions to other historians to sharing the files he built up over the decades to memorable meals at the Century Club in New York. High on the same list are literary agents Howard Yoon and Dara Kaye at RossYoon in Washington, DC. They patiently listened as I tossed out ideas for my next book, offering wise counsel. Dara later offered much-appreciated editorial rudder. I was of course thrilled when Howard reached out to Peter Hubbard at William Morrow (and now Mariner) to cement the project. I could not have asked for a better editor than Peter. Working along with author and editor were assistant editor Molly Gendell, copyeditor Mark Steven Long, and production editor Laura Brady. All made much-appreciated contributions. Sharyn Rosenblum, vice president and senior director for media relations at HarperCollins, has been uniquely supportive. She has also cheerfully endured more of my book talks than just about anyone else alive.

Early on I had the good fortune to encounter long-lost comrade and fellow historian Chalmers Hood at the Library of Congress. I benefited from his knowledge of the sources and especially his deep knowledge of Vichy France. Chalmers was kind enough to introduce me to Gillian Bennett, formerly chief historian of the Foreign and Commonwealth Office, who suggested various British collections to search. David Schaefer of King's College London performed yeoman service by sifting through files at the British National Archives in Kew. From Cambridge University, my good friend Dr. Peter Martland generously shared material when our research interests overlapped, as did Dr. Dan Larsen, a scholar at work on World War I intelligence. The incomparable Hayden Peake was kind enough to point me in the direction of other scholars and their works.

On this side of the Atlantic, I am grateful to Richard Busick for sharing his expert knowledge of FBI organization and history. Professor Sarah-Jane Corke and her colleagues at the University of New Brunswick were kind enough to read an early chapter and serve as a focus group on the concept for the book. Thanks to Sarah-Jane and our mutual friend Professor Mark Stout of Johns Hopkins, I was able to continue the process online at a session hosted by the North American Society for Intelligence History. Mark remains a tremendous resource for any intelligence scholar, myself included. Rick Schroeder, Tom Ahern, and Bryan Lintott listened to me carry on about my ideas and shared their reactions. Rick's book on this period was near at hand while I was writing about the end of the war. The same was true of David F. Rudgers's *Creating the Secret State.* Liza Mundy, the author of *Code Girls,* was kind enough to respond to this author's queries, as was Professor Richard Breitman, the eminent scholar of the Holocaust. Christopher Andrew's *For the President's Eyes Only* and Michael Warner's *The Rise and Fall of Intelligence* were both sure guides to the context in which my story is set, to say nothing of the relevant volumes of *The Foreign Relations of the United States* produced by the Department of State's incomparable historians of diplomacy.

Much of this book was written during the pandemic, which meant that I had to rely on the kindness of librarians and archivists who did what they could from afar. The standout has been Robert J. Simpson of the National Cryptologic Museum Library. Rob made a series of documents available in drop boxes without which my work would have ground to a halt until the archives reopened (which, as of this writing in the spring of 2022, has yet to happen). Likewise, Melissa Davis, the director of library and archives at the Marshall Foundation, readily responded to my queries and sent me files when she could. A world-class expert on World War II records, Paul Brown of the National Archives in College Park was kind enough to answer my queries, in person before the pandemic and after that online. I salute the archivists at the FDR Library in Hyde Park, New York, who have set the standard by digitizing hundreds of thousands of pages of files.

Perhaps my most far-flung correspondent was Zdisław Kapera, of the Enigma Press in Krakow, Poland, who unearthed an out-of-date monograph that was important to my work. Annette Amerman and Fred Allison, of my old home team at the Marine Corps' History Division, responded to my searches for digitized records, as did the US Naval Institute's Janis Jorgenson. Sarah Holcomb generously shared records of her father's accomplishments in World War II. CIA Museum experts Toni Hiley, Ann Todd, and Rob Byer were in my corner when I needed them.

I have benefited enormously from readers who gave of their time and expertise, helping me to sharpen my arguments and saving me from a good bit of embarrassment. They include Paul Nevin, who must have been an editor in another life, he is so good at it in this life, and Dr. Jay Ridler, the Swiss Army knife of our profession, equally at home in the fields of writing and military history. Jay Venables was, thankfully, on hand for months on end to apply her expertise to a long manuscript; more than once I would have been lost without Jay. Nor could I have done without the expertise of Chris Buehler, Gil Barndollar, and Thomas Boghardt, who read parts of the manuscript and kept me from straying too far from the truth. The same is true of Jean Bartholomew, Katie Sanders, and Jeff Rogg, who did what only good friends will do—offer constructive criticism. Karen Jensen of *World War II Magazine* expertly edited articles that I spun off from this work; her probing questions prompted me to go back and improve the text itself. So did the discussions I had with fellow writer and historian Andy Kutler, who was writing about World War II at the same time.

Lifelong friends Tom Sancton and Mark Bradley, both accomplished authors in their own right, offered support and encouragement. My uniquely talented wife and partner, Becky, graciously put up with yet another literary enterprise and all of the travail that it entailed. I owe heartfelt thanks to all, but especially to her.

ABBREVIATIONS

AFHQ—Allied Force Headquarters, the regional command responsible for operations in the Mediterranean theater, including Italy

ASA—Army Security Agency, from September 1945, the successor agency to the SSA

BSC—British Security Coordination, a covert office established by MI6 in New York in 1940

CBI—China-Burma-India Theater

COI—Coordinator of Information, title for Donovan and his first intelligence organization, the precursor to OSS

FBI—Federal Bureau of Investigation

G-2—Common abbreviation for US military intelligence function or officer

GC&CS—Government Code and Cypher School, British codebreaking operation at Bletchley Park

IJN—Imperial Japanese Navy

JIC—Common abbreviation for the British Joint Intelligence Sub-Committee, later duplicated by Washington

JCS—Joint Chiefs of Staff

MI5—The British internal security service, primarily responsible for running the Double Cross Program

MI6—The British external intelligence service, responsible during World War II for both espionage and codebreaking

MID—Military Intelligence Division, part of the Department of War in Washington, more or less synonymous with G-2

MO—Morale Operations, a branch of OSS

NARA—US National Archives and Records Administration

NCML—National Cryptologic Museum Library, Ft. Meade, Maryland

NKVD—People's Commissariat for Internal Affairs, Soviet spy and security agency, precursor of KGB

ONI—Office of Naval Intelligence

Op-20-G—Navy Department office responsible for codes

OSS—Office of Strategic Services, Donovan's wartime intelligence organization

PHA—Pearl Harbor Attack, a reference to the proceedings of the Joint House-Senate Committee that met in 1945 and 1946

R&A—Research and Analysis, OSS branch

SB—Special Branch, wartime part of G-2 responsible for processing Magic, headed by Clarke and McCormack, first titled Special Service Branch

SHAEF—Supreme Headquarters, Allied Expeditionary Force. Eisenhower's headquarters from late 1943 to mid-1945, responsible for operations in Northern Europe

SI—Secret Intelligence, OSS branch responsible for espionage

SIS—Signal Intelligence Service, US Army organization responsible for making and breaking codes from 1930 to 1943, part of the Signal Corps

SIS—Special Intelligence Service, wartime FBI organization in Latin America, also a title sometimes used for MI6

SO—Special Operations, OSS branch responsible for irregular warfare

SOE—Special Operations Executive, wartime British agency responsible for special operations

SSA—Signal Security Agency, from 1943 to 1945 the successor organization to the Army's Signal Intelligence Service

SSU—Strategic Services Unit, part of the War Department, successor to OSS

TNA—The British National Archives

X-2—OSS counterintelligence branch

PRINCIPAL PRIMARY SOURCES

Sources consulted mostly or partly online are marked with an asterisk.

Central Intelligence Agency*
Freedom of Information Act, Electronic Reading Room
Miscellaneous OSS Files

Department of State*
Foreign Relations of the United States

Federal Bureau of Investigation*
History of the SIS Division (at https://vault.FBI.gov)
William J. Donovan File (at https://vault.FBI.gov)

Franklin D. Roosevelt Presidential Library, Hyde Park, NY (FDRL)*
Executive Orders and Presidential Proclamations
Map Room Papers
President's Secretary's File
Presidential Press Conferences

George C. Marshall Foundation, Lexington, VA
Pogue Collection Interviews

Georgetown University Library Special Collections, Washington, DC
Anthony Cave Brown Papers

Library of Congress, Washington, DC
Frank Knox Papers

National Archives and Records Administration II, College Park, MD (NARA II)

Record Group 51 (Bureau of the Budget)
- Emergency and War Agencies (1939–1949)

Record Group 218 (Joint Chiefs of Staff)
- Donovan Organization

Record Group 226 (OSS)
- Personnel Files
- Donovan Office Papers (M 1642)

Record Group 263 (CIA)
- Thomas Troy Papers

The National Archives, Kew, UK (TNA)

Admiralty
Air Ministry
Cabinet Office
Foreign Office
Government Code & Cypher School
Prime Minister's Office
Security Service
Special Operations Executive

National Cryptological Museum Library, Ft. Meade, MD (NCML)

David Kahn Papers
Oral History Collection
Special Research History Collection

University of Maryland Library, College Park, MD

Gordon Prange Papers

US Army Heritage and Education Center, Carlisle, PA

William J. Donovan Papers

US Naval Institute, Annapolis, MD

Oral History Collection

Yale University Library, New Haven, CT

Henry L. Stimson Diary

NOTES

Introduction

1. Depending on context, "intelligence" can refer either to the product, the process, or the organization that collects and analyzes secret information. See for example Michael Warner, "Wanted: A Definition of 'Intelligence,'" *Studies in Intelligence*, vol. 46, no. 3 (2002).
2. The roots of American intelligence can be traced back to George Washington. See for example Christopher Andrew, *For the President's Eyes Only: Secret Intelligence and the American Presidency from Washington to Bush* (New York: HarperCollins, 1995) and Jeffrey Rogg, "The Spy and the State: The History and Theory of American Civil-Intelligence Relations" (PhD diss., Ohio State University, 2020). Michael Warner, *The Rise and Fall of Intelligence: An International Security History* (Washington, DC: Georgetown University Press, 2014) is an excellent overview of US and foreign intelligence.

I: Friends in Desperate Need

1. George Axelsson, "Reich Flag Raised Over Versailles," *New York Times*, Jun. 16, 1940. The flag signaled who had conquered whom, but the goose step was apparently a mark of respect for the enemy's war dead.
2. Clare Boothe, *Europe in the Spring* (New York: Knopf, 1940), 176; Orville H. Bullitt, ed., *For the President, Personal and Secret: Correspondence between Franklin D. Roosevelt and William C. Bullitt* (Boston: Houghton Mifflin, 1972), 469–70; Robert Murphy, *Diplomat among Warriors* (New York: Pyramid Books, 1965), 56; and more generally, Herbert R. Lottman, *The Fall of Paris: June 1940* (New York: HarperCollins, 1992).
3. John H. Godfrey, *The Naval Memoirs of Admiral J. H. Godfrey*, vol. 5, part 1 (London: privately published, ca. 1964), 27.

4. Still the best-known of the many titles of this agency, "MI6" stands for "Military Intelligence 6," even though it was part of the Foreign Office in 1939. For the sake of clarity, MI6 is the title I will use throughout this text.
5. Malcolm Muggeridge, *Chronicles of Wasted Time: An Autobiography* (Vancouver, British Columbia: Regent, 2006), 396.
6. Nigel West, *MI6: British Secret Intelligence Service Operations 1909–45* (New York: Random House, 1983), 70–76. For a more recent account, see Keith Jeffery, *MI6: The History of the Secret Intelligence Service, 1909–1949* (London: Bloomsbury, 2010), 382–6. For the contemporary sense of the German reaction, see C. Brooke Peters, "Nazis Doubt Diplomatic Good-Will of Dutch over Border 'Incident,'" *New York Times*, November 25, 1939. Stevens and Best spent the war in captivity in Germany, much of the time in concentration camps.
7. West, *MI6*, 74.
8. See Lynne Olson, *Last Hope Island: Britain, Occupied Europe, and the Brotherhood That Helped Turn the Tide of War* (New York: Random House, 2017), 145–6. Erskine Childers's novel *The Riddle of the Sands* is an early example.
9. Amanda Smith, ed., *Hostage to Fortune: The Letters of Joseph P. Kennedy* (New York: Viking, 2001), 418.
10. See for example Ernest May, *Strange Victory: Hitler's Conquest of France* (New York: Hill and Wang, 2000), 459.
11. The total number of evacuees approached 340,000, including French and Belgian soldiers, many of whom would soon return home.
12. Winston S. Churchill, *Their Finest Hour* (Boston: Houghton Mifflin, 1949), 114.
13. David Dilks, ed., *The Diaries of Sir Alexander Cadogan O.M., 1938–1945* (London: Cassell, 1971), 285, 299.
14. Churchill, *Finest Hour*, 231. Since their 1931 invasion of Manchuria, the Japanese had been expanding on the Asian mainland.
15. Gabriel Gorodetsky, ed., *The Maisky Diaries: Red Ambassador to the Court of St. James's 1932–1943* (New Haven: Yale University Press, 2015), 105–6.
16. Quoted in David Nasaw, *The Patriarch: The Remarkable Life and Turbulent Times of Joseph P. Kennedy* (New York: Penguin, 2012), 431–2. Kennedy was likely remembering the social and political consequences of World War I.
17. Nasaw, *Patriarch*, 417.
18. Edward Wood, First Earl of Halifax, Diary, entry for October 10, 1940, Hickleton Papers, Borthwick Institute, University of York.
19. Quoted in Nasaw, *Patriarch*, 431.
20. John F. Kennedy, "Remarks Upon Signing a Proclamation Conferring Honorary Citizenship on Sir Winston Churchill, 9 April 1963," John F. Kennedy Presidential Library and Museum, audio, accessed Jul. 18, 2020, https://www.jfklibrary.org/asset-viewer/archives/JFKWHA/1963/JFKWHA-175-003/JFKWHA-175-003.

21. Churchill, *Finest Hour*, 225–6. For the argument that Churchill and Britain saved western civilization, see Robin Prior, *When Britain Saved the West: The Story of 1940* (New Haven: Yale University Press, 2015).
22. This was a complicated issue, with a range of opinions that are explored in various books. See Lynne Olson, *Those Angry Days: Roosevelt, Lindbergh, and America's Fight over World War II, 1939–1941* (New York: Random House, 2014). William K. Klingaman, *The Darkest Year: The American Home Front 1941–1942* (New York: St. Martin's Press, 2019), 9, offers this detail: in 1939, "public opinion polls revealed that more than 80 percent of the nation's voters opposed entry into the war—a number that would remain remarkably stable over the next two years." This was true even though, in 1940, a majority favored aiding Britain. Prior, *When Britain Saved the West*, 285. The modern American polling industry traces its roots to George Gallup's startup in the mid-1930s.
23. Winston S. Churchill, *The Gathering Storm* (New York: Houghton Mifflin, 1948), 440–1.
24. Richard J. Aldrich and Rory Cormac, *The Black Door: Spies, Secret Intelligence and British Prime Ministers* (London: William Collins, 2016), esp. 90–91.
25. Gill Bennett, *Churchill's Man of Mystery: Desmond Morton and the World of Intelligence* (London: Routledge, 2007), 33–35.
26. Churchill, *Gathering Storm*, 80.
27. Joseph E. Persico, *Roosevelt's Secret War: FDR and World War II Espionage* (New York: Random House, 2001), 9.
28. On the Room, see Jeffery M. Dorwart, *Conflict of Duty: The U.S. Navy's Intelligence Dilemma, 1919–1945* (Annapolis, MD: Naval Institute Press, 1983), 161–7. During World War II, Roosevelt would make Astor a coordinator of intelligence in the New York metropolitan area.
29. James Roosevelt and Sidney Shalett, *Affectionately, F.D.R.: A Son's Story of a Lonely Man* (New York: Harcourt, Brace, 1959), 277. Astor inherited the yacht from his father, who named it *Nourmahal*, which means "palace of light or pleasure" in Arabic and Hindi.
30. Eager to please his friend, Astor set out on this journey with Kermit Roosevelt, but wisely decided not to test the Japanese after they denied his request to sail into waters under their control.
31. Steven T. Usdin, *Bureau of Spies: The Secret Connections between Espionage and Journalism in Washington* (New York: Prometheus, 2018), esp. chapter 5.
32. US Congress, *Report of the Joint Committee on the Investigation of the Pearl Harbor Attack*, part 2 (Washington, DC: US Government Printing Office, 1946), 899. The "above board" remark was by General Sherman Miles, the army's chief intelligence officer in 1941. Though each service had its own codebreakers in the 1930s, their work would not bear much fruit until the end of the decade.
33. The army and navy considered fighting more important than spying. MID was not a strong organization, but it generally fared better than ONI. A wartime

intelligence staffer described it as "a static and bureaucratic organization ... that ... worked on ... humdrum tasks [like] ... the silhouettes of alien ships and ... biographical data.... It kept tabs on ship movements." Ladislas Farago, *The Tenth Fleet* (New York: Paperback Library, 1964).

34. See chapter 4 for a discussion of coordination among these agencies and the Department of State.
35. Rogg, "The Spy and the State," makes the argument that it was the American tradition of liberalism that was the limiting factor as much as or more than the Puritanical tradition of the city on the hill, as others have suggested. See for example David Kahn, *The Reader of Gentlemen's Mail: Herbert O. Yardley and the Birth of American Codebreaking* (New Haven: Yale University Press, 2004), 101–3.
36. While he was reporting the war from Madrid in 1937, Ernest Hemingway wrote the first draft of a play about fighting subversion that he called *The Fifth Column*. The play opened on Broadway in March 1940 and ran through eighty-seven performances. He was just one of many writers to publicize the concept.
37. Boothe, *Europe in the Spring*, 250–1.
38. Edgar Ansel Mowrer, *Triumph and Turmoil: A Personal History of Our Time* (London: Allen and Unwin, 1970), 309; Col. William J. Donovan and Edgar A. Mowrer, *Fifth Column Lessons for America* (Washington, DC: American Council on Public Affairs, 1941), 5–9.
39. Churchill, *Finest Hour*, 285.
40. See for example Patrick Howarth, *Intelligence Chief Extraordinary: The Life of the Ninth Duke of Portland* (London: Bodley Head, 1986), 135.
41. See, for example, George H. Lobdell, Jr., "A Biography of Frank Knox" (Ph.D. diss., University of Illinois, Champagne-Urbana, 1954).
42. Henry L. Stimson and McGeorge Bundy, *On Active Service in Peace and War* (New York: Octagon Books, 1971), 332.
43. Nigel Nicolson, ed., *Harold Nicolson, Diaries and Letters, 1939–1945* (London: Collins, 1967), 324. Greg Robinson, *By Order of the President: FDR and the Internment of Japanese Americans* (Cambridge, MA: Harvard University Press, 2001), 77, notes that as early as 1933, Knox called for the internment of Japanese Americans in case of war.
44. Frank Knox, "Memo of Conference with President Roosevelt dated 12/12/39," Box 8, Knox Papers, Library of Congress.
45. Frank Knox to Annie Knox, Jun. 15, 1940, Box 3, Knox Papers.
46. Knox admitted that he knew little about the navy but was willing to learn and work hard. This he did. He helped to energize the most traditional and hidebound of America's armed services. Lobdell, "A Biography of Frank Knox," 328–56.
47. Sir John Wheeler-Bennett, *Special Relationships: America in Peace and War* (London: Macmillan, 1975), 156.
48. Edmond Taylor, *Awakening from History* (Boston: Gambit, 1969), 344.

49. Douglas Waller, *Wild Bill Donovan: The Spymaster Who Created the OSS and Modern American Espionage* (New York: Free Press, 2011), 200, cites the claim by OSS officer Rolfe Kingsley that "you knew to have women at the receptions Donovan attended. He'd take care of the rest of it."
50. Mary Bancroft, *Autobiography of a Spy* (New York: William Morrow, 1983), 198–9.
51. For a contemporary portrait of Donovan, see Elizabeth R. Valentine, "Fact-finder and Fighting Man," *New York Times*, May 4, 1941.
52. A nuanced description of Donovan is in Taylor, *Awakening from History*, 344: a "moderately successful Republican politician and... prosperous, able corporation lawyer... the perhaps slightly too candid blue of... [his] eyes [and] the bland courtesy of his smile sometimes suggested the bridge club shark or even, in cases of extreme provocation, the shady stock promoter."
53. William J. Donovan, "Should Men of 50 Fight Our Wars?" *New York Herald Tribune*, Apr. 14, 1940.
54. According to two secondary sources, the actual words were "fighting and fucking." Anthony Cave Brown, *The Last Hero: Wild Bill Donovan* (New York: Times Books, 1982), 352, and Rick Atkinson, *The Day of Battle: The War in Sicily and Italy, 1943–1944* (New York: Henry Holt, 2007), 104. Waller, *Wild Bill Donovan*, 175, places the conversation in a prewar setting. The original source is a postwar anecdote recorded by an aide who was considerably more subdued in his own memoirs but conveyed a similar idea: Donovan was like a fire horse waiting for the fire bell to ring. William J. vanden Heuvel, *Hope and History: A Memoir of Tumultuous Times* (Ithaca, NY: Cornell University Press, 2019), 42.
55. One wartime colleague found Donovan to be a warrior "spoiling for a general's star and a gun," but one who was willing to leave the technical details of the military profession to others. Thomas F. Troy, ed., *Wartime Washington: The Secret OSS Journal of James Grafton Rogers 1942–1943* (Frederick, MD: University Publications of America, 1987), 15.
56. General Sherman Miles quoted in Mark Riebling, *Wedge: The Secret War between the FBI and the CIA* (New York: Knopf, 1994), 14.
57. See for instance William L. Langer, *In and Out of the Ivory Tower* (New York: Neal Watson Academic Publications, 1977), 181.
58. The house would later be owned by Katharine Graham, the publisher of the *Washington Post*.
59. These events are described in: Frank Knox to Annie Knox, Jul. 6 and 14, 1940, Box 3, Knox Papers.
60. See for example Knox to Roosevelt, Dec. 15, 1939, Box 4, Knox Papers.
61. Roosevelt to Knox, Dec. 29, 1939, Box 4, Knox Papers.
62. Theodore Roosevelt and his son, Theodore Jr., a member of Astor's circle, followed Donovan's military exploits in World War I. The elder Roosevelt complimented Donovan, "as about the finest example of the American fighting gentleman." Roosevelt to Donovan, October 25, 1918, facsimile in

Corey Ford, *Donovan of OSS* (Boston: Little, Brown, 1970), 47. That the TRs were Republicans occasionally made relations tense with FDR and his wife.

63. "Mrs. F. D. Roosevelt Assails Donovan," *New York Times*, Oct. 27, 1932. While running for governor of New York, Donovan campaigned against the record of the outgoing FDR administration.
64. Quoted at length in Thomas F. Troy, *Wild Bill and Intrepid: Donovan, Stephenson, and the Origin of CIA* (New Haven: Yale University Press, 1996), 45.
65. Troy, *Wild Bill and Intrepid*, 45–46. On the same day, July 11, both Hull and Knox prepared letters of recommendation for Donovan to carry to London. The date—just three days before Donovan's departure—reflects the last-minute nature of this initiative. Facsimiles in Ford, *Donovan*, 92–93.
66. Roosevelt may have done so in passing through a third party or during a meeting around the time Knox was sworn in.
67. Troy, *Wild Bill and Intrepid*, 46.
68. Waller, *Wild Bill Donovan*, 59.
69. "Colonel Donovan Leaves on the Atlantic Clipper," *New York Times*, Jul. 15, 1940.
70. Howarth, *Intelligence Chief Extraordinary*, 130–131.
71. The king mentioned these facts to Lord Halifax, who had been granted the rare privilege of walking through the palace gardens. Halifax asked, a little anxiously, for the king to let him know the schedule for target practice.
72. A. M. Sperber, *Murrow: His Life and Times* (New York: Fordham University Press, 1998), 161–2; Lynne Olson, *Citizens of London: The Americans Who Stood with Britain in Its Darkest, Finest Hour* (New York: Random House, 2010), 41–52.
73. Mowrer, *Triumph and Turmoil*, 314. There is more than one story to explain how Donovan got his nickname. The best explanation seems to come out of his dash and energy on the battlefield in World War I.
74. Jeffery, *MI6*, 442.
75. Jeffery, *MI6*, 442.
76. Vansittart to Churchill, Jul. 23, 1940, quoted in Troy, *Wild Bill and Intrepid*, 52.
77. Mowrer, *Triumph and Turmoil*, 315.
78. No one person handled Donovan's schedule. The many phone messages and scheduling notes from the time of the trip in the Donovan Papers attest to the impossibly high tempo of the visit—and the conflicting demands on Donovan's time. Jeffery's conclusion that "Menzies oversaw all the arrangements" puts too much emphasis on intelligence. Jeffery, *MI6*, 446. Troy, *Wild Bill and Intrepid*, 55, offers a cautionary note: in July 1940, Donovan was still a generalist.
79. For more on the fifth column as a recurring theme during the Donovan-Mowrer visit, see Nigel West, ed., *The Guy Liddell Diaries*, vol. 1, *1939–1942:*

MI5's Director of Counter-Espionage in World War II (London: Frank Cass, 2005), 86–87.

80. James Leutze, ed., *The London Observer: The Journal of General Raymond E. Lee 1940–1941* (London: Hutchinson, 1971), 26, 34.
81. Cooper to Donovan, Jul. 30, 1940, and Metcalfe to Donovan, dated "Saturday," Box 81B, Donovan Papers, US Army Heritage and Education Center (USAHEC), Carlisle, PA.
82. Andrew Roberts, *"The Holy Fox": A Biography of Lord Halifax* (London: Weidenfeld & Nicolson, 1991), 302.
83. Halifax Diary, entry for Aug. 1, 1940.
84. Kim Philby, *My Silent War* (New York: Ballantine, 1983), 66.
85. Jeffery, *MI6*, 478. Philby, *My Silent War*, 74, describes Broadway in similar terms.
86. One topic that Donovan asked about repeatedly was conscription. Leutze, ed., *London Observer*, 21, and Donovan to Menzies, Aug. 27, 1940, Box 81A, Donovan Papers. Menzies remembered stressing the need for equipment, especially ships. Bennett, *Churchill's Man of Mystery*, 369.
87. In addition to Godfrey, see Donald McLachlan, *Room 39: A Study in Naval Intelligence* (New York: Atheneum, 1968), 232.
88. Godfrey, *Naval Memoirs*, vol. 5, part 1, 128.
89. Leutze, ed., *London Observer*, 27–28.
90. Lee to Donovan, Aug. 8, 1940, Box 81B, Donovan Papers.
91. Bracken to Donovan, Undated, Box 2, Anthony Cave Brown Papers, Georgetown University Library.

2: The British Come Calling

1. "British Plane Here on Regular Flight," *New York Times*, Aug. 5, 1940.
2. Untitled Article ("Col. William J. Donovan of New York"), *Washington Post*, Aug. 6, 1940, 5.
3. Stark to Roosevelt, Apr. 25, 1941, Box 62, PSF, FDR Library (FDRL).
4. Frank Knox to Annie Knox, Aug. 8, 1940, Box 3, Knox Papers.
5. Henry L. Stimson Diary, entry for Aug. 6, 1940, Yale University Library.
6. This stems from his famous remark: "Gentlemen do not read each other's mail," an attitude that led him to close America's first, semiprofessional codebreaking bureau.
7. Much of the evidence for this trip is indirect. On August 6, Roosevelt stated that Donovan would be coming up to Hyde Park with Knox and would join in on the inspection tour. Troy, *Wild Bill and Intrepid*, 57. Roosevelt's log shows times of departure and arrival: "Franklin D. Roosevelt: Day by Day," entries for Aug. 9 and 10, 1940, http://www.fdrlibrary.marist.edu/daybyday/daylog/august-9th-1940/. The Associated Press reported that Donovan was only on part of the trip. "Col. Donovan rode the Presidential train from Hyde Park to Portsmouth, and the Potomac from Portsmouth

to Boston to report on a confidential mission to Europe on which Knox had sent him." Associated Press, "President Tours New England Navy Yards and Arsenal and 'Is Very Well Pleased'; Keeps Silence on Donovan Report," *Washington Post*, Aug. 11, 1940.

8. The Pullman Company built a series of private cars named after explorers.
9. Donovan to Godfrey, Aug. 27, 1940, Box 81 B, Donovan Papers.
10. Troy, *Wild Bill and Intrepid*, 59.
11. Donovan to Vansittart, Sept. 26, 1940, Box 81 B, Donovan Papers. This was presumably a reference to the Republican nomination for the Senate seat that was being contested in 1940.
12. Stimson Diary, entry for Aug. 6, 1940.
13. Mowrer, *Triumph and Turmoil*, 317–8. One version of their work is William J. Donovan and Edgar Ansel Mowrer, *Fifth Column Lessons for America* (Washington, DC: American Council on Public Affairs, 1941). The British ambassador, Lord Lothian, reported to London that, being "rather slight," the articles lacked substance. Troy, *Wild Bill and Intrepid*, 59. They also wildly overplayed the importance of the fifth column. See for example Col. William J. Donovan and Edgar Mowrer, "French Debacle Held Masterpiece of Fifth Columnists Under Hitler," *New York Times*, Aug. 22, 1940.
14. "Donovan Backs Conscription Bill," *New York Times*, Aug. 18, 1940.
15. Under the agreement, the US was "trading" the destroyers for land rights to British bases. The more formal Lend-Lease Act would not be signed until March 1941.
16. J. R. M. Butler, *Lord Lothian* (New York: Macmillan, 1960), 296, quotes a letter from Lothian to Lady Astor, Aug. 16, 1940; Godfrey to Donovan, Sept. 14, 1940, answers Donovan to Godfrey, Aug. 27, 1940, both in Box 81 B, Donovan Papers.
17. Donovan to Godfrey, Aug. 27, 1940. See also Waller, *Wild Bill Donovan*, 62.
18. Quoted in Jeffery, *MI6*, 443. Menzies told a slightly different version of this story in 1941, using the same phrase and taking credit for supporting the Donovan mission that had produced the destroyers. Bennett, *Churchill's Man of Mystery*, 370n. Stephenson's comment was puffery. Donovan played a supporting, not a pivotal, role in the destroyer deal.
19. See for example Henry Hemming, *Agents of Influence: A British Campaign, A Canadian Spy, and the Secret Plot to Bring America into World War II* (New York: Public Affairs, 2019).
20. Stephenson wrote to Tunney from London, enclosing a letter for Hoover, which Tunney then "arranged to get . . . into the hands of Mr. Hoover, having known him quite well." Tunney to Troy, 6, Aug. 18 and Sept. 1969, quoted in Troy, *Wild Bill and Intrepid*, 39.
21. Jeffery, *MI6*, 439.
22. Jeffery, *MI6*, 439–40.
23. Jeffery, *MI6*, 440; Bennett, *Churchill's Man of Mystery*, 254. Bennett quotes from the original memorandum.

24. The nature of their relationship is clear from the notes between them in "Justice, J. Edgar Hoover," Folder, Box 57, PSF, FDRL.
25. Stephenson would claim that it was Churchill who sent him to the US. See for example: Troy, *Wild Bill and Intrepid*, 5; Jeffery, *MI6*, 440; and Bennett, *Churchill's Man of Mystery*, 253, 369n, which offer compelling firsthand evidence that it was "C," not Churchill, who sent Stephenson to New York, and that Stephenson reported only to MI6.
26. Troy, *Wild Bill and Intrepid*, 35, is a facsimile of the landing card.
27. Godfrey, *Naval Memoirs*, vol. 5, part 1, 7.
28. At the time, MI6's official designation was MI 1(c). As noted prior, I will use the designation MI6 throughout this book. The mission for the New York office was to spy on and counter the many anti-British revolutionaries and saboteurs active in the United States. Irish expatriates supported revolution against the empire at home while German spies sought to interrupt the flow of munitions from America to the Allies. The German ambassador to the United States was an intermediary between his government and the Irish revolutionary Sir Roger Casement. In July 1916, German agents caused the massive explosion at Black Tom, a marshaling yard in Jersey City for shipments to Europe through the port of New York. See generally Richard Spence, "Englishmen in New York: The SIS American Station, 1915–21," *Intelligence and National Security* 19:3 (2004), 511–37.
29. Arthur C. Willert, *The Road to Safety: A Study in Anglo-American Relations* (London: Derek Verschoyle, 1952), 64. Willert was the *London Times* correspondent in the United States during the war, and kept in regular touch with Wiseman.
30. Willert, *Road to Safety*, 65.
31. House to Wilson, Jan. 16, 1917, quoted in W. B. Fowler, *British-American Relations, 1917–1918: The Role of Sir William Wiseman* (Princeton, NJ: Princeton University Press, 1969), 8. This book is both a narrative and a reference, containing fifty-three pages of original documents, mostly notes, letters, and cables written by Wiseman to London. Wiseman's reporting is of exceptional quality, capturing detail and nuance.
32. Fowler, *British-American Relations*, 243, 254, 258, 259.
33. Fowler, *British-American Relations*, 22, 25. See also Jeffery, *MI6*, 116.
34. Wiseman to Arthur C. Murray, Aug. 30, 1918, reproduced in Fowler, *British-American Relations*, 280–3. Murray was the assistant military attaché at the British embassy in Washington.
35. Edith Bolling Wilson, *My Memoir* (New York: Arno Press, 1980), 286. She wrote that she had never liked "this plausible little man" who was "a secret agent of the British government."
36. Willert, *Road to Safety*, 62.
37. Quoted in Charles E. Neu, *Colonel House: A Biography of Woodrow Wilson's Silent Partner* (Oxford and New York: Oxford University Press, 2014), 357.
38. Wiseman to Arthur C. Murray, Aug. 30, 1918.

39. Willert, *Road to Safety*, 17, describes a meeting between Wiseman and David Lloyd George, who in turn described Wiseman as a remarkable young diplomat. See also *War Memoirs of David Lloyd George: 1917–1918* (Boston: Little, Brown, 1936), 575.
40. "Fifth Av. Penthouse Is Leased by Banker," *New York Times*, Mar. 12, 1937.
41. Alexander Cadogan, Untitled Memorandum, Oct. 11, 1939, FO 1093–139, The National Archives, Kew (TNA).
42. [Wiseman], Untitled Memorandum, Jun. 3, 1940, FO 1093–140, TNA. Although the document is unsigned and untitled, the next document in the file identifies it as Wiseman's.
43. Untitled Memo, Jun. 5, 1940, FO 1093–140, TNA. Cadogan edited, initialed, and dated the memo, though he does not appear to have written it. Both Cooper and Vansittart had recently met with Donovan.
44. Untitled Memo, Jun. 5, 1940. See also Jeffery, *MI6*, 440; Bennett, *Churchill's Man of Mystery*, 369n; and Halifax Diary, entry for Jun. 6, 1940: "Sir William Wiseman . . . is occupying himself on Philip Lothian's behalf with possible publicity action in the States."
45. Cadogan to Lothian, Jun. 10, 1940, FO 1093–140, TNA.
46. Jeffery, *MI6*, 440. Stephenson did not hesitate to reach out to Wiseman for advice and support. See also H. Montgomery Hyde, *Room 3603: The Story of the British Intelligence Center in New York during World War II* (New York: Farrar, Straus, 1962), esp. 62–63. Hyde, who worked with Stephenson in New York, notes that Wiseman "had been in charge of the same security job a quarter of a century before."
47. See for example Jeffery, *MI6*, 442.

3: Gentleman Headhunters Make a Placement

1. Jeffery, *MI6*, 439–441, is an overview of BSC based on access to still-secret MI6 files. The other works that cover BSC are less authoritative: William S. Stephenson, ed., *British Security Coordination: The Secret History of British Intelligence in the Americas, 1940–1945* (New York: Fromm International, 1999) (hereinafter *BSC*); William Stevenson, *A Man Called Intrepid* (New York: Harcourt, Brace, Jovanovich, 1976); and Hyde, *Room 3603*. In his introduction to *The Secret History*, ix–xix, respected historian Nigel West reviews the unusual backstory of these three works, concluding that there is considerable overlap among them and that none is particularly accurate. He notes that Stephenson ordered BSC's files burned. The most recent entry in the field is Hemming, *Agents of Influence*.
2. Nigel West comments in his introduction to the "official" BSC report that it is worth reading for its "shocking" revelations of BSC's influence on American radio commentators, who were manipulated into peddling British propaganda. He also praises the ingenuity with which Stephenson outmaneuvered the Nazis in North and South America. It was, he argues,

an extraordinary achievement, both for the organization and for its chief. Stephenson, ed., *BSC,* xix. To similar effect, read the excellent overview of BSC and its report by William Boyd, "The Secret Persuaders," *Guardian,* Aug. 19, 2006.

3. Both COI and OSS would occupy space in Rockefeller Center. Much later, Donovan's law firm would move to the thirty-ninth floor of the RCA Building in Rockefeller Center, three stories above the space BSC had occupied.
4. Stephenson claimed that they met earlier. Troy, *Wild Bill and Intrepid,* 42–44, reviews this matter carefully, and includes a facsimile of a note by Donovan stating that they met after he returned from his first trip to London.
5. Jeffery, *MI6,* 446. This is a partial quote from a cable dated November 1940.
6. Jeffery, *MI6,* 444, is a facsimile of the cable as received in London. To the same effect, see Knox to Ghormley, Nov. 16, 1940, Box 4, Knox Papers.
7. Stimson wrote in his diary that Donovan was once again traveling for Knox with the approval of the British and American governments. Stimson Diary, entry for Dec. 2, 1940. See also Troy, *Wild Bill and Intrepid,* 78.
8. Vladimir Petrov, *A Study in Diplomacy: The Story of Arthur Bliss Lane* (Chicago: Regnery, 1971), 137. Cordell Hull, in *The Memoirs of Cordell Hull, Vol. II* (New York: Macmillan, 1948), 200, found the administration's practice of dispatching unofficial emissaries disturbing. "Col. Donovan Flies over Atlantic on Secret Mission Tied to France," *New York Times,* Dec. 7, 1940.
9. Halifax to London, Nov. 27, 1940; Halifax to Sinclair, Dec. 5, 1940, Air 8/368, TNA. Halifax announces Donovan's trip, mentioning Knox as "the prime mover." See also Troy, *Wild Bill and Intrepid,* 78–79; Jeffery, *MI6,* 447.
10. Troy, *Wild Bill and Intrepid,* 87.
11. See for example "Spain, Central," February 12, 1941, FO 371–26966, TNA. Bureaucrats at the Foreign Office circulated cables about Donovan's travels, debating policy in the margins.
12. Jeffery, *MI6,* 447, is a facsimile of the first page of a memorandum by "C," dated December 7, 1940, summarizing the cable from Stephenson for the prime minister. Emphasis in the original.
13. "Col. Donovan Flies over Atlantic on Secret Mission Tied to France"; Troy, *Wild Bill and Intrepid,* 79–80.
14. Richard Dunlop, *Donovan: America's Master Spy* (Chicago: Rand McNally, 1982), 233.
15. Edward R. Murrow, *This Is London* (New York: Simon and Schuster, 1941), 228.
16. Ghormley to Knox, Oct. 11, 1940, Box 4, Knox Papers.
17. Troy, *Wild Bill and Intrepid,* 80, quotes Cadogan to Halifax, Dec. 17, 1940.
18. Halifax Diary, entry for Dec. 17, 1940; "In London on a Mission for the U.S.," *New York Times,* Dec. 19, 1940, shows a photo of Donovan departing

10 Downing Street after lunch. There is no evidence that Stephenson attended.

19. Morton to "C," Dec. 1940, quoted in Bennett, *Churchill's Man of Mystery*, 258.
20. Morton to "C," Dec. 1940.
21. Troy, *Wild Bill and Intrepid*, 81.
22. Alex Danchev, ed., *Establishing the Anglo-American Alliance: The Second World War Diaries of Brigadier Vivian Dykes* (London: Brassey's, 1990), 2–5. See also Dykes, "Diary upon Arrival in Jerusalem," n.d. [1941], Box 94A, Donovan Papers.
23. The estimates and an overview of the schedule are in the preface of Dykes, "Diary upon Arrival in Jerusalem." See also Various, "Balkan Trip 1941 of William J. Donovan," Box 94A, Donovan Papers.
24. Malcolm Muggeridge, *Chronicles of Wasted Time* (Vancouver, British Columbia: Regent College Publishing, 2006), 453. The details of the Donovan-Clarke meeting are lost to history, but it is fair to surmise that Donovan was taken by this unusual and imaginative man who promoted irregular warfare. He was also a transvestite who would be arrested in Madrid a few months later for dressing as a woman in clothes that seemed to fit all too well. Ben Macintyre, *Rogue Heroes: The History of the SAS, Britain's Secret Special Forces Unit That Sabotaged the Nazis and Changed the Nature of War* (New York: Crown, 2017), 27. He is the star of Thaddeus Holt, *The Deceivers: Allied Military Deception in the Second World War* (New York: Skyhorse, 2007).
25. John O. Iatrides, ed., *Ambassador MacVeagh Reports: Greece, 1933–1947* (Princeton, NJ: Princeton University Press, 1980), 281–3.
26. London to Cairo, Jan. 30, 1941, FO 371–29792, TNA.
27. Cairo to London, Feb. 8, 1941, FO 371–29792.
28. Churchill to Foreign Office, Feb. 12, 1941, FO 371–26966, TNA.
29. Nicholas Rankin, *Ian Fleming's Commandos* (New York: Faber & Faber, 2011), 101–2; Andrew Lycett, *Ian Fleming: The Man behind James Bond* (Atlanta: Turner Publishing, 1995), 124–6. Gibraltar's prime features are its harbor and a 1,200-foot hill known as the Rock.
30. See for example John Colville, *The Fringes of Power: Downing Street Diaries, 1939–1955* (London: Hodder & Stoughton, 1985), 359, 364. Max Hastings, *The Secret War: Spies, Ciphers, and Guerrillas, 1939–1945* (New York: HarperCollins, 2016), 98, recounts the impression of Major General John Kennedy, who found Donovan likable but was put off by the "fat & prosperous lawyer" who wanted "to lay down the law so glibly about what we... should" do. Troy, *Wild Bill and Intrepid*, 89, details the visit with SOE.
31. Anon., "Churchill Dines Winant, Col. Donovan and Gen. de Gaulle are Received by the King," *New York Times*, March 5, 1941.
32. Martin Gilbert, ed., *The Churchill War Papers* vol. 3, *The Ever-Widening War, 1941* (New York and London: W. W. Norton, 2000), 337.

33. "Col. Donovan Back; Has Seen 'a Lot'," *New York Times,* Mar. 19, 1941.
34. "Franklin D. Roosevelt: Day by Day," entry for March 19, 1941, accessed July 4, 2019, http://www.fdrlibrary.marist.edu/daybyday/daylog/march-19th-1941/; Frank L. Kluckhohn, "President on Way to Florida Vacation; Hears Donovan Report on Europe's Trend," *New York Times,* Mar. 20, 1941.
35. Stimson Diary, entry for Mar. 19, 1941.
36. This is not the standard interpretation. See for example Danchev, ed., *Establishing the Anglo-American Alliance,* 23: "He was undeniably impressed with the benefits of Anglo-American intelligence cooperation . . . fired with the need for . . . some kind of central intelligence agency."
37. Balfour to Lothian, Nov. 28, 1940, cited in Thomas F. Troy, *Donovan and the CIA: A History of the Establishment of the Central Intelligence Agency* (Frederick, MD: Aletheia Books, University Publications of America, 1981), 34.
38. Stimson Diary, entry for Dec. 2, 1940. In 1940, the secretary of war portfolio included the army and Army Air Corps.
39. Robert E. Sherwood, *Roosevelt and Hopkins: An Intimate History* (New York: Harper, 1948), 230. Roosevelt believed that it would be disastrous for these tensions to come to a head; it would risk turning millions of Irish Americans against Britain. At issue was neutral Ireland's role in the war.
40. Troy, *Wild Bill and Intrepid,* 118, 123; Stimson Diary, entry for Apr. 17, 1941, 169–70. Fiorello La Guardia was appointed to head civil defense. "La Guardia to Head Home Defense," *New York Times,* May 20, 1941. Treasury Secretary Morgenthau offered a defense savings position.
41. Knox spoke to Justice Frankfurter about Donovan. Troy, *Wild Bill and Intrepid,* 123. Frank L. Kluckhohn, "Relations with Reich Under Greater Strain," *New York Times,* Apr. 6, 1941.
42. Stephenson to Menzies, Jun. 18, 1941, quoted in Stephenson, ed., *BSC,* 25.
43. Donovan to Knox, Apr. 26, 1941, reproduced in part in Troy, *Donovan and the CIA,* 417–8. A copy of the original is in Box 119, Donovan Papers, USAHEC.
44. Stephenson, ed., *BSC,* 24; Jeffery, *MI6,* 448, both quoting from a cable with the same date.
45. He would complain that he took a job in intelligence because Stephenson had intrigued on his behalf and pressured him into taking the job. Stephenson to Menzies, Jun. 18, 1941, quotes Donovan's complaint.
46. Joint Intelligence Sub-Committee (JIC), "Minutes," Jun. 17, 1941, CAB 81–88, TNA.
47. JIC, "Minutes," Jun. 17, 1941, refer to the Godfrey mission and the desire "to stimulate interest amongst United States Intelligence Departments in joint intelligence machinery." JIC, "Minutes," Jun. 6, 1941, CAB 81–88, TNA, state that Admiral Godfrey proceeded to Washington to examine, "on an inter-service basis, the whole problem of the exchange of intelligence with the United States." See also Howarth, *Intelligence Chief Extraordinary,* 148, to the effect that the various American intelligence functions should fall under one single body.

48. "Mme. Schiaparelli Arrives on Clipper," *New York Times*, May 26, 1941; Lycett, *Ian Fleming*, 127.
49. Jeffery, *MI6*, 448–9; New York to Foreign Office, Jun. 25, 1941, FO 371-26231, TNA, both stating that Godfrey stayed with Donovan.
50. Troy, *Donovan and the CIA*, 59. McLachlan, *Room 39*, 234, discusses the extent to which Stephenson, Godfrey, and Fleming contributed to the memorandum. It is impossible to pinpoint who contributed what, but highly likely that Donovan accepted British input.
51. Jeffery, *MI6*, 449. To the same effect, see: J. H. Godfrey, "Intelligence in the United States," Jul. 7, 1941, quoted in Nelson MacPherson, *American Intelligence in War-Time London: The Story of the OSS* (London: Frank Cass, 2003), 46–47: "Colonel Donovan was persuaded to increase his personal interest in Intelligence."
52. Jeffery, *MI6*, 449. Jeffery quotes extensively from this cable but does not give its date.
53. See for example Aldrich and Cormac, *Black Door*, 110–111.
54. Lycett, *Ian Fleming*, 129; Godfrey, *Naval Memoirs*, vol. 5, part 1, 132–3.
55. Godfrey, *Naval Memoirs*, vol. 5, part 1, 133.
56. Winston S. Churchill, *The Grand Alliance* (Boston: Houghton Mifflin, 1951), 307.
57. "Capital Stunned by Hood's Sinking," *New York Times*, May 25, 1941.
58. Sherwood, *Roosevelt and Hopkins*, 295.
59. Franklin D. Roosevelt, "Proclamation 2487," The American Presidency Project, May 27, 1941, accessed July 14, 2019, https://www.presidency.ucsb.edu/documents/proclamation-2487-proclaiming-that-unlimited-national-emergency-confronts-this-country. Roosevelt gave himself almost unlimited authority to deal with the crisis as he saw fit, including the power to create government agencies. See for example: Frank L. Kluckhohn, "Wide Executive Powers in Reserve, Proclamation Makes Them Available If and When Needed," *New York Times*, June 1, 1941.
60. Sherwood, *Roosevelt and Hopkins*, 298.
61. Donovan to Roosevelt, May 28, 1941, quoted in Troy, *Wild Bill and Intrepid*, 123.
62. Godfrey, *Naval Memoirs*, vol. 5, part 1, 134.
63. Godfrey, *Naval Memoirs*, vol. 5, part 1, 133. Sulzberger was an occasional guest at the White House. See for example: "Franklin D. Roosevelt: Day by Day," entry for May 26, 1941, http://www.fdrlibrary.marist.edu/daybyday/daylog/may-26th-1941/.
64. Godfrey, *Naval Memoirs*, vol. 5, part 1, 135–7, is the principal source for this meeting. Jeffery, *MI6*, 449, puts the meeting on June 11, quoting from Godfrey's cable of June 12. The White House logs put the meeting on June 10, and list Godfrey as an attendee. "Franklin D. Roosevelt: Day by Day," entry for Jun. 10, 1941, accessed Jul. 12, 2019, http://www.fdrlibrary.marist.edu/daybyday/daylog/june-10th-1941/. Godfrey's memoirs, written in the

1960s, appear remarkably accurate. For example, he remembered watching a film after dinner about members of the Roosevelt family in the Far East. The logs show that dinner guests watched *Wheels Across India*, produced by Armand Denis and Leila Roosevelt.

65. Jeffery, *MI6*, 449. A flywheel is a mechanical device that both restrains and distributes energy.
66. The original memorandum is William J. Donovan, "Memorandum of Establishment of Service of Strategic Information," Box 128, Coordinator of Information Subject File, PSF, FDRL. It is signed but undated, except for a note in pencil on the first page: "Memorandum for the Pres. Of the United States, 6–10–41." Robert Sherwood mentioned "your memo of June 10 to the President" in a note to Donovan on Oct. 20, 1941, Roll 123, Frame 92, Microfilm 1642, NARA II, affirming that he had roughed out the organization of an intelligence agency like COI. Troy is probably wrong that Roosevelt requested the memorandum. Troy, *Wild Bill and Intrepid*, 125.
67. Donovan's ideas dovetailed with his belief that the fifth column was a form of psychological warfare. By now, governments on both sides of the Atlantic were using radio broadcasts to convey information and propaganda.
68. Troy, *Donovan and the CIA*, 61. See generally: William Lasser, *Benjamin V. Cohen: Architect of the New Deal* (New Haven: Yale University Press, 2002).
69. Henry Morgenthau Jr. Diary, Book 411, 67, accessed August 12, 2019, http://www.fdrlibrary.marist.edu/_resources/images/morg/md0552.pdf: "I just wanted to tell you myself that along the lines you and I talked, the President accepted in totem."
70. Donovan, "Memorandum," Jun. 10, 1941. Donovan told Cohen that he had persuaded the president that his organization should be distinct from the civilian Office of Emergency Information (OEM). B. L. Gladieux, "Conversation with Ben Cohen on Strategic Information," Jun. 19, 1941, Box 37, Entry 107A, RG 51. As noted, the president claimed broad new powers to create government bodies in an emergency and made liberal use of them.
71. Edward A. Tamm, "Memorandum for the Director," Jun. 27, 1941, Part 6, William J. Donovan File, accessed August 11, 2020, https://vault.fbi.gov/William%20J%20Donovan%20/. See also Donovan to William D. Whitney, Aug. 19, 1941, Folder 2, Box 1, Troy Papers, RG 263, NARA II, College Park.
72. These are largely the same conditions as those in Donovan's June 10 "Memorandum," with the exception that the president's office would secretly fund the new agency. Tamm was clearly not impressed, pointing out that there was already "complete coordination" among G-2, ONI, and the FBI.
73. Jeffery, *MI6*, 449. The same cable is quoted at greater length in Stephenson, ed., *BSC*, 25. Jeffery dates the cable June 19, and Stephenson June 18, which may be the difference between time sent and time received.
74. New York to Foreign Office, Jun. 25, 1941. Donovan apparently used similar language in conversations with Knox and Stimson. Stimson Diary, entries

for Jun. 20 and 22, 1941. In any event, the British continued to believe what they wanted to believe.

4: J. Edgar Hoover

1. See their correspondence in Part 6, William J. Donovan File, accessed Sept 1, 2020, https://vault.fbi.gov/William%20J%20Donovan%20/.
2. Hoover to Tolson and Tamm, Nov. 29, 1940, Part 6, Donovan File.
3. Tamm to Hoover, Jun. 11, 1941, Part 6, Donovan File.
4. Tamm to Hoover, Jun. 27, 1941, Part 6, Donovan File. Donovan said he was worried that Hoover would misinterpret the press reports about the COI's powers.
5. Kenneth D. Ackerman, *Young J. Edgar: Hoover and the Red Scare, 1919–1920* (New York: Da Capo, 2007) is a detailed narrative.
6. The ACLU grew out of the National Civil Liberties Bureau, founded in 1917 to protect the rights of antiwar advocates, especially free speech, which were then threatened by the Palmer Raids.
7. He also took a commission in the army reserves, serving in the Military Intelligence Corps from 1922 to 1942.
8. See for example Brown, *Last Hero*, 95.
9. One letter, from Hoover to Missy LeHand, informed the presidential intimate that sitting next to her at dinner made for "one of the most delightful evenings [he] . . . had ever spent." Hoover to LeHand, Feb. 12, 1936, Hoover Folder, Box 10, Grace Tully Archive, FDRL. Presidential logbooks show the first one-on-one meeting between Hoover and Roosevelt as occurring on August 24, 1936, and additional personal meetings over lunch on February 5, 1937, and June 16, 1938. See "Franklin D. Roosevelt: Day by Day," http://www.fdrlibrary.marist.edu/daybyday/ for these dates (accessed Aug. 23, 2020). Especially after 1940, Hoover also performed small favors for the president, sending him tidbits of information—often political gossip—and occasionally small gifts. See for example Richard Gid Powers, *Secrecy and Power: The Life of J. Edgar Hoover* (New York: Free Press, 1987), 233–4.
10. Hoover, "Confidential Memorandum," Aug. 24, 1936, quoted extensively in US Senate, *Supplementary Detailed Staff Reports on Intelligence Activities and the Rights of Americans, Book III* (Washington, DC: GPO, 1976), 394. See also Raymond J. Batvinis, *The Origins of FBI Counterintelligence* (Lawrence, KS: University Press of Kansas, 2007), 48.
11. Hoover, "Confidential Memorandum," Aug. 25, 1936.
12. A persistent story spread by historians and journalists has it that Hull responded, "Go ahead and investigate the cock-suckers." See for example Powers, *Secrecy and Power*, 229; Don Whitehead, *The FBI Story: A Report to the People* (New York: Random House, 1956), 158.

13. US Senate, *Intelligence Activities*, 396. The director stated in 1939 that the General Intelligence Division had "compiled extensive indices of individuals, groups, and organizations engaged in . . . subversive activities, in espionage activities, or any activities that are possibly detrimental to the internal security of the United States." US Senate, *Intelligence Activities*, 408–9. In May 1940, FDR would authorize the FBI to add wiretaps to its repertoire. This violated US law at the time.
14. These suspects were not confined. When Turrou told them that they would have to appear before a grand jury, they fled the country and did not return. Only four out of eighteen appeared in the dock.
15. "Hoover and Hardy Clash in Spy Case," *New York Times*, June 2, 1938; "Hoover Jealous, Turrou Declares," *New York Times*, July 2, 1938.
16. Press Conference #469, Jun. 24, 1938, Press Conferences of President Franklin D. Roosevelt, 1933–1945, FDRL. Roosevelt noted that he was not advocating for the US to enter the field of foreign intelligence—that is, spying by Americans on foreigners. Rhodri Jeffreys-Jones, *The Nazi Spy Ring in America: Hitler's Agents, the FBI, and the Case That Stirred the Nation* (Washington, DC: Georgetown University Press, 2020) is the most up-to-date, detailed treatment of the Rumrich case.
17. Press Conference #489, Oct. 7, 1938, Press Conferences of President Franklin D. Roosevelt, 1933–1945, FDRL. September had brought the Munich Agreement whereby Great Britain and France allowed Hitler to seize part of Czechoslovakia, which he did on October 1. See also Troy, *Donovan and the CIA*, 11, to the effect that the FBI was now investigating hundreds instead of tens of espionage cases.
18. US Senate, *Intelligence Activities*, 397.
19. US Senate, *Intelligence Activities*, 397–8.
20. US Senate, *Intelligence Activities*, 409. The exact timing is not clear, but it appears that Hoover added the offices prior to appearing before a Senate appropriations committee in 1939.
21. US Senate, *Intelligence Activities*, 402–3.
22. Beatrice Bishop Berle and Travis Beal Jacobs, eds., *Navigating the Rapids, 1918–1971: From the Papers of Adolf A. Berle* (New York: Harcourt Brace Jovanovich, 1973), 320.
23. Troy, *Donovan and the CIA*, 17.
24. Raymond J. Batvinis, *Hoover's Secret War against Axis Spies: FBI Counterespionage during World War II* (Lawrence, KS: University Press of Kansas, 2014), 8–10; Cartha D. DeLoach, *Hoover's FBI* (Washington, DC: Regnery, 1995), 12, 14, 88. Hoover did not design the New Deal–era building; he merely occupied part of it. He did much of his work in a smaller adjoining office decorated with knickknacks from his mother's estate. His desk was raised on a dais, which meant that he literally looked down on his visitors.

25. Berle, ed., *Navigating the Rapids*, 320. Preliminary talks occurred between Berle and Hoover earlier in May; it was Berle who first suggested that the FBI take on the responsibility of running the new service. FBI, *History of the S.I.S. Division* part 1 (Washington, DC: unpublished, 1947), 46–47, available at https://vault.fbi.gov/special-intelligence-service (accessed Sept 1, 2020). Berle had a number of portfolios, including Latin America.
26. G. Gregg Webb, "Intelligence Liaison Between the FBI and State," *Studies in Intelligence* 49, no. 3 (2005): 25–38.
27. Berle, ed., *Navigating the Rapids*, 321, 323.
28. FBI, *S.I.S. History*, part 1, 42; Adolf A. Berle, Jr., "Memorandum Prepared by Assistance Secretary of State Berle," June 24, 1940, reproduced in G. Gregg Webb, "New Insights into J. Edgar Hoover's Role," *Studies in Intelligence* 48, no. 1 (2004): 48.
29. Berle, "Memorandum," June 24, 1940.
30. SIS would be funded by White House discretionary funds. Roosevelt would not sign a more formal order until December 23, 1941. Even then, it was a DOJ initiative; Hoover and the attorney general drafted a suitable document for his signature. FBI, *S.I.S. History*, part 1, 53.
31. FBI, *S.I.S. History*, part 1, 2.
32. Berle, ed., *Navigating the Rapids*, 337. Miles also referred to Hoover by his rank in the army reserves—a mere lieutenant colonel at the time. Batvinis, *Origins*, 72. Hoover maneuvered Miles into retracting his comment.
33. Quoted in Webb, "New Insights," 49.
34. Berle, ed., *Navigating the Rapids*, 298.
35. FBI, *S.I.S. History*, part 1, 92, 141–3. The discussion to follow relies on this contemporary official history, especially part 1, 1–10, 44–47. Batvinis, *Origins*, 207–25, is a more recent, comprehensive discussion of SIS, as is Batvinis, *Hoover's Secret War*, 106–25.
36. Batvinis, *Origins*, 217.
37. One estimate has the number of Special Agents on duty at the beginning of 1940 as 898, then doubling during that year. Powers, *Secrecy and Power*, 218.
38. FBI, *S.I.S. History*, part 1, 5.
39. FBI, *S.I.S. History*, part 1, 4–7.
40. Batvinis, *Origins*, 220–1.
41. Batvinis, *Origins*, 222–3.
42. Webb, "New Insights," 52. A few months later, he even appeared open to transferring SIS to Donovan's control.
43. Webb, "New Insights," 52.
44. FBI, *S.I.S. History*, part 1, 8.
45. FBI, *S.I.S. History*, part 1, 8. As much as anyone else, Berle would keep the faith, defending SIS from potential raiders and even bolstering FBI morale. Almost Pollyannaish, he wrote in September 1941 that it had done "an excellent job, with such great efficiency, completely without friction." Quoted in Webb, "New Insights," 54.

5: The Oil Slick Principle

1. Lycett, *Ian Fleming*, 130.
2. Ian Fleming, "Memorandum to Colonel Donovan," Jun. 27, 1941, Box 2, Folder 19, Troy Papers.
3. Robert L. Benson, *A History of U.S. Communications Intelligence during World War II: Policy and Administration* (Ft. Meade, MD: National Security Agency, 1997), esp. 21. Donovan was never particularly security conscious, but even he would not have allowed a foreign national to direct his communications.
4. Fleming to Menzies, Jul. 19, 1941, quoted at length in Bennett, *Churchill's Man of Mystery*, 257.
5. Troy, *Wild Bill and Intrepid*, 11, 160. In September 1941, Berle would describe Ellis in almost exactly the same terms—but intending to complain. Nicholas J. Cull, *Selling War: The British Propaganda Campaign against American "Neutrality" in World War II* (New York and Oxford: Oxford University Press, 1995), 174. In a strange twist, Ellis was accused after the war of having betrayed MI6 secrets to the Nazis before the war—and then to the Soviets after the war. He confessed to the first charge in writing, but denied the second. Nigel West, *The Friends: Britain's Postwar Secret Intelligence Operations* (London: Coronet Books, 1988), 211–5.
6. The Bureau of the Budget was the forerunner of the Office of Management and Budget (OMB). Gladieux was chief of the War Organization Section of the Bureau, which some detractors viewed as "a sort of parens patriae, Gestapo, school principal, and scold for all the Government" that pretended to know administrative principles but actually wallowed in "gossip and espionage." Troy, ed., *Wartime Washington*, 16, 62n.
7. The drafts are in Box 37, Entry 107A, RG 51, NARA II.
8. Stimson Diary, entry for Jun. 24, 1941.
9. See for example H. D. Smith to Roosevelt, Jul. 3, 1941, Box 37, Entry 107A, RG 51.
10. Stimson Diary, entry for Jun. 25, 1941.
11. Stimson Diary, entry for Jul. 3, 1941.
12. B. L. Gladieux, "Conference with Colonel J. [*sic*] Donovan and Ben Cohen on Coordinator of Defense Information," Jul. 3, 1941, Box 37, Entry 107A, RG 51.
13. "Dictated by Ben Cohen's Secretary," n.d., Box 80A, Donovan Papers.
14. Federal Register, vol. 6, no. 136, Jul. 15, 1941. The president signed the designation on July 11. The BoB history would refer to it as "a presidential order," which appears to have been a lesser life-form. Bureau of the Budget, *The United States at War: Development and Administration of the War Program by the Federal Government* (Washington, DC: US Government Printing Office, 1946), 523, suggesting a rough hierarchy of orders: Executive, Military, Presidential, and Administrative.

15. Was this an oversight, or was it intentional in the interests of discretion? The drafters might have thought about spying, sabotage, and propaganda, but might not have wanted to put it in writing. Or—just as likely—no American, including Donovan, was entirely sure just what the COI *should* do.
16. "Donovan is Named Information Head," *New York Times*, Jul. 12, 1941. This document and a letter about COI, sent by Roosevelt to members of his cabinet, were drafted by BoB. Box 37, Entry 107A, RG 51.
17. The BoB described the original intent in the same terms: the proposal, as approved by Roosevelt in 1941, called for a small staff to carefully review, analyze, and collate strategic information and data collected by other agencies. "Analysis of the 1942 Budget, Coordinator of Information," Oct. 12, 1941, Box 37, Entry 107A, RG 51.
18. Hitler had always been interested in Eastern Europe where vast territories could supply the *Lebensraum*—room for expansion—that he believed Germany needed. He would have preferred to defeat Britain first, but had tired of waiting for that to happen. Hitler and Stalin were unnatural allies, but they had had an agreement that was mutually beneficial. Olson, *Those Angry Days*, 95.
19. Edwin O. Reischauer, *My Life between Japan and America* (New York: Harper & Row, 1986), 88.
20. Quoted in Riebling, *Wedge*, 31.
21. Stanley P. Lovell, *Of Spies and Stratagems: Incredible Secrets of World War II Revealed by a Master Spy* (New York: Pocket Books, 1964), 4. Riebling, *Wedge*, 32, describes its "secret boys-only-after-school treehouse charm," where you could hide out from your wife, drink bad coffee, and think of crazy ways to save the world.
22. Mowrer, *Triumph and Turmoil*, 323.
23. Troy, *Donovan and the CIA*, 78.
24. James B. Reston, "Rebutting Dr. Goebbels," *Washington Post*, Oct. 6, 1941. For this purpose the COI had set up ten shortwave transmitters in the United States.
25. Langer was serving as the chairman of the History Department at Harvard in 1941, and would succeed Baxter as head of OSS research and analysis. He was also the brother of psychoanalyst Walter C. Langer who would create a psychological profile of Hitler for OSS. Barry M. Katz, *Foreign Intelligence: Research and Analysis in the Office of Strategic Services 1942–1945* (Cambridge, MA: Harvard University Press, 1989) is an excellent overview of R&A.
26. Alden Whitman, "James P. Baxter 3d Dies," *New York Times*, Jun. 19, 1975.
27. FDR to WJD, Jul. 23, 1941, Box 37, Entry 107A, RG 51. See also: T. G. Early to H. D. Smith, Jan. 23, 1942, Box 2, Folder 21, Troy Papers. Working for the government as well as a law firm or a newspaper, as Knox continued to do, was not considered to be a conflict of interest at the time.

28. Brig. Gen. James Roosevelt, USMCR, "Record," Reference Branch, History Division, Marine Corps University, Quantico. During the second half of 1941, Roosevelt was attached to the office of the Commandant of the Marine Corps, which in turn seconded him to COI. He left COI in 1942 to fight in the Pacific Theatre, where he distinguished himself in combat.
29. James Roosevelt to Grace Tully, Dec. 1, 1941, "Coordinator of Information," Box 128, PSF, FDRL.
30. William O. Hall, "Conference with Robert Sherwood," Jul. 16, 1941, Box 37, Entry 107A, RG 51. Hall went on to a distinguished career with the State Department, rising to the level of director general of the Foreign Service.
31. William O. Hall to Bernard L. Gladieux, "Functional Confusion in the Office for Coordination of Information," Aug. 28, 1941, Box 37, Entry 107A, RG 51. Hall was not reassured when, on November 2, Donovan stated that "the only way to develop an effective team, when individualists are being utilized in a program, is to let the organization break down through a lack of direction. Then all the individuals will recognize the need for supervision." W. O. Hall, "Conference with Colonel Donovan," Nov. 2, 1941, Box 37, Entry 107A, RG 51.
32. Langer, *In and Out of the Ivory Tower*, 182. A report on R&A's initial accomplishments is James Baxter, "Report on the Research and Analysis Branch," Oct. 20, 1941, Box 128, PSF, FDRL.
33. For examples see "Office of Strategic Services—Reports—'The War This Week,' March 26, 1942- January 7, 1943" in Box 154, PSF, FDRL.
34. W. O. Hall, "Conference with Colonel Donovan, Coordinator of Information, and Staff," Jul. 16, 1941, Box 37, Entry 107A, RG 51, reflects the early discussions about COI. Later developments are reflected in W. O. Hall, "Scope and Function of the Office of the Coordinator of Information," Sept. 11, 1941, Box 1, Folder 3, Troy Papers; Donovan to V. Stefanson, Oct. 28, 1941, Box 1, Folder 2, Troy Papers.
35. In the fall of 1941, both services were committed to a combination of business as usual—mostly drawing conclusions about enemy capabilities from attaché reports and open sources—as well as signals intelligence, a field they wanted very much to protect and deny to the upstart COI. See chapters 6 and 7 below.
36. See for example Stark to Knox, Sept. 25, 1941, Box 1, Folder 2, Troy Papers: "the secret intelligence of the services should be consolidated under the Coordinator of Information. . . . [This function would be] much more effective if under one [civilian] head rather than three."
37. Wyman H. Packard, *A Century of U.S. Naval Intelligence* (Washington, DC: Navy Department, 1996), 45.
38. W. O. Hall, "Scope and Function," Sept. 11, 1941.
39. Donovan to Roosevelt, Oct. 10, 1941, Folder 2, Box 1, Troy Papers. Roosevelt noted his approval with a characteristic "OK"—so long as Donovan also obtained the approval of the Department of State.

40. See for example "Hearing on Office of Coordinator of Information Second Quarter Budget," Oct. 20, 1941, Box 1, Folder 3, Troy Papers.
41. FDR's reactions are recorded in T. G. Early to Rex Johnson, Nov. 8, 1941, Folder 2, Box 1, Troy Papers. He was reacting to: Director, BoB to FDR, "Budget Request for the Coordinator of Information," Nov. 5, 1941, Box 1, Folder 3, Troy Papers.
42. The president did not, however, involve Donovan in the Arcadia Conference with Churchill in August, when the two leaders met and crafted a joint declaration known as the Atlantic Charter.
43. Frames 101–123, Roll 123, M1642 Microfilm, NARA II, contain the documentation for this event, starting with [William J. Donovan], "Memo on Matters to Take Up at Conference with President 3:30 pm," Oct. 21, 1941. Troy, *Donovan and the CIA*, 115, also lists agenda items.
44. [Roosevelt], "President Roosevelt's Navy Day Address on World Affairs," *New York Times*, Oct. 28, 1941. This text is a transcription from an on-site recording.
45. Frank L. Kluckhohn, "Nazi Ire over 'Secret Map' is a 'Scream' to Roosevelt," *New York Times*, Oct. 29, 1941.
46. Fireside Chat No. 17, Box 62, Master Speech File, FDRL. This speech includes mention of a Nazi plot in Bolivia reported in another secret BSC forgery that found its way into Roosevelt's inbox. See also Christopher Andrew, *For the President's Eyes Only: Secret Intelligence and the American Presidency from Washington to Bush* (New York: Harper Perennial, 1992), 102.
47. Who was lying to whom? Was London involved? Did Stephenson tell Donovan that the map was a forgery? Did Donovan alert the president or string him along? The answers are not clear. See for examples Bennett, *Churchill's Man of Mystery*, 255; Hemming, *Agents of Influence*, 202–65; and John F. Bratzel and Leslie B. Rout Jr., "FDR and the 'Secret Map'," *Wilson Quarterly* 9, no. 1 (1985), 171. An undated translation of the map's legend (but not the map itself) is in the folder for "Germany July 1941–1944," Box 3, PSF, FDRL, where it is sandwiched in with notes from Donovan. A memo dated November 3, 1941, shows signs that someone investigated the origins of the map. Donovan's name appears in the upper right-hand corner. Henry S. Sterling to Preston E. James, Nov. 3, 1941, Germany 1940–1941, Box 31, PSF, FDRL.
48. Desmond Morton to Ian Jacob, Sept. 18, 1941, PREM 3/463, TNA; also quoted in Troy, *Wild Bill and Intrepid*, 132. There is scant evidence for the claim that Stephenson was in direct touch with Roosevelt. Although he overstates Stephenson's influence, Troy lays out the evidence. Troy, *Wild Bill and Intrepid*, 186–8.
49. Quoted in Cull, *Selling War*, 174. Berle would go on to attempt to limit BSC's unilateral powers.
50. See Jordan A. Schwarz, *Liberal: Adolf A. Berle and the Vision of an American Era* (New York and London: Free Press, 1987), 164–74. See also quotations

from Berle's diary in Andrew, *For the President's Eyes Only*, 128. Andrew, 101–2, states the case: "Never before had one power [GB] had so much influence in the development of the intelligence community of another independent state [the US]."

51. The president's schedule is online at "Franklin D. Roosevelt: Day by Day," https://www.fdrlibrary.marist.edu/daybyday/; Box 1, Folder 2, Troy Papers, contains copies of Donovan's agenda and supporting documentation. Roosevelt's copies of some of the same documents appear in Box 128, Folder for "Coordinator of Information, 1941," PSF, FDRL.
52. Copy of Donovan, Untitled Note, Nov. 28, 1941, Folder 2, Box 1, Troy Papers.
53. Whether Roosevelt was masterminding a conspiracy to land the United States in the war is a matter that has sparked lasting controversy. The overwhelming weight of evidence is against any such conspiracy theory. See for example Andrew, *For the President's Eyes Only*, 103. Andrew points out that a conspiracy would have required a degree of organization that the Roosevelt administration did not have when it came to intelligence.
54. L. Effrat, "Tuffy Leemans Receives the Gifts, but Two Dodgers Steal Spotlight," *New York Times*, Dec. 8, 1941.
55. An excellent firsthand description of the day at the White House is in Sherwood, *Roosevelt and Hopkins*, 432–4. Another equally good description comes from the highly observant secretary of labor Frances Perkins. Frances Perkins, *The Roosevelt I Knew* (New York: Penguin, 2011), 362–4.
56. Frank K. Kluckhohn, "Guam Bombed; Army Ship Is Sunk," *New York Times*, Dec. 8, 1941.
57. Charles Hurd, "1 Battleship Lost," *New York Times*, Dec. 9, 1941. Connally was proud of his temper tantrum. Senator Tom Connally with Alfred Steinberg, *My Name Is Tom Connally* (New York: Thos. Y. Crowell, 1954), 249–50.
58. Perkins, *The Roosevelt I Knew*, 363.
59. Langer, *In and Out of the Ivory Tower*, 184.
60. On December 2, Roosevelt sent Murrow a telegram acknowledging the debt that millions of Americans owed him for his reporting, and inviting him and his wife to dinner on the seventh. Olson, *Citizens of London*, 142.
61. Edward Bliss Jr., ed., *In Search of Light: The Broadcasts of Edward R. Murrow 1938–1961* (New York: Knopf, 1967), 108–9, is Murrow's 1945 memory of December 7. The White House logs show the meeting ran from 12:00 to 12:25. "Franklin D. Roosevelt: Day by Day," entry for Dec. 8, 1941, http://www.fdrlibrary.marist.edu/daybyday/daylog/december-8th-1941/. See also Grace Tully, *F.D.R: My Boss* (New York: Scribner's, 1949), 258, and Troy, *Donovan and the CIA*, 116. Troy has Donovan remembering, almost eight years later, that on that night the president said he was glad Donovan had gotten him started on intelligence. This comment seems too good to be true.

6: Spying or Riding to the Sound of the Guns?

1. This was not an idle worry in 1941. The army officers dispatched to investigate their service's role in the Pearl Harbor disaster would die in a plane crash on December 12.
2. Frank E. Beatty, "Secretary Knox and Pearl Harbor: The Background of the Secret Report," *National Review* (December 13, 1966), 1261–1265, is the firsthand account written by Knox's aide. For an earlier version, see [Gordon Prange], "Excerpts from Beatty's Report on Knox's Visit, Written in 1953," Box 57, Pearl Harbor Collection, Gordon Prange Papers, University of Maryland Special Collections.
3. The *Arizona, California,* and *West Virginia* were listed as "sunk at berth"; the *Nevada* was severely damaged but beached; the *Oklahoma* had capsized.
4. Beatty, "Secretary Knox and Pearl Harbor," 1263.
5. Edwin T. Layton with Roger Pineau and John Costello, *"And I Was There": Pearl Harbor and Midway—Breaking the Secrets* (New York: Quill / William Morrow, 1985), 328–333, is Layton's account of the Knox visit. Kimmel was both commander in chief of the Pacific Fleet and commander in chief of the US Fleet, but is remembered most for the former title.
6. [Knox], "Report of the Secretary of the Navy to the President," *Pearl Harbor Attack: Hearings Before the Joint Committee on the Investigation of the Pearl Harbor Attack* (Washington, DC: US Government Printing Office, 1946), part 24, 1749–1756. The "alibi" remark is on page 1753. Hereinafter, this voluminous document will be cited as: *PHA*, part:page.
7. The actual number was higher. There was still a good bit of confusion among American officials at this point.
8. Charles Hurd, "Knox Reports One Battleship Sunk at Hawaii," *New York Times*, Dec. 16, 1941. The *Times* also printed this handout in the same edition: Associated Press, "Knox Statement on Hawaii."
9. There would be a series of wartime inquiries and investigations as well as seemingly unending controversy after the war. The most comprehensive examination was the PHA that would begin on November 15, 1945. A mercifully clear summary of the many investigations is in Craig Nelson, *Pearl Harbor: From Infamy to Greatness* (New York: Scribner's, 2016), 437–54. A comprehensive narrative is in the magisterial book by Gordon W. Prange, *At Dawn We Slept: The Untold Story of Pearl Harbor* (New York: McGraw-Hill, 1981), esp. 582–738.
10. Kimmel and Short were relieved on the same day, December 16, and entered that purgatory that attends government investigations. For the rest of their lives, they had to endure the agony of having their every action and motive questioned in the public eye. By most accounts, they were competent, dedicated officers who made errors of omission but did not deserve to be vilified. Nor were they the only leaders who should have been blamed. Douglas MacArthur's performance in the Philippines on December 8 was

not unlike that of Kimmel or Short, except that Pearl Harbor occurred first and he had the benefit of that foreknowledge. MacArthur was never held to account.

11. Quoted in Nelson, *Pearl Harbor*, 443. Such crude phrases were common in 1941 and 1942, and suggest that for the United States, Pearl Harbor was a failure of imagination as much as a failure to collect, analyze, and disseminate information. The November 27 message did not specifically point to Pearl Harbor but, more generally, to the possibility of war in the Pacific. Under these circumstances, a reasonable precaution would have been to aggressively patrol the air and sea approaches to Oahu.
12. Kimmel's and Short's defenses boiled down to the charge that Washington had withheld information that would have made a difference. Revisionist historians, and longtime Roosevelt haters, would take this farfetched defense further, alleging that the administration knowingly withheld information, and even provoked the Japanese into attacking. See Prange, *At Dawn We Slept*, 839–850.
13. Stimson Diary, entry for Dec. 22, 1941.
14. What appear to be 1941-vintage carbons of these reports are in Box 2, Anthony Cave Brown Papers, Georgetown University Library.
15. Donovan to Roosevelt, Dec. 22, 1941, Folder 2, Box 1, Troy Papers. At this point Donovan's strategic vision included propaganda, fifth column subversion, guerrilla warfare, and finally, conventional operations—in that order.
16. Donovan, Memorandum on British Commandos, n.d. (sent to Roosevelt Dec. 22, 1941), COI, 1941 Folder, Box 128, PSF, FDRL.
17. Roosevelt to Donovan, Dec. 23, 1941, OSS Reports-Donovan-1941–3, Box 153, PSF, FDRL.
18. Stimson Diary, entry for Jan. 12, 1942, describes a dinner at the British embassy with Churchill, Knox, and Donovan in attendance. Stimson was irked at Knox for supporting Churchill's ideas without having done his homework. He added that Donovan supported Knox during the conversation.
19. This officer was one Colonel Robert Solborg, who reported back to Donovan on the state of the Special Operations Executive (SOE), now eighteen months old. Solborg to Donovan, Jan. 12, 1942, Box 2, Folder 19, Troy Papers. As early as the fall of 1941, Donovan had dispatched a similar mission to explore the potential for cooperation between COI and SOE. See R. M. J. Fellner to Donovan, Nov. 2, 1941, Box 2, Anthony Cave Brown Papers.
20. King to Holcomb, Jan. 8, 1942, reproduced in Robert E. Mattingly, *Herringbone Cloak: GI Dagger Marines of the OSS* (Washington, DC: USMC History and Museums Division, 1989), 237.
21. King to Holcomb, Jan. 8, 1942.
22. T. Holcomb to S. W. Meek, Jan. 19, 1942; facsimile in Mattingly, *Herringbone Cloak*, 254.

23. T. Holcomb to C. B. Vogel, Feb. 10, 1942; facsimile in Mattingly, *Herringbone Cloak*, 265. These units would ultimately be known as Marine Raiders and include James Roosevelt among their officers.
24. Donovan to Knox, Feb. 6, 1942, Folder 21, Box 2, Troy Papers.
25. Donovan to Roosevelt, Feb. 21, 1942, Folder 21, Box 2, Troy Papers.
26. Marshall to Donovan, Feb. 27, 1942, Folder 21, Box 2, Troy Papers. Stimson also heard about Donovan's idea of "guerrilla fighting in the Philippines to support our present forces," and invited the colonel over for lunch to discuss them before concluding that his ideas were "mostly wind." Stimson Diary, entry for Feb. 21,1942.
27. Donovan to Roosevelt, Jan. 3, 1942, Box 2, Folder 21, Troy Papers.
28. Thomas Moon and Carl F. Eifler, *The Deadliest Colonel* (New York: Vantage Press, 1975), 324.
29. Yasutaro Soga, *Life behind Barbed Wire: The World War II Internment Memoirs of a Hawai'i Issei* (Honolulu: University of Hawaii, 2008), 31–32. See also: Gail Honda, ed., *Family Torn Apart: The Internment Story of the Otokichi Muin Osaka Family* (Honolulu: Japanese Culture Center of Hawaii, 2012), 21–23.
30. Moon and Eifler, *Deadliest Colonel*, 339.
31. Moon and Eifler, *Deadliest Colonel*, 31.
32. Before the war, the US armed forces had a group known as "the Joint Board" that dealt with interservice matters. The Joint Chiefs of Staff, a far more robust structure, emerged in the early months of 1942, at first as a semiofficial body. The JCS breathed life into the American JIC that had been a dead letter before the war.
33. There were memos on March 4, 7, 16, and 30. See Troy, *Donovan and the CIA*, 124 and 137.
34. Sherwood to Roosevelt, Mar. 19, 1942, Box 128, PSF, FDRL.
35. Hall to Gladieux, Mar. 21, 1942, Folder 23, Box 3, Troy Papers.
36. W. B. Smith to H. Hopkins, Mar. 26, 1942, Box 37, RG 51. The same was now true of Stimson, who worried that Donovan would upset the work of his G-2. Stimson Diary, entry for May 26, 1942.
37. W. B. Smith to King, Mar. 23, 1942, Folder 23, Box 3, Troy Papers.
38. A different version blames a drunk driver who turned out to be an FBI employee. This colorful story, allegedly told by Donovan to a British friend, is unsupported. Sir John Wheeler-Bennett, *Special Relationships: America in Peace and War* (London: Macmillan, 1975), 167–8.
39. Wheeler-Bennett, *Special Relationships*, 168. Donovan was only an occasional drinker, and unlike so many generals and government officials during the war, he did not smoke.
40. Waller, *Wild Bill Donovan*, 114, describes these events and offers general source notes, which can make it difficult to pinpoint specific facts.
41. See for example "New Information Unit Reported Delayed by Status of Donovan, Coordinator Insists on Being Responsible Only to President,"

Washington Evening Star, Apr. 3, 1942, and Raymond P. Brandt, "Coordinator of Wartime News Being Sought by White House," *Washington Evening Star*, Apr. 5, 1942.

42. Copies of the notes from both Roosevelts can be found in Folder 23, Box 3, Troy Papers.
43. Donovan to Roosevelt, Apr. 14, 1942, COI 1942 Folder, Box 128, PSF, FDRL.
44. Carter to Roosevelt, Jan. 19, 1942, Box 98, PSF, FDRL.
45. Berle, ed., *Navigating the Rapids*, 396–7.
46. The charge was that Donovan was operating in Latin America despite the presidential directive reserving that continent for others. Donovan was indeed interested in expanding to Latin America but hoped to do so in tandem with the FBI. See Webb, "New Insights," 54.
47. Donovan to Roosevelt, May 9, 1942, discussed and quoted in Troy, *Donovan and the CIA*, 142, and Brown, *Last Hero*, 221.
48. "Franklin D. Roosevelt: Day by Day," entry for May 15, 1942, http://www.fdrlibrary.marist.edu/daybyday/daylog/may-15th-1942/.
49. Donovan to H. D. Smith, Jun. 9, 1942, Folder 22, Box 3, Troy Papers, citing COI's current strength as 1,796.
50. OSS would still be able to beam "black" propaganda at the enemy overseas—that is, false or misleading information—while OWI would focus on "white" propaganda, that is, the dissemination of accurate information, generally favorable to the war effort. See Clayton D. Laurie, *The Propaganda Warriors* (Lawrence, KS: University Press of Kansas, 1996).
51. Donovan to Wavell, Jun. 13, 1942, Box 3, Folder 22, Troy Papers.
52. OWI was created by executive order, OSS by military order—both more powerful instruments than the hybrid instrument that had created COI.
53. See for example: Maj. J. K. Woolnough to Col. A. C. Wedemeyer, Jun. 24, 1942, Folder 24, Box 3, Troy Papers, which proposes quick action to implement the president's order before Donovan's return from overseas on the grounds that "he might upset the apple cart." Donovan had informed Roosevelt (and presumably others) that he would be out of town. Donovan to FDR, Jun. 8 and 10, 1942, Box 149, PSF, FDRL. Journalist Irving Pflaum remembered Donovan's surprise at hearing the news: Pflaum to Edward P. Lilly, Jan. 26, 1949, Box 3, Folder 22, Troy Papers.
54. [Hambro], "Note on Conversation with G.50.000 [Donovan]," Jun. 15, 1942, HS 8/13, TNA. Donovan knew that by early 1942, SOE had grown to a strength of approximately five hundred officers and one thousand enlisted personnel. It had its own laboratories, training camps, and dedicated aircraft, and claimed to have conducted thirty-two successful and three unsuccessful operations on the continent. Solborg to Donovan, Jan. 12, 1942, Box 2, Folder 19, Troy Papers.
55. "Collaboration between British and American S.O.E," Jun. 26, 1942; "Relations with the American S.O.E.—Now Known as O.S.S.," Jul. 27, 1942, both in HS 8/13, TNA. In the latter document, the writer notes that "there

is no doubt that Donovan intends to be very active in setting up subversive organizations."

56. Chiefs of Staff Committee, Minutes, Jun. 16, 1942, CAB 79/21, TNA.
57. To the same effect, Churchill's intelligence advisor Desmond Morton reported that Donovan had "a mandate covering all SOE and SIS work in all parts of the world other than North and South America and the Pacific Islands." Director, SOE Circular, "G.50,000," Jun.10, 1942, HS 8-13, TNA.

7: Army Cipher Brains

1. One explanation for the term is that Army Signal Corps general Mauborgne referred to his SIS codebreakers as magicians. Frank Rowlett, quoted in Layton et al., *"And I Was There,"* 80. David Kahn offers a more general explanation in *The Codebreakers* (New York: Macmillan, 1967), 3.
2. Miles to Marshall, Jul. 14, 1941, Folder 2, Box 1, Troy Papers.
3. *PHA,* 40:253.
4. King Directive, Jun. 20, 1942, quoted in Stephen Budiansky, *Battle of Wits: The Complete Story of Codebreaking in World War II* (New York: Simon and Schuster, 2002), 261. "Radio intelligence" is a more general term, akin to signals intelligence or communications intelligence.
5. Individual letters could also be transposed (rearranged). If numbers were substituted for letters, the process could get infinitely more complicated. "Additives," for example, were numbers that could be tacked onto codes. They would have to be stripped away before anyone could read the code. Perhaps the most accessible explanation of basic codemaking and -breaking for the layman is in John Keegan, *Intelligence in War* (New York: Knopf, 2003), 146–9. Two other very good explanations are in Kahn, *Codebreakers,* xiii–xxvi, and Budiansky, *Battle of Wits,* 62–88.
6. A recent scholarly overview of Friedman's life and work is: Rose Mary M. Sheldon, "William F. Friedman: A Very Private Cryptographer and His Collection," *Cryptologic Quarterly* 34, no. 1 (2015): 4–30. The only full biography of Friedman is somewhat dated and does not contain source notes: Ronald W. Clark, *The Man Who Broke Purple: The Life of Colonel William F. Friedman, Who Deciphered the Japanese Code in World War II* (Boston: Little, Brown, 1977).
7. An excellent, recent biography of Elizebeth is Jason Fagone, *The Woman Who Smashed Codes: A True Story of Love, Spies, and the Unlikely Heroine Who Outwitted America's Enemies* (New York: Dey Street, 2017). Her mother insisted on spelling her daughter's name with three *e*'s because she did not like the nickname "Eliza," a usage that the conventional spelling invited.
8. William F. Friedman, "A Brief History of the Signal Intelligence Service," June 29, 1942, Special Research History (SRH) 029, National Cryptologic Museum Library (NCML), Ft. Meade, MD. This history also contains basic

facts about Friedman's career. The SRH series is also available at NARA II in College Park, MD.

9. See for example [William Friedman], *The Friedman Legacy: A Tribute to William and Elizebeth Friedman* (Ft. Meade, MD: National Security Agency, Center For Cryptologic History, 2006), 108.
10. For much of its existence, MID was virtually synonymous with G-2, the highest level of the intelligence branch of the army. In the 1930s and '40s, despite various reorganizations, G-2's assistant chief of staff remained the army's chief intelligence officer and head of MID.
11. Here the US was following the British model for organizing military intelligence. One of the British officers advising the startup was Lieutenant Colonel C. G. Dansey, who would conflict with his American colleagues in the next world war. Bruce W. Bidwell, *History of the Military Intelligence Division, Department of the Army General Staff: 1775–1941* (Frederick, MD: University Publications of America, 1986), 121.
12. Herbert O. Yardley, *The American Black Chamber* (Annapolis, MD: Naval Institute Press, 2013), 241.
13. The Federal Communications Act of 1934 would go further, imposing penalties for intercepted messages between foreign countries and the United States. Kahn, *Codebreakers*, 11. The way around the prohibition was to make private arrangements with the telegraph companies, usually through their directors.
14. Yardley, *Black Chamber*, 313. Pages 273–317 cover this period and include the actual messages decrypted and passed to Washington.
15. Yardley, *Black Chamber*, 317.
16. Yardley, *Black Chamber*, 318.
17. Yardley, *Black Chamber*, 323; Kahn, *Reader of Gentlemen's Mail*, 81–82.
18. Friedman, "A Brief History of the Signal Intelligence Service," 9; Yardley, *Black Chamber*, 369–70.
19. Stimson Diary, entry for Jun. 1, 1931.
20. Stimson and Bundy, *On Active Service in Peace and War*, 188. See also Louis Kruh, "Stimson, the Black Chamber, and the 'Gentlemen's Mail' Quote," *Cryptologia* 12, no. 2 (April 1988): 65–89.
21. Kahn, *Reader*, 129.
22. Kahn, *Reader*, 102–3.
23. Friedman, "A Brief History of the Signal Intelligence Service," 11.
24. Yardley, *Black Chamber*, 372.
25. Yardley, *Black Chamber*, 247.
26. Both Friedman and Yardley would be buried in Arlington, but in different sections.
27. Rowlett would later say, "We knew it was illegal and therefore we better keep quiet about it." Frank Rowlett, Oct. 1, 1976, NSA Oral History, 350.
28. Friedman, "A Brief History of the Signal Intelligence Service," is a useful overview.

29. Rowlett, NSA Oral History, 37–38.
30. Solomon Kullback, Aug. 26, 1982, NSA Oral History, 16–17; Frank Rowlett, *The Story of Magic: Memoirs of an American Cryptologic Pioneer* (Laguna Hills, CA: Aegean Park Press, 1998), 34–36.
31. Rowlett, *Story of Magic*, 36. Also described in Kullback, NSA Oral History, 16–17.
32. When generated by hand from a codebook, a code might have a one-to-one solution (as in "Golfplatz" meaning "Britain"); a machine could generate a vast number of possible substitutions according to its design, with the letters for the word "Britain" becoming numbers or letters that might vary. Perhaps the best-known machine was the Enigma, invented in Germany in 1923.
33. The words "code" and "cipher" are not interchangeable, but overlap and are often used imprecisely. Usually machine generated, a cipher is a kind of code, but not all codes are ciphers. Nevertheless, codebreakers remain codebreakers whether they are breaking old-fashioned codes or modern ciphers.
34. Thomas H. Dyer, "The Reminiscences of Capt. Thomas H. Dyer, USN (Ret.)," August–September 1983, US Naval Institute Oral History, Annapolis, MD, 228. Although Dyer was a navy officer, his comments apply to most codebreakers in the 1930s and early 1940s.
35. Yardley, *Black Chamber*, 120–1.
36. Dyer, USNI Oral History, 278.
37. Rowlett, *The Story of Magic*, 88–89.
38. William F. Friedman, "Expansion of the Signal Intelligence Service from 1930 to 7 December 1941," 1945, SRH 134, NCML, 7.
39. Rowlett, NSA Oral History, 343; Rowlett to Kahn, Feb. 21, 1987, David Kahn Papers, NCML.
40. Friedman, "Expansion of the Signal Intelligence Service from 1930 to 7 December 1941," 7.
41. Friedman, "Expansion," 7–8. This is an approximation. Exact comparisons are difficult given different accounting methods.
42. Rowlett, NSA Oral History, 450. William F. Friedman, "Preliminary Historical Report of the Solution of the 'B' Machine," Oct. 14, 1940, SRH 159, NCML 9, puts the date of his discussion with Mauborgne in August 1939. The word "it" apparently refers to Friedman's work as a manager.
43. Rowlett, NSA Oral History, 452.
44. Rowlett, *Story of Magic*, 114.
45. Yardley's employees had reported disturbing themes: trying to catch a bulldog that had gotten into their bedroom and had the word "code" on his side, or carrying an enormous sack of pebbles on a lonely beach and looking for matches. Yardley, *Black Chamber*, 320–1.
46. Rowlett states that the event occurred "shortly before Labor Day," which was September 2 in 1940. Rowlett, *The Story of Magic*, 151–2. Friedman,

"A Brief History," writes "by September." Kahn argues for September 20. Kahn, "Pearl Harbor and the Inadequacy of Cryptanalysis," *Cryptologia* 15, no. 4 (October 1991): 46. Grotjan would stress teamwork, and that her insight was just one step in a long process. Genevieve Grotjan Interview, Jun. 25, 1995, David Kahn Papers.

47. Rowlett, *Story of Magic*, 151–3.
48. She would not receive full recognition until after her death. In 2011 Grotjan was inducted into NSA's Cryptologic Hall of Honor. In 1946 she received a far lesser, though not unwelcome, award for exceptional civilian performance. Ann Whitcher Gentzke, "An American Hero," *At Buffalo*, Spring 2018, University at Buffalo, accessed Oct. 3, 2020, http://www.buffalo.edu/atbuffalo/article-page-spring-2018.host.html/content/shared/www/atbuffalo/articles/Spring-2018/features/an-american-hero.detail.html.
49. This was irregular but apparently not forbidden; security regulations lagged behind the technology. Rowlett himself was hired without any formal clearances.
50. Their prototypes would morph into the army's SIGABA and the navy's Electric Cipher Machine II, which served the United States well in World War II. Timothy J. Mucklow, *The SIGABA/ECM II Cipher Machine: "A Beautiful Idea"* (Ft. Meade, MD: National Security Agency, 2015).
51. Yardley, *Black Chamber*, 318; Kahn, *Codebreakers*, 22–23. Rowlett, NSA Oral History, 422–6, 457–65 makes the point that it was a combination of factors, not just the pressure of breaking Purple, that caused Friedman's breakdown. Colin MacKinnon, "William Friedman's Bletchley Park Diary," *Intelligence and National Security* 20, no. 4 (2005): 657–8, fn. 9.
52. Rowlett, NSA Oral History, 465.
53. MacKinnon, "William Friedman's Bletchley Park Diary," 657–8; Fagone, *Woman Who Smashed Codes*, 219–21.
54. Edward A. Tamm to Director, Jul. 5, 1940, File 62-9798-88, FBI, Reply to 2019 FOIA Request By Author. Here Tamm describes a meeting with Miles and Mauborgne, during which Mauborgne described his intervention with Watson.
55. Tamm to Director, Jul. 5, 1940. At the same meeting, ONI director Walter S. Anderson reported that he had personally discussed "what the Navy Department was doing" and that the president had "indicated his desire that this program continue."
56. Stimson Diary entry for Sept. 25, 1940.
57. Rowlett, *Story of Magic*, 172–3.
58. Stimson Diary, entry for Oct. 23, 1940.
59. Japanese Army messages were hard to acquire in the first place—being relatively weak signals typically sent over land in Asia—let alone decrypt.
60. Kahn, "Roosevelt, MAGIC, and ULTRA," *Cryptologia* 16, no. 4 (October 1992). Friedman dates the first complete decrypts to November 27,

1940. Quoted in Frederick D. Parker, *Pearl Harbor Revisited: U.S. Navy Communications Intelligence 1924–1941* (Ft. Meade, MD: National Security Agency, Center for Cryptologic History, 2013), 23–24.

61. See for example: L. G. Safford, Untitled Memorandum, July 20, 1940; Safford and S. B. Akin, "Directive to Joint Army-Navy Committee," July 31, 1940, SRH 200-001, NCML.
62. S. Miles and J. James, "Handling and Dissemination of Certain Special Material," Jan. 25, 1941, SRH 200-001.
63. There are many discussions of this arrangement. Bidwell, *History of the Military Intelligence Division*, 446–8 hews closely to the original sources, as does Benson, *History of U.S. Communications Intelligence*, 13–14.
64. Frank Rowlett to David Kahn, Feb. 21, 1987, David Kahn Papers. Bratton's full title was chief of the Far Eastern Section of the Intelligence Branch in the MID. He was dual-hatted: responsible for processing Magic and for producing analyses of the situation in the Pacific—in other words, hopelessly overworked.
65. Joseph P. Rochefort Interview, August 26, 1964, Prange Collection.
66. Smith would serve as chief of staff to Dwight Eisenhower in Europe in World War II, ambassador to the Soviet Union, and director of the CIA; Taylor would command the 101st Airborne during the war and eventually serve as chairman of the Joint Chiefs of Staff.
67. *PHA*, 29:2450–1. This was Bratton's testimony about the procedure in 1944. It is consistent with S. Miles and J. James, "Handling and Dissemination of Certain Special Material," January 25, 1941, SRH 200-001.
68. Bratton commented in 1944 that, by late 1941, General Marshall had decreed that customers would see the actual text. *PHA*, 29:2450. Secretary Hull and the president himself expressed the same preference. Bidwell, *History of the Military Intelligence Division*, 449. Bratton had included cryptic references to Magic in more general analyses before Marshall's decree. Department of Defense, *The "Magic" Background of Pearl Harbor* (Washington, DC: US Government Printing Office, 1977), hereinafter *MBPH*, contains many, if not most, of the messages that customers saw in 1941.
69. Other offices in G-2 did produce and circulate analyses, but they were not always worth reading, since many of their authors were not cleared to see Magic.
70. *PHA*, 2:447–8.
71. William F. Friedman, "Certain Aspects of 'Magic' in the Cryptological Background of the Various Official Investigations into the Attack on Pearl Harbor," 1957, SRH 125, NCML, 55.
72. Rowlett, NSA Oral History, 315. See also *PHA*, 29:2450. In other countries the situation was not far different. Stalin insisted on seeing raw reports and famously relied on his own intuition, with disastrous results in the spring of 1941 when he dismissed at least eighty-four separate warnings from many different kinds of sources that the Germans were about to attack. Churchill,

too, was, as often as not, his own analyst. He wanted to see the raw reports and make up his own mind. But he also had the wisdom to listen to his intelligence officers, and to support a system that produced joint analyses.

73. *PHA*, 11:5373.
74. *PHA*, 2:447.
75. Cordell Hull, *Memoirs of Cordell Hull*, vol. 2 (New York: Macmillan, 1948), 1068.
76. Stimson Diary, entry for Jan. 2, 1941. Discussed in Andrew, *For the President's Eyes Only*, 108. See also Carl Boyd, *Hitler's Japanese Confidant: General Ōshima Hiroshi and MAGIC Intelligence, 1941–1945* (Lawrenceville, KS: University Press of Kansas, 1993).
77. Alfred McCormack, "Memorandum for General Lee, Colonel Bratton," Feb. 12, 1942, SRH 141, NCML.
78. Kahn, "Roosevelt, MAGIC, and ULTRA." By comparison, Churchill would seize on particular Enigma intercepts and demand "action this day."
79. Churchill insisted on receiving raw decrypts in a locked box to which only he had the key. He would take the documents out, read them, and replace them. Most days the box was delivered by the chief of MI6 in person.
80. *PHA*, 11:5474-6. See also Kahn, "Roosevelt, MAGIC, and ULTRA."
81. Accordingly, the army, navy, and White House agreed on a new routine on Nov. 10. *PHA*, 11:5475.
82. *PHA*, 11:5476.
83. See for example: *MBPH*, vol. 4, 132.
84. *MBPH*, vol. 4, 208–9.
85. *MBPH*, vol. 4, 209–12.
86. See Roger B. Jeans, *Terasaki Hidenari, Pearl Harbor, and Occupied Japan: A Bridge to Reality* (Lanham, MD, and Boulder, CO: Lexington Books, 2009), esp. 123–5.
87. Terasaki's widow, Gwen, is the secondhand source for the president's pledge. Gwen Terasaki, *Bridge to the Sun: A Memoir of Love and War* (Chapel Hill: University of North Carolina Press, 1957), 68–69. Jeans suggests that the quote is too good to be true, but that the other details are plausible: Jeans, *Terasaki Hidenari*, esp. 123–5.
88. Dispatched on December 6, the appeal differed more in tone than in substance from American policy over the past few months. See *MBPH*, vol. 4, 98.
89. Quoted in Layton et al., *"And I Was There,"* 270. A December 5 message to Tokyo praised Terasaki's work as the embassy's intelligence coordinator. *MBPH*, vol. 4, 139.
90. Jeans, *Terasaki Hidenari*, 61, 122, 129.
91. Andrew, *For the President's Eyes Only*, 115. The message went to Joseph Grew, the American ambassador in Tokyo who requested an audience with the emperor but settled for the foreign minister's pledge that he would deliver the message.

92. The events of the following few days are well known. An integral part of any history of intelligence in World War II, I retell them here in abbreviated form. See Kahn, *Codebreakers*, esp. chapter 1; Prange, *At Dawn We Slept*; and Andrew, *For the President's Eyes Only*. Most rely on the voluminous contemporary records of the 1946 congressional investigation of the Pearl Harbor attack.
93. Rowlett, NSA Oral History, 325.
94. Hull, *Memoirs*, vol. 2, 1096–7.
95. Friedman, "Expansion," 29.

8: More Wall Street Lawyers

1. Stimson and Bundy, *On Active Service*, 393. This was not an unusual reaction. Admiral William H. Standley, a member of the Roberts Commission, commented that three thousand casualties was not a high price to pay for national unity. Quoted in Joseph J. Rochefort, *The Reminiscences of Capt. Joseph J. Rochefort, U.S. Navy (Retired)*, 1983, US Naval Institute Oral History, Annapolis, MD, 172.
2. Stimson Diary, entry for Dec. 31, 1941.
3. This line of reasoning is embodied in Alfred McCormack, "Origin, Functions and Problems of the Special Branch, M.I.S.," April 15, 1943, SRH 116, NCML, 6.
4. Kai Bird, *The Chairman: John J. McCloy & the Making of the American Establishment* (New York: Simon and Schuster, 1992), 106.
5. Stimson Diary, entry for Dec. 31, 1941.
6. McCormack, "Origin, Functions and Problems." See also David Kahn, "Roosevelt, MAGIC, and ULTRA" in *Selections from Cryptologia*, eds. Cipher A. Deavours, David Kahn, Louis Kruh, Greg Mellen, and Brian J. Winkel (Boston and London: Artech House, 1998), esp. 128–30; Bird, *Chairman*, esp. 142; and Bruce Lee, *Marching Orders: The Untold Story of World War II*. (New York: Crown, 1995), 8.
7. Stimson Diary, entry for Jan. 19, 1942.
8. McCormack, "Memorandum for General Lee," Feb. 12, 1942, SRH 141, NCML.
9. McCormack, "Memorandum for General Lee."
10. McCormack, "Draft of Memorandum," Feb. 5, 1942, SRH 141, NCML. McCormack was one of the few to claim that Pearl Harbor could have been predicted through logical analysis. Today there is general agreement that nothing in Magic specifically pointed to Pearl Harbor. Kahn, "Pearl Harbor and the Inadequacy of Cryptanalysis," 35–51. Andrew, *For the President's Eyes Only*, 119–120, observes that, *if* the navy had been able to decrypt and translate all the IJN traffic that it had intercepted, it would have found messages that pointed to the impending attack. However, the relevant Japanese Navy Codes were not broken until years later.

11. McCormack, "Memorandum for General Lee."
12. Alfred T. McCormack, "Origin, Functions and Problems of the Special Branch," Apr. 15, 1943, SRH 116, NCML, 9.
13. McCormack, "Memorandum for General Lee."
14. McCormack, "Origin, Functions and Problems," 9.
15. McCormack, "Origin, Functions and Problems," 34.
16. [Alfred T. McCormack], "War Experience of Alfred McCormack," July 31, 1947, SRH 185, NCML, 6; McCormack, "Origins, Functions and Problems," 8. French prime minister Georges Clemenceau commented that "war was too important to be left to the generals."
17. McCormack, "Organization of Work in Hand," April 21, 1942, SRH 141, NCML.
18. [McCormack], "War Experience," 7.
19. Thomas Ervin Interview, March 11, 1986, David Kahn Papers.
20. Kahn, "Roosevelt, MAGIC, and ULTRA," 128; Telford Taylor Oral History, Jan. 22, 1985, NSA-OH-1-85, 21; Thomas Ervin Interview.
21. George V. Strong, "Reorganization and Expansion of Special Service Branch," May 27, 1942, SRH 141, NCML. The title would be shortened to "Special Branch" (SB).
22. Carter W. Clarke Interview, July 6, 1959, George C. Marshall Library and Archives, VMI, 9.
23. "History of the Special Branch, MIS, War Department, 1942–1944," n.d. [ca. 1944], SRH 035, NCML, 8.
24. Benjamin R. Shute Interview, Mar. 8, 1986, David Kahn Papers.
25. [McCormack], "War Experiences," 7–8.
26. Thomas Ervin Interview. See also [McCormack], "War Experience," 9. "History of the Special Branch," 60, which puts the number in June 1944 at 140 officers, 142 civilians, and 100 enlisted.
27. Edwin O. Reischauer, *My Life between Japan and America* (New York: Harper & Row, 1986), 98. Reischauer would one day become the US ambassador to Japan.
28. They initially failed to bring SIS under SSB, which, as McCormack had pointed out, would have improved communication between these two components. See McCormack to McCloy, Mar. 1, 1942, SRH 141-1, NCML.
29. Kahn, "Roosevelt, MAGIC, and ULTRA." There would be more specialized summaries later in the war, as well as longer studies on particular subjects.
30. Magic was also the only source on Japanese-Soviet relations, and a rich source of information on such diverse topics as Japanese Army supply lines and German fortifications on the French coast. For its value as perceived by George Marshall, see chapter 20 to follow.
31. See for instance James L. Gilbert and John P. Finnegan, eds., *U.S. Army Signals Intelligence in World War II: A Documentary History* (Washington, DC: US Army, Center of Military History, 1993), esp. 134.
32. To optimize his time, G-2 and ONI agreed that they would limit Roosevelt's reading to big-picture topics from Magic. Kahn, "Roosevelt, MAGIC, and

ULTRA," 127. They did not know that he was wasting a good deal of time reading the reports that J. F. Carter continued to produce during the war. Usdin, *Bureau of Spies*, 185–200.

9: Navy Cipher Brains

1. The operators are identified with Room 40, where they worked when the war began under the aegis of a younger Winston Churchill, then first lord of the Admiralty. See Thomas Boghardt, *The Zimmermann Telegram: Intelligence, Diplomacy, and America's Entry into World War I* (Annapolis, MD: Naval Institute Press, 2012).
2. Quoted in Kevin Wade Johnson, *The Neglected Giant: Agnes Meyer Driscoll* (Ft. Meade, MD: National Security Agency, Center for Cryptologic History, 2015), 8. See also Laurance F. Safford, "A Brief History of Communications Intelligence in the U.S.," March 1952, SRH 149, NCML, 2–3.
3. Safford, "A Brief History," 3–4. To the same effect, "The Birthday of the Naval Security Group," ca. 1968, SRH 150, NCML.
4. Quoted in Johnson, *Neglected Giant*, 8. The "more by accident" phrase is from the 1935 official navy history. See also Safford, "A Brief History," 4.
5. See Elliot Carlson, *Joe Rochefort's War: The Odyssey of the Codebreaker Who Outwitted Yamamoto at Midway* (Annapolis, MD: Naval Institute Press, 2011), 31–32, for a description of Safford and his career.
6. Johnson, *Neglected Giant*, 3–12. A British officer would describe her as a difficult interlocutor. "Interview with Brigadier [John] Tiltman," Dec. 17, 1978, Oral History Collection, NCML, 9.
7. Safford, "A Brief History," 4–5.
8. Johnson, *Neglected Giant*, 3, 20–21. As a woman in the 1920s, Driscoll was placed in clerical positions and consistently paid at lower rates than comparable male codebreakers like Friedman. Being embittered about pay and status might have led her to sell a US device to German intelligence before World War II, an allegation made by a German spy in a confession reported in a 1938 FBI document. Jeffreys-Jones, *Nazi Spy Ring in America*, 6, 244n; Tamm to Hoover, Apr. 5, 1938, FBI FOIA Release (courtesy of Prof. Jeffreys-Jones).
9. Op-20-G was divided into the following sections: Cryptographic, Communications Security, Radio Intercept and Tracking, Cryptanalytic, and Translation. "The Birthday of the Naval Security Group."
10. Carlson, *Joe Rochefort's War*, 33.
11. Rochefort, "The Reminiscences of Capt. Joseph J. Rochefort, USN (Ret.)," 53.
12. Rochefort, USNI Oral History, 255–6.
13. See for example: Durwood G. Rorie, NSA Oral History Interview, Oct. 5, 1984, NCML, 17.
14. Rochefort, USNI Oral History, 90.

15. Quoted in Johnson, *Neglected Giant*, 20. Lieutenant Commander Wenger conducted the study. He would only go so far as to suggest adding two more civilian positions, hardly a full-throated expression of support.
16. See for example Carlson, *Joe Rochefort's War*, 50.
17. Written Japanese is made up of some two thousand "ideographs," or characters, which were hard to send electronically. One solution was to come up with syllables for the sounds that the characters represented.
18. Rowlett, *Story of Magic*, 112. Wenger was head of Op-20-G from 1935 to 1938.
19. This was the machine that the Unites States built to decode Japanese messages. William F. Friedman, "Preliminary Historical Report on the Solution of the B Machine," Oct. 14, 1940, SRH 159, NCML.
20. See Carlson, *Joe Rochefort's War*, 93–95, as well as W. J. Holmes, *Double-Edged Secrets* (New York: Berkeley Books, 1981), esp. 143. Though published decades later, Holmes's memoir is based on an official history that he wrote in 1946. See W. J. Holmes, "Narrative of the Combat Intelligence Center, JICPOA," Nov. 8, 1945, SRH 306, NCML, 33–36.
21. See Edwin T. Layton, NSA Oral History, 1983, NCML, 110–111, for a description of Rorie. For Rorie's own testimony: Rorie, NSA Oral History, NCML.
22. The word "solve" implies that the enemy code has been made completely transparent, while "break" suggests a more limited compromise, like insight into part of the target system. Estimates vary on the degree to which the US Navy had made inroads into JN-25 before December 1941, but most sources agree that they were limited, in part because the navy devoted resources to Magic. Carlson, *Joe Rochefort's War*, 204.
23. Edwin T. Layton, "The Reminiscences of Rear Adm. Edwin T. Layton, U.S. Navy (Retired)," 1975, US Naval Institute Oral History, Annapolis, MD, 93–94.
24. Holmes estimated that only one in ten intercepted messages yielded information. Holmes, *Double-Edged Secrets*, 79.
25. Quoted in Carlson, *Joe's Rochefort's War*, 138. Parker, *Pearl Harbor Revisited*, 66–74, includes a more comprehensive listing and abstracts of summaries. Combat Intelligence, War Diary, 1942, SRH 279, NCML, is a slightly abbreviated version of daily findings.
26. Up until this point, the senior combatant commander and his staff had been based on the battleship *Pennsylvania*, which was based at Pearl, ready in case the admiral made the traditional decision to deploy with the fleet. Thus, although he served on shore, the fleet intelligence officer was considered a deployable staff member.
27. One measure of ONI's weakness was that it had four different directors in 1941. This turbulence almost guaranteed lack of effectiveness.
28. Ellis M. Zacharias, *Secret Missions: The Story of an Intelligence Officer* (Annapolis, MD: Naval Institute Press, 1946), 254.
29. Carlson, *Joe Rochefort's War*, 123.

30. Carlson, *Joe Rochefort's War*, 137.
31. The IJN had a deception plan, but Hypo apparently did not intercept the signals generated by that plan. Carlson, *Joe Rochefort's War*, 213.
32. Layton, et al., *"And I Was There,"* 244.
33. Layton, USNI Oral History, 79.
34. Rochefort, USNI Oral History, 235.
35. Holmes, *Double-Edged Secrets*, 61.
36. Dyer, USNI Oral History, 233.
37. Dyer, USNI Oral History, 220, 206.
38. Dyer took phenobarbital, which was a "downer," and benzedrine sulphate to counteract its effects. Rochefort himself helped to perpetuate the myth. Rochefort, USNI Oral History, 124.
39. Rochefort admitted this in his oral history, calling it a practical matter rather than an expression of eccentricity. Rochefort, USNI Oral History, 125.
40. Carlson, *Joe Rochefort's War*, 262.
41. Rorie, NSA Oral History, 35. According to Layton, Nimitz personally authorized the band's assignment. Layton, NSA Oral History, 88.
42. Dyer, USNI Oral History, 233. Rochefort stated that his work was usually not like breaking out the enemy's secrets from a book, but "more like finding a bit of information here, a little more there and then deciding what he might do and hope you are right." Prange, Rochefort Interview, August 26, 1964, Prange Papers.
43. Prange, Rochefort Interview, September 1, 1964, Prange Papers.
44. Edward Van Der Rhoer, *Deadly Magic: A Personal Account of Communications Intelligence in World War II in the Pacific* (New York: Scribner's, 1978), 83. The title of the book is misleading; the author spent the war in Washington as a civilian at Op-20-G.
45. Layton, NSA Oral History, 143–4.
46. Layton, USNI Oral History, 124–5; Carlson, *Joe Rochefort's War*, 310–2. He meant that they had become part of the same radio net.
47. Layton later claimed the navy was working from the decrypted enemy operations order. Layton, USNI Oral History, 125. But no such order was ever found in Hypo's files. A contemporary analysis conveys the impression that Layton and Rochefort assembled the picture themselves from the rich assortment of messages at their disposal. Op-20-G, "The Role of Radio Intelligence in the American-Japanese Naval War," vol. 1, Sept. 1, 1942, SRH 012, NCML.
48. King had replaced Kimmel as commander in chief in December 1941, and Stark as chief of naval operations in March 1942, becoming "dual-hatted."
49. See Jonathan Parshall and Anthony Tully, *Shattered Sword: The Untold Story of the Battle of Midway* (Sterling, VA: Potomac Books, 2007), 433.
50. J. W. Holmes made a plausible claim to have been the first to propose the plan. Holmes, *Double-Edged Secrets*, 101.

51. See for example Gordon W. Prange, Donald M. Goldstein, and Katherine V. Dillon, *Miracle at Midway* (New York: Penguin, 1982), 77. There was no one person who made the broadcasts, but rather a number of very good English speakers who were women.
52. Japanese ships were more flammable than American ships because they were, amazingly, built partly of wood, and because their fire control procedures were inferior. Parshall and Tully, *Shattered Sword*, 244–8.
53. Van Der Rhoer, *Deadly Magic*, 98. The author remembered reading distress signals from the Japanese carriers after they had been hit.
54. Combat Intelligence Unit, War Diary, entry for June 6.
55. Holmes, *Double-Edged Secrets*, 112–4. Nimitz eventually visited Gay and sat by his hospital bed, listening to his story. He was the only survivor from his squadron.
56. When he decided to move his staff from its confined spaces on a flagship (which happened to be the USS *Pennsylvania*) to offices on land, Kimmel picked Facility 661 because it was available. A new headquarters was soon under construction and would be occupied after the Battle of Midway.
57. See Carlson, *Joe Rochefort's War*, 362–82; Layton et al., *"And I Was There,"* esp. 448; and Rochefort, USNI Oral History, esp. 234.
58. Rochefort, USNI Oral History, 235.
59. Rochefort, USNI Oral History, 265–6. The venue might have been J. W. Holmes's home at 4009 Black Point Road. Holmes does not, however, mention the party in his memoirs. Holmes, *Double-Edged Secrets*.
60. See Carlson, *Joe Rochefort's War*, esp. 442–56; Holmes, *Double-Edged Secrets*, 120–1.
61. Although he was junior to most officers who typically received the award, Rochefort otherwise met the criteria. The DSM and the Navy Cross were at the same level, one usually awarded for noncombat service, the other for combat service.
62. Carlson, *Joe Rochefort's War*, 65; Willson considered cryptology a dead-end career choice.
63. Willson to King, Jun. 22, 1942, facsimile in Layton et al., *"And I Was There,"* 528. Whether he dictated or drafted the letter, Willson signed it—and made liberal use of the pronoun "I."

10: Reorganizing Naval Intelligence

1. Secretary of the Navy, *Annual Report of the Secretary of the Navy for the Fiscal Year 1941* (Washington, DC: US Government Printing Office, 1941), 26, 31.
2. See for example the comparative production tables in Williamson Murray and Allan R. Millett, *A War to Be Won: Fighting the Second World War, 1937–1945* (Cambridge, MA, and London: Belknap Press of Harvard University Press, 2000), 535.

3. US Army Center of Military History, "Mobilization," accessed Mar. 18, 2020, https://history.army.mil/documents/mobpam.htm; Naval History and Heritage Command, "U.S. Navy Personnel in World War II," accessed Mar. 18, 2020, https://www.history.navy.mil/research/library/online-reading-room/title-list-alphabetically/u/us-navy-personnel-in-world-war-ii-service-and-casualty-statistics.html.
4. Van Der Rhoer, *Deadly Magic*, 48–49.
5. Walter Muir Whitehill, "A Postscript to 'Fleet Admiral King, A Naval Record,'" *Proceedings of the Massachusetts Historical Society*, Third Series, vol. 70 (1950–1953), 219, 222. Roosevelt enjoyed hearing that King had received a model of a blowtorch—supposedly what he used to shave every morning. This in turn led to other jokes, like giving King an iron bar and calling it a toothpick.
6. Frank O. Hough, Verle E. Ludwig, and Henry I. Shaw Jr., *Pearl Harbor to Guadalcanal*, History of U.S. Marine Corps Operations in World War II, vol. 1 (Washington, DC: US Government Printing Office, 1958), 235.
7. Holmes, *Double-Edged Secrets*, 123, remembers Holcomb as the first to advance the idea, writing that it was "well in advance of anyone else's at that time." See also Benson, *History of U.S. Communications Intelligence*, 63.
8. Holcomb to King, "Establishment of Advanced Joint Intelligence Centers," Mar. 24, 1942, quoted in Carlson, *Joe Rochefort's War*, 385.
9. Packard, *A Century of U.S. Naval Intelligence*, 229–30.
10. Packard, *A Century of U.S. Naval Intelligence*, 230.
11. See the discussion in Carlson, *Joe Rochefort's War*, 387–91. As noted, the Fourteenth Naval District was headquartered at Pearl Harbor along with the Pacific Fleet and was already the administrative home for Hypo. The Pacific theater was divided into the Pacific Ocean Areas under Nimitz, and the Southwest Pacific Area under MacArthur, who opted out of any system he could not control.
12. Packard, *A Century of U.S. Naval Intelligence*, 230.
13. Packard, *A Century of U.S. Naval Intelligence*, 230; Holmes, *Double-Edged Secrets*, 124.
14. Holmes, *Double-Edged Secrets*, 123–4.
15. Rochefort, USNI Oral History, 254. "Clowns" is the word he used in his postwar oral history. He spoke as if there was one message that amounted to this declaration of independence after Midway. Rochefort's biographer did not find a copy of the actual message. Carlson, *Joe Rochefort's War*, 527.
16. Benson, *History of U.S. Communications Intelligence*, 43.
17. Dyer, USNI Oral History, 269. The talented British cryptographer and historian F. H. Hinsley agreed. On a visit to Washington, he found Redman "was chock full of grievances largely because he likes grievances for their own sake." Monthly Letter (Washington to London), May 28, 1944, HW 14/142, TNA.

18. John R. Redman to Vice Admiral F. J. Horne, June 20, 1942, SRH 268, NCML. Redman noted that both King and Nimitz had endorsed the basic concept.
19. Redman to Horne, June 20, 1942.
20. Dyer, USNI Oral History, 263; Holmes, *Double-Edged Secrets*, 126–8; Alva B. Lasswell, USMC Oral History, 33, Marine Corps History Division, 1968. Nimitz objected to the intrigue against Rochefort; for two weeks he was so angry that he refused to speak to Redman.
21. Redman would stay in the job until March 1945.
22. Rochefort, USNI Oral History, 258. "Flatly refused" is likely an overstatement. The navy culture allowed regulars some leeway in negotiating assignments.
23. Rochefort, USNI Oral History, 268–72. Rochefort did serve in one more intelligence assignment toward the end of the war.
24. Holmes, *Double-Edged Secrets*, 126–8; Dyer, USNI Oral History, 265–6.
25. Rorie, NSA Oral History, 18, NCML.
26. Kahn, *Codebreakers*, 563, 573. Kahn put its strength at approximately 1,000 in 1945. Holmes estimates the number was 1,300 or more in January 1945. Holmes, *Double-Edged Secrets*, 211.
27. Dyer, USNI Oral History, 286. In addition to the codes, Dyer calculated that Hypo defeated something like twenty keys. Dyer would wind up establishing another record by serving from July 1936 to December 1945 in the same job at Pearl.
28. Holmes, *Double-Edged Secrets*, 137. For most of 1942, protecting the source was easy for a simple reason. American torpedoes were defective and sank few Japanese ships; the Japanese had no reason to be suspicious.
29. Dyer, USNI Oral History.
30. Layton et al., *"And I Was There,"* 471–2.
31. Layton, et al., *"And I Was There,"* 472. Force commander Vice Admiral Charles A. Lockwood also refers to "special internal codes" carried only by submarines which, one can only hope, were more sophisticated. Lockwood to CNO, "Contribution of Communication Intelligence to the Success of Submarine Operations against the Japanese in World War II," June 17, 1947, SRH 306, NCML.
32. Holmes, *Double-Edged Secrets*, 138.
33. Op-20-G-7, "The Role of Communication Intelligence in Submarine Warfare in the Pacific (January 1943–October 1943)," vols. 1–8, November 1945–January 1946, SRH 011, NCML; Ronald H. Spector, *Eagle against the Sun: The American War with Japan* (New York: Free Press, 1985), esp. 452–3, 478–85. Spector follows the argument laid out in SRH-011, esp. vol. 1, ii–ix. The more positive testimony by Vice Admiral Lockwood lacks the precision of the Op-20-G-7 analysis. Lockwood to CNO, June 17, 1947, SRH 306, NCML.
34. Dyer's judgment was that "the information [about convoy routings] was useful but not critical." Dyer, USNI Oral History, 287.

35. J. N. Wenger, "Admiral Yamamoto, the Death of," May 23, 1945, SRH 306, NCML.
36. Lasswell, USMC Oral History, 39. The multiple addressees on the message suggested that it might be worth the effort to decrypt.
37. Holmes, *Double-Edged Secrets*, 149; Layton et al., *"And I Was There,"* 474.
38. Wenger, "Admiral Yamamoto," May 23, 1945. Hypo and Negat routinely shared intercepts and decrypts.
39. Layton et al., *"And I Was There,"* 475, states that Nimitz obtained presidential approval before acting. John Prados, *Combined Fleet Decoded: The Secret History of American Intelligence and the Japanese Navy in World War II* (Annapolis, MD: Naval Institute Press, 1995), 460, points out that Roosevelt was out of town for much of the time between April 13 and 18.
40. Van Der Rhoer, *Deadly Magic*, 144. Prados, *Combined Fleet Decoded*, 460, mentions a similar memory by Ladislas Farago.
41. This was quite a feat of airmanship and navigation given the technology of the day.
42. Adonis C. Arvanitakis, "Killing a Peacock: A Case Study of the Targeted Killing of Admiral Isoroku Yamamoto," U.S. Army Command and General Staff College Master's Thesis, 2015, 1–3. The author cites a postwar interview with the officer who led the patrol. Yamamoto's seat became a museum artifact.
43. Prados, *Combined Fleet Decoded*, 459–63, is a good discussion of the operation and its aftermath.
44. Wenger, "Admiral Yamamoto," May 23, 1945.
45. FDR to "Bill," May 24, 1943, Box 11, PSF, FDRL.
46. Layton et al., *"And I Was There,"* 475.
47. Layton et al., *"And I Was There,"* 476. In 1950, Captain Roger Pineau, USNR, conducted an interview with one of Yamamoto's aides who was convinced that the breach had come from a message transmitted by the Imperial Japanese Army. British intelligence officers were aghast at the risk the Americans took with their equivalent of the Ultra secret. Richard W. Cutler, *Counterspy: Memoirs of a Counterintelligence Officer in World War II and the Cold War* (Washington, DC: Potomac Books, 2004), 15.
48. Hough et al., *Pearl Harbor to Guadalcanal*, 242–6. The Solomons were a British protectorate between the Coral Sea and the South Pacific. The Marines discovered that "not a single accurate or complete map of Guadalcanal or Tulagi existed in the summer of 1942." Most of what the Marines knew of the Japanese forces they learned from "coastwatchers," the mostly Australian, rugged individualists, brave beyond measure, who operated on their own in the jungle behind Japanese lines. See Martin W. Clemens, *Alone on Guadalcanal: A Coastwatcher's Story* (Annapolis, MD: Naval Institute Press, 2012).
49. See Prados, *Combined Fleet Decoded*, 411–5.
50. Holmes, *Double-Edged Secrets*, 135–6.

51. The situation changed as the war progressed and more Japanese were captured. Prados, *Combined Fleet Decoded*, 495–8. Paradoxically, Japanese who were captured through no fault of their own often proved willing to talk, like the survivors from the doomed aircraft carrier *Hiryu* at the Battle of Midway. No one had thought to call down to the engine room when the order to abandon ship had been given; its crew emerged from the bowels of the ship onto the deserted hangar deck minutes before she sank—managing to save themselves only to be captured by the US Navy. Parshall and Tully, *Shattered Sword*, 357–9.
52. Packard, *A Century of U.S. Naval Intelligence*, 231–5. The postwar official history found JICPOA to be unique among field intelligence organizations for its jointness and outstanding strategic studies.
53. Layton et al., *"And I Was There,"* 470; Packard, *A Century of U.S. Naval Intelligence*, 231. As noted prior, Nimitz commanded the Pacific Ocean Areas while MacArthur held sway in the Southwest Pacific. The focus here is on POA and institutions that represent broad trends. SWP had its own codebreakers in Central Bureau but did not develop its own fusion center. See Prados, *Combined Fleet Decoded*, 422–4, and Edward J. Drea, *MacArthur's Ultra: Codebreaking and the War against Japan, 1942–1945* (Lawrence, KS: University Press of Kansas, 1992), 20–22.
54. Layton et al., *"And I Was There,"* 470.

11: Army and Navy Codebreakers in Washington

1. Steven E. Maffeo, *U.S. Navy Codebreakers, Linguists, and Intelligence Officers against Japan, 1910–1941: A Biographical Dictionary* (New York: Rowman & Littlefield, 2015), 178–186 et seq., is a short biography of Wenger relying on original sources. Bradley F. Smith, *The Ultra-Magic Deals and the Most Secret Special Relationship, 1940–1946* (Shrewsbury, UK: Airlife, 1993), 126, describes Wenger as a man with a sensitive stomach given to worry but not, unlike his brother officers, to drink. According to Liza Mundy, *Code Girls: The Untold Story of the American Women Code Breakers of World War II* (New York: Hachette, 2018), 279, Wenger would eventually suffer a nervous breakdown and be sidelined for six months, not unlike Yardley and Friedman. A British officer commented on "the draught of Wenger's illness" at that time. Director Monthly Letter (Washington to London), Oct. 28, 1943, HW 14/142, TNA.
2. Joseph N. Wenger, "Future Cooperation between Army and Navy," June 1, 1943, SRH 403, NCML.
3. The father of the field was becoming too senior for working-level assignments and was apparently considered too eccentric for higher command.
4. To the same effect, see VCNO to Fourteenth Naval District, "Communication Intelligence Activities," April 19, 1943, SRH 279, NCML, a document

drafted by Wenger or one of his subordinates: "After more than a year of war operation, it has become clear that consolidation of effort should be the aim whenever possible."

5. Joseph N. Wenger, "Reorganization of Section 20-G," January 26, 1942, SRH 279. Wenger literally underlined his main point: "There must be a central coordinating authority for all intelligence communication activities." Another requirement, enabled by advances in technology, was excellent and quick communication among the various stations.
6. Quoted in Mundy, *Code Girls*, 12–13. Comstock has been celebrated as one of the pioneers of women's education in the United States.
7. Op-20-G, "Historical Review of Op-20-G," February 17, 1944, SRH 152, NCML.
8. Quoted in Mundy, *Code Girls*, 13.
9. Op-20-G, "Historical Review."
10. Redman to Horne, June 20, 1942. Benson, *History of U.S. Communications Intelligence*, 45, puts the number of personnel at 475 in April 1942, 750 by June 1942, and 1,000 by the end of the year. See also Op-20-G, "Historical Review of Op-20-G."
11. VCNO to Fourteenth Naval District, "Communication Intelligence Activities," April 19, 1943, SRH 279. WAVES were Women Accepted for Volunteer Emergency Service. Op-20-G, "Historical Review," puts the number of female officers at 406 and WAVE enlisted at 2,407 as of February 1944.
12. According to Maffeo, Wenger played a leading role in the initiatives to acquire machines and bring women on board. Maffeo, *U.S. Navy Codebreakers*, 183–4.
13. See for example David Brinkley, *Washington Goes to War* (New York: Knopf, 1988), 117. Sources differ somewhat on the additional amount the navy paid.
14. Wenger to Redman, "Future Cooperation between Army and Navy," June 1, 1943, SRH 403. In this document Wenger begins by addressing past cooperation.
15. John R. Redman to VCNO, "Cryptanalytical and Decryption Operations on Diplomatic Traffic," June 25, 1942, SRH 200–1, NCML.
16. Wenger to Redman, "Future Cooperation between Army and Navy." See also Wenger to John Redman, January 27, 1945, SRH 200–2, NCML, specifically referring to the "gentlemen's [*sic*] agreement." ONI, which was conspicuous by its almost total absence from Wenger's plans, does not appear to have played an important role in the negotiations. See Packard, *A Century of U.S. Naval Intelligence*, 24.
17. This sometimes-complicated round of negotiations is described by Benson, *History of U.S. Communications Intelligence*, 49–53. John Redman would later take credit for suggesting the agreement. See Redman to Wenger, Feb. 9, 1945, SRH 200–2.
18. W. O. Hall to B. L. Gladieux, July 20, 1942, Box 6, RG 51, NARA II.
19. See for example John Redman to Frederick J. Horne, June 16, 1942, SRH 200–1, calling for "an executive memorandum or executive order."

20. JCS, "Memorandum for the President," July 6, 1942, SRH 403.
21. FDR, "Memorandum for the Director of the Budget," July 8, 1942, SRH 403.
22. Hall to Gladieux, July 20, 1942. See also Benson, *History of U.S. Communications Intelligence*, 54.
23. Op-20-G, "Report of Meeting of Standing Committee for Coordination of Cryptanalytical Work," [August 25, 1942], SRH 200–1.
24. Benson, *History of U.S. Communications Intelligence*, 34. The army went through a complicated reorganization of its intelligence in March 1942 and again in 1944. Throughout the war, the push-pull between Communications and G-2 would continue; G-2 did not run SIS, but the two maintained cordial relations, especially compared to Op-20-G and ONI.
25. Assistant Chief of Staff, G-2, "The Achievements of the Signal Security Agency in World War II," Feb. 20, 1946, 22, SRH 349, NCML. As noted, the enemy's army signals, typically beamed over distant land masses, were more difficult to intercept than his navy signals, and were largely unbroken until mid-1943.
26. G-2, "Achievements of the Signal Security Agency," 4.
27. G-2, "Achievements of the Signal Security Agency," 6. A court apparently forced the army to accept this price. Thomas L. Burns, *The Quest for Cryptologic Centralization and the Establishment of NSA: 1940–1952* (Ft. Meade, MD: National Security Agency, Center for Cryptologic History, 2005), 13.
28. SIS would be renamed three times between 1942 and 1944. Benson, *History of U.S. Communications Intelligence*, 40. For the sake of clarity, this text uses SIS. The analysts under McCormack and Clarke stayed at the Pentagon.
29. SIS had thirteen machines and twenty-one operators at the beginning of the war, and 407 machines with 1,275 operators by the spring of 1945, reckoning that the machines could do the work of thousands of codebreakers. G-2, "Achievements of the Signal Security Agency," 16. The prime contract for building the army machines was with AT&T.
30. G-2, "Achievements of the Signal Security Agency," 6.
31. Mundy, *Code Girls* is a recent exploration.
32. Brinkley, *Washington Goes to War*, 243. Sally F. Reston wrote a series of articles on the role of women in the war. See for example: Sally Reston, "Women Officials Aid U.S. in London," *New York Times*, Oct. 5, 1943. Reston was married to another accomplished journalist, James B. "Scotty" Reston.

12: Jeeping into Action

1. Robin Winks, who studied OSS for years, concluded that haste and chance were the original drivers for recruitment. Robin W. Winks, "Getting the Right Stuff: FDR, Donovan, and the Quest for Professional Intelligence," in George C. Chalou, ed., *The Secrets War: The Office of Strategic Services in World War II* (Washington, DC: National Archives and Records

Administration, 1992), 26. As the war went on, more sophisticated processes would be introduced, some of them scientific and innovative.

2. Katz, *Foreign Intelligence*, 13, citing John H. Herz.
3. Troy, ed., *Wartime Washington*, xx, xxi, 8. This underappreciated book contains the diary Rogers kept while he was with OSS. It is a contemporaneous source, written by a relatively disinterested party.
4. Alpine Club Library to the author of this book, May 20, 2020.
5. The writer Malcolm Cowley, who worked for Archibald MacLeish during the war, quipped that "Washington in wartime is a combination of Moscow (for overcrowding) . . . and Hell (for its livability)." Quoted in Brinkley, *Washington Goes to War*, vii.
6. Troy, ed., *Wartime Washington*, 3.
7. Troy, ed., *Wartime Washington*, 4.
8. Troy, ed., *Wartime Washington*, 4, 5.
9. Troy, ed., *Wartime Washington*, 5.
10. Troy, ed., *Wartime Washington*, 8.
11. Troy, ed., *Wartime Washington*, 9; Langer, *In and Out of the Ivory Tower*, 181–2.
12. Quoted in Chalou, ed., *Secrets War*, 20.
13. Originally "Special Activities/Bruce," the branch was known for most of its existence as Secret Intelligence.
14. The Army Air Corps became the Army Air Forces in 1941 and the US Air Force in 1947.
15. Elizabeth Peet McIntosh, *Sisterhood of Spies: The Women of the OSS* (Annapolis, MD: Naval Institute Press, 2009), 41.
16. The original name of the branch was Special Activities/Goodfellow. For most of its existence, it was known as Special Operations. Morale Operations, responsible for black propaganda, would eventually split off from Special Operations.
17. Troy, ed., *Wartime Washington*, 4n; "Col. M. P. Goodfellow, 81 Dies," *New York Times*, September 6, 1973.
18. Troy, ed., *Wartime Washington*, 9, 14. Goodfellow considered employing renegade codebreaker Herbert O. Yardley and John V. Grombach, a right-wing West Point graduate with an uneven record, probably to establish an OSS black chamber. Donovan vetoed Goodfellow's plans on the grounds that these two men were not trustworthy. Donovan to Goodfellow, April 28, 1942, Folder 19, Box 2, Troy Papers, RG 263. Compare Timothy Naftali, "X-2 and the Apprenticeship of American Counterespionage, 1942–1944" (PhD diss., Harvard University, 1993), 84–85. In the first half of 1942, Grombach was on temporary loan from the army to COI. See Mark Stout, "The Pond: Running Agents for State, War, and the CIA: The Hazards of Private Spy Operations," *Studies in Intelligence* 48, no. 3 (2004).
19. See for example Brian M. Hayashi, *Asian American Spies: How Asian Americans Helped Win the Allied Victory* (New York and Oxford: Oxford

University Press, 2021), esp. 38–39, focusing on Asian Americans in the general context of OSS personnel policies.

20. OSS management considered formally employing Ernest Hemingway but decided against the proposal on the grounds that he was too independent. Leicester Hemingway, who lived in Ernest's shadow, applied to COI/OSS but wound up doing radio intelligence for the FCC. He would eventually found an independent country on a coral reef in the Caribbean in order to mint and sell unique postage stamps. Looking for more glory than he was finding in the military police, John (Jack) Hemingway chanced upon an OSS unit in Northern Africa and became a full-fledged member of SO. See Nicholas Reynolds, *Writer, Sailor, Soldier, Spy: Ernest Hemingway's Secret Adventures, 1935–1961* (New York: William Morrow, 2017).
21. Donovan was willing to employ women but had old-fashioned notions of where they should work and what they should do. For the conclusion that his approach was "not regressive," see Katherine Breaks, "Ladies of the OSS: The Apron Strings of Intelligence in World War II," *American Intelligence Journal* 13, no. 3 (Summer 1992): 91–96. There would ultimately be some 4,500 women in OSS, dozens of whom would see overseas service, but only a few of whom would work in enemy or neutral countries. Perhaps the most well-known female veteran of OSS was Julia McWilliams, who became Julia Child when she married another OSS veteran. See Jennet Conant, *A Covert Affair: The Adventures of Julia Child and Paul Child in the OSS* (New York: Simon and Schuster, 2011). See generally McIntosh, *Sisterhood of Spies*, and Ann Todd, *OSS Operation Black Mail: One Woman's Covert War against the Imperial Japanese Army* (Annapolis, MD: Naval Institute Press, 2017).
22. Counting the members of OSS is tricky given the various types of affiliation and transfers in and out. It appears that on any given day, the number of employees and detailees never exceeded thirteen thousand. However, a good many more, perhaps as many as twenty thousand, were affiliated with OSS at some point. See Michael Warner, *The Office of Strategic Services: America's First Intelligence Agency* (Washington, DC: Central Intelligence Agency, 2000), 4. SOE, the wartime British paramilitary service, was of comparable size to OSS. SIS, or MI6, the British intelligence service, peaked at well over ten thousand during the war but included both codebreakers and spies. By comparison, a US Army division numbered about fifteen thousand.
23. See for example a 1943 request by OSS for 454 army officers, 1,805 enlisted soldiers, and 89 members of the Women's Army Auxiliary Corps. Director, OSS to Chief of Staff, US Army, May 20, 1943, Box 375, Entry UDI, RG 218, NARA II.
24. Roger Hall, *You're Stepping on My Cloak and Dagger* (Annapolis, MD: Naval Institute Press, 2004), 16. This is one of the better-known—and most irreverent—OSS memoirs.
25. William J. Morgan, *The O.S.S. and I* (New York: Norton, 1957), 6. This book rivals Hall's for a sense of what it was like to sign up for OSS.

26. The JIC produced a weekly roundup, but it did not contain any reporting from signals intelligence, while McCormack's Magic Summaries focused primarily on Magic intercepts. G-2 and ONI produced other, very specific reports such as summaries of ship sinkings. See for example John Patrick Finnegan and Romana Danysh, *Military Intelligence* (Washington, DC: US Army, Center of Military History, 1998), 63.
27. See "OSS—Reports" in Box 154, PSF, FDRL.
28. Barry M. Katz, *Foreign Intelligence* (Cambridge, MA: Harvard University Press, 1989), 55.
29. See Richard Breitman with Norman J. W. Goda, "OSS Knowledge of the Holocaust," in Breitman et al., *U.S. Intelligence and the Nazis* (Cambridge, UK: Cambridge University Press, 2005), esp. 11–13. From 1943 on, OSS had a one-man desk gathering information on the Holocaust. S. Aronson, "Preparations for the Nuremberg Trial: The OSS, Charles Dwork, and the Holocaust," *Holocaust and Genocide Studies* 12, no. 2 (1998). Even though J. F. Carter and others reported accurately on the subject as early as 1942, the Roosevelt administration did not take action until January 1944 when it established the War Refugee Board. Usdin, *Bureau of Spies,* 185–6; Rebecca Erbelding, *Rescue Board* (New York: Doubleday, 2018). See also Richard Breitman, *Official Secrets, What the Nazis Planned, What the British and Americans Knew* (New York: Hill and Wang, 1998), esp. 116–120.
30. WJD to FDR, January 9, 1943, "OSS—Reports," Box 154, PSF, FDRL, advised that the digest was being discontinued by order of the JCS; Langer, *In and Out of the Ivory Tower,* 182–3. A "PW [Psychological Warfare] Weekly" with a narrower focus would be a partial replacement. Arthur M. Schlesinger Jr., *A Life in the 20th Century* (Boston: Houghton Mifflin, 2000), 297–8. Schlesinger lamented that OSS reports went unread.
31. Rogers was also a member of a bewildering variety of other committees which were part of the staffs of JCS and OSS. See Troy, ed., *Wartime Washington,* xxiii–xxv.
32. Troy, ed., *Wartime Washington,* 21.
33. Troy, ed., *Wartime Washington,* 18.
34. His daily routine appears in various places in Rogers's diary. A good summary is at Troy, ed., *Wartime Washington,* 181.
35. Troy, ed., *Wartime Washington,* 21.
36. The OSS mandate included the dissemination of black propaganda behind enemy lines—an operation usually associated with the Special Operations or Morale Operations Branches—which was split off from SO in March 1943. Complicating matters was a tendency by the military to define the term "psychological warfare" more broadly to include other types of unconventional warfare.
37. Troy, ed., *Wartime Washington,* 14–15.
38. Troy, ed., *Wartime Washington,* 112.

39. Troy, ed., *Wartime Washington*, 156, 42. Lovell was largely free to invent as he pleased. However, both he and his staff lamented the general disorder, claiming that OSS was "wasting the time of scientists by unplanned orders for research." Troy, ed., *Wartime Washington*, 157. Lovell's own memoir is *Of Spies and Stratagems*.
40. Troy, ed., *Wartime Washington*, 80.
41. Troy, ed., *Wartime Washington*, 66. See also p. 81: "neither Bill [Donovan] nor his loyal operators welcome planning or interference by authority, and they are by-passing the board."
42. Troy, ed., *Wartime Washington*, 49.
43. For general information on Strong, see his personnel file in Box 1763, Entry NM 418, RG 165 (Records of the War Department General and Specific Staffs), NARA II; Rogers's and Troy's comments in Troy, ed., *Wartime Washington*, 17, 17n, and 21; and Holt, *Deceivers*, 250.
44. It is possible that Strong and Donovan met during the First World War; they both fought in the Battle of Saint-Mihiel.
45. This is the thrust of Strong's official complaints, as laid out in Troy, *Donovan and the CIA*, esp. 165–8.
46. Troy, ed., *Wartime Washington*, 17, 21. The date of their conversation was November 14, 1942.
47. Donovan Memo to Members of JPWC, Oct. 31, 1942, Box 371, RG 218, NARA II.
48. Donovan, Memoranda to JIC, October 21 and 22, 1942, quoted in Benson, *History of U.S. Communications Intelligence*, 54. At this point, Strong was chairman of the JIC.
49. See Stout, "The Pond," *Studies in Intelligence*.
50. Troy, ed., *Wartime Washington*, 80.
51. Troy, ed., *Wartime Washington*, 19. Rogers would work on a special plan for "military psychological warfare" in case Germany invaded Spain, a good example of a contingency plan. Troy, ed., *Wartime Washington*, 62, 64, 68.
52. Under what was known as the Murphy-Weygand Accord, the US would permit the French to buy goods on deposit. Robert Murphy, *Diplomat among Warriors* (New York: Pyramid Books, 1965), 99.
53. Troy, ed., *Wartime Washington*, 20. Rogers was off the mark here. OSS played a role in the political planning—but it was not the principal role, which was played by Roosevelt, Murphy, and Eisenhower.
54. Murphy, *Diplomat among Warriors*, 109–10. Murphy was grateful for the help but noted the danger in Donovan's repeated failure to coordinate with other agencies. Another—and very casual—plan that Donovan hatched was to empower an American soldier named Charles Sweeney to go to French Morocco and explore the potential for organizing resistance against the Axis. Donovan asked the president's approval to launch Sweeney. Roosevelt wrote "OK, Go Ahead" and initialed the document. Donovan to Roosevelt, January 9, 1942, Box 128, PSF, FDRL.

55. Thomas W. Lippman, *Arabian Knight: Colonel Bill Eddy USMC and the Rise of American Power in the Middle East* (Vista, CA: Selwa Press, 2008), 7; "Colonel William A. Eddy, USMC (Retired)," ca. 1947, Biographical Files, Marine Corps Historical Division, Quantico, VA.
56. Multiple sources mention a prosthesis. See for example Robin W. Winks, *Cloak and Gown: Scholars in the Secret War, 1939–1961* (New York: William Morrow, 1987), 190. Still, his biographer Lippman does not. Amputation would not normally be indicated in ankylosis, Eddy's condition.
57. C. A. Prettiman, "The Many Lives of William Alfred Eddy," *Princeton University Library Chronicle* 53, no. 2 (Winter 1992): 206–7.
58. Lippman, *Arabian Knight*, 59.
59. Murphy, *Diplomat among Warriors*, 110. Murphy was accredited to the government in Vichy but spent much of 1942 moving around French North Africa.
60. Lippman, *Arabian Knight*, 62.
61. See Lippman, *Arabian Knight*, 64, quoting Eddy; Anthony Cave Brown, ed., *The Secret War Report of the OSS* (New York: Berkley, 1976), esp. 135–42.
62. See for instance Carleton S. Coon, *A North Africa Story: The Anthropologist as OSS Agent, 1941–1943* (Ipswich, MA: Gambit, 1980).
63. Murphy, *Diplomat among Warriors*, 110. Eddy described Murphy as "the policy man." Lippman, *Arabian Knight*, 65.
64. Waller, *Wild Bill Donovan*, 134–5; Lippman, *Arabian Knight*, 81–82. Dunlop, *Donovan*, 369, portrays Buxton as the host of the dinner in London, and the other guests as Generals George S. Patton and Jimmy Doolittle.
65. Mark W. Clark, *Calculated Risk* (New York: Enigma Books, 2007), 85.
66. Germany and Italy placed armistice commissions on French territory to monitor compliance with the terms of the French surrender in 1940. Clark, *Calculated Risk*, 85, quotes a lengthy passage from Eddy's memo. Richard Harris Smith, *OSS: The Secret History of America's First Central Intelligence Agency* (Berkeley, CA: University of California Press, 1972), 57, reports on Eddy's proposal, and adds a citation to a postwar letter from Eddy.
67. William D. Leahy, *I Was There* (New York: McGraw-Hill, 1950), 112.
68. Murphy, *Diplomat among Warriors*, 110. This would have upset the French, not to mention the market for subversion. The French did not want anyone to interfere in the internal affairs of their colonies. $50,000 was more than Murphy's entire budget.
69. Murphy, *Diplomat among Warriors*, 127.
70. Troy, ed., *Wartime Washington*, 20, records Rogers's belief that OSS played an important role in deceiving the Germans. Similarly, the War Report credits OSS with effecting this "large scale deception plan." Brown, ed., *Secret War Report*, 145. An exhaustive study of Allied deception in World War II shows that the OSS initiative was but a small part of a complicated (and more successful) British campaign. Holt, *Deceivers*, 254–70.

71. Frank Holcomb had received a direct commission into the Marine Corps Reserves. Like Eddy, he suffered from a leg injury that made him unfit for duty on the front lines. Nicholas Reynolds, "The 'Scholastic' Marine Who Won a Secret War: Frank Holcomb, the OSS, and American Double-Cross Operations in Europe," *Marine Corps History* 6, no. 1 (Summer 2020): 18–29. Malavergne kept a diary of his wartime experiences. Leon B. Blair, "Rene Malavergne and His Role in Operation Torch," *Proceedings of the Meeting of the French Colonial Historical Society* 4 (1979): 206–12.
72. Blair, "Rene Malavergne," 210.
73. Brown, ed., *Secret War Report*, 144–5.
74. Lippman, *Arabian Knight*, 95; Brown, ed., *Secret War Report*, 140.
75. Brown, ed., *Secret War Report*, 142. He also received the Silver Star. Frank Knox signed the citation for Malavergne's award.
76. Quoted in Rick Atkinson, *An Army at Dawn: The War in North Africa, 1942–1943* (New York: Henry Holt, 2002), 123. Robert L. Melka, "Darlan between Britain and Germany 1940–41," *Journal of Contemporary History* 8, no. 2 (April 1973): 57–80, is a review of Darlan's views and policies highlighting his Anglophobia.
77. Murphy, *Diplomat among Warriors*, 162–3.
78. See for example Olson, *Citizens of London*, esp. 195, and Lynne Olson, *Last Hope Island: Britain, Occupied Europe, and the Brotherhood That Helped Turn the Tide of War* (New York: Penguin Random House, 2017), 224. Roosevelt's attitude was pragmatic; Darlan had given him Algiers, and that was all that mattered.
79. Brown, *Last Hero*, 254. Rogers's diary reflects the tension that he and others felt. Troy, ed., *Wartime Washington*, 19–20.
80. Waller, *Wild Bill Donovan*, 141; Brown, ed., *Secret War Report*, 144–5.
81. Quoted at length in Brown, *Last Hero*, 261–2. Brown notes that it is unclear for whom the memorandum, dated December 7, 1942, was intended, or even if it was delivered.
82. Troy, ed., *Wartime Washington*, 30, 28. What Rogers had in mind is not clear. During a meeting of the Planning Group, he used the phrase "disposing of Darlan," which appears to cover a range of possibilities from sidelining him to killing him. Taken together with Donovan's earlier memorandum, it is more likely that he was thinking of a political solution.
83. Brown, ed., *Secret War Report*, 145.
84. See for example Brown, ed., *Secret War Report*, 140.
85. Quoted in Troy, *Donovan and the CIA*, 191.
86. Quoted in Troy, *Donovan and the CIA*, 191.
87. Since Murphy found Darlan useful, he had no motivation to sanction his killing. Nor did Roosevelt, to whom almost all US government officials deferred on life-or-death decisions. The president might have been guarded about Darlan's political legacy and future, but his chief of staff, Leahy, appreciated the benefits that Darlan conferred. Roosevelt was sympathetic

to a man whose son suffered from polio, the same disease that had crippled him. Roosevelt had even written to Darlan about his son, and would arrange for the young man to spend time in Warm Springs, Georgia, the spa that he owned and ran for polio victims. Murphy, *Diplomat among Warriors*, 164; Leahy, *I Was There*, 132.

88. One theory is that he was part of a conspiracy to restore the French monarchy in North Africa. The conspiracy might have been real even though the goal was not realistic. See for example Taylor, *Awakening from History*, 329.
89. Taylor, *Awakening from History*, 330–1.
90. It was Cordier who heard Bonnier de la Chapelle's confession and gave him absolution *before* he murdered Darlan.
91. Coon, *A North Africa Story*, 47–48, 61–62. According to Brown, *Last Hero*, 269–70, 847n, Coon wrote a memorandum to Donovan in 1943 arguing for OSS to develop a capability for political assassination.
92. Troy, ed., *Wartime Washington*, 32.
93. Taylor, *Awakening from History*, 318.
94. Troy, ed., *Wartime Washington*, xxvi, 57; Dunlop, *Donovan*, 384–7.
95. Troy, *Donovan and the CIA*, 197. Roosevelt's calendar shows a thirty-five-minute meeting with Strong on February 18, 1943: "Franklin D. Roosevelt: Day by Day," http://www.fdrlibrary.marist.edu/daybyday/daylog/february-18th-1943.
96. Troy, ed., *Wartime Washington*, 57.
97. Donovan to Roosevelt, February 23, 1943, OSS Reports-Donovan-1941–3, Box 153, PSF, FDRL.
98. Troy, *Donovan and the CIA*, 201–3, is an excellent, carefully documented reconstruction of the bidding.

13: Traveling the World

1. The lines between OSS and military intelligence were always blurred. Generally speaking, the military used physical and aerial reconnaissance to gather information about the short-term and near distance, while OSS used spies and foreign contacts to look further out.
2. Moon and Eifler, *Deadliest Colonel*. This is something like an autobiography written in collaboration with Moon, who served with Detachment 101. See also Carl F. Eifler Personnel File, Box 214, Entry 224, RG. 226, NARA II; Troy James Sacquety, "The Organizational Evolution of OSS Detachment 101 in Burma, 1942–1945" (PhD diss., Texas A&M University, 2008); and Troy Sacquety, *The OSS in Burma: Jungle War against the Japanese* (Lawrence, KS: University Press of Kansas, 2014).
3. War Department Travel Order for Major Carl F. Eifler, May 6, 1942, Eifler Personnel File.
4. Moon and Eifler, *Deadliest Colonel*, 58.

5. Maochun Yu, *OSS in China: Prelude to Cold War* (Annapolis, MD: Naval Institute Press, 2011), 25.
6. Eifler to Goodfellow, September 28, 1942, quoted in Smith, *OSS*, 248.
7. Quoted in W. R. Peers, "Intelligence Operations of OSS Detachment 101," n.d., accessed July 13, 2020, CIA, https://cia.gov/static/d09cbe41d73881b12671a029186525fc/Operations-of-OSS-Detachment101.pdf. Stilwell's orders to the Detachment called for operations in the vicinity of the airbase at Myitkyina in northern Burma.
8. Mentioned in Eifler to Stilwell, November 11, 1942, Box 74, Entry 99, RG 226.
9. Brown, *Last Hero*, 412, quotes this report authored by an officer on Stilwell's staff, who apparently had limited expectations for the Detachment.
10. There are numerous accounts of this operation. The best is Eifler to Donovan, April 6, 1943, Folder for "CBI-OSS Under Stilwell," Box 74, Entry 99, RG 226. See also Moon and Eifler, *Deadliest Colonel*, 115–9.
11. Citation for Legion of Merit, July 14, 1943, Eifler Personnel File. Moon and Eifler, *Deadliest Colonel*, 332, is a contemporary letter written by Eifler about the incident. Sacquety, "Organizational Evolution," 77, discusses the group's fate.
12. Eifler to Donovan, April 6, 1943.
13. Moon and Eifler, *Deadliest Colonel*, 120; Mark A. Bradley, *A Very Principled Boy: The Life of Duncan Lee, Red Spy and Cold Warrior* (New York: Basic Books, 2014), 81. Sources vary on timing. Lee was with the Detachment in July and then again at the end of August, most likely into September.
14. Dunlop, *Donovan*, 421.
15. Lee to Donovan, October 20, 1943, quoted in Brown, *Last Hero*, 413. To similar effect, Peers, "Intelligence Operations of OSS Detachment 101," and Sacquety, "Organizational Evolution," esp. 186, 203, 209–10. Under Peers, the Detachment would improve the quality of its intelligence reporting—which was primarily tactical—about enemy strength and location, roads and trails, or bombing targets. Later in the war, an R&A detachment would produce reports and briefings about Burma.
16. Moon and Eifler, *Deadliest Colonel*, 120, 266–7. Eifler left little doubt that he and his men used harsh methods.
17. For the complete text of the October 1943 letter from Wingate, see Tom Moon, *This Grim and Savage Game: O.S.S. and the Beginning of U.S. Covert Operations in World War II* (Los Angeles: Burning Gate Press, 1991), 163–4. See also Hayashi, *Asian American Spies*, 115–6; Moon and Eifler, *Deadliest Colonel*, 107–8. Another member of Detachment 101, Roger Hilsman, *American Guerrilla: My War behind Japanese Lines* (Washington, DC: Brassey's, 1990), esp. 137–40, 187 discusses the related topic of tolerating torture by America's Kachin allies.
18. R. H. Oliver to "Miss Smith," June 20, 1944, Eifler Personnel File.
19. Troy, ed., *Wartime Washington*, 153. Their intent was also to generally streamline operations.

20. Magruder to Donovan, September 11, 1943, "Letter to William J. Donovan," Electronic Reading Room (ERR), accessed January 19, 2021, CIA, https://www.cia.gov/readingroom/docs/CIA-RDP83-01034R000200090004-7.pdf.
21. Troy, ed., *Wartime Washington*, 153.
22. Troy, ed., *Wartime Washington*, 108.
23. Troy, ed., *Wartime Washington*, 156.
24. Troy, ed., *Wartime Washington*, 160.
25. Donovan to Magruder, November 8, 1943, quoted in Waller, *Wild Bill Donovan*, 203.
26. Waller, *Wild Bill Donovan*, 174–5. According to Rogers, OSS played no part in the operation. Troy, ed., *Wartime Washington*, 131.
27. Troy, ed., *Wartime Washington*, 141; Waller, *Wild Bill Donovan*, 178–9.
28. Troy, ed., *Wartime Washington*, 170.
29. Waller, *Wild Bill Donovan*, 79.
30. Troy, ed., *Wartime Washington*, 61.
31. Troy, ed., *Wartime Washington*, 170. Rogers would leave Washington and OSS on December 3, 1943.
32. Taylor, *Awakening from History*, 343–4.
33. Waller, *Wild Bill Donovan*, 204–5. Leahy, *I Was There*, 189, 491, mentions Donovan's briefing to Roosevelt. Dai Li, by most accounts a ruthless operator often likened to Heinrich Himmler, exchanged threats with Donovan. In the summer of 1943, Stilwell was still enamored with Eifler but did not fully trust Donovan, who, he told subordinates, was "out to screw us." "Meeting held Wed. June 16, 1943 at Gen. Stilwell's Home/Subject: OSS," reproduced in *NAPKO Project of OSS* (Seoul: South Korean Patriots and Veterans Administration Agency, 2001), 87–90.
34. Hilsman, *American Guerrilla*, 70. See also William R. Peers and Dean Brelis, *Behind the Burma Road* (New York: Avon Books, 1963), 26.
35. Dunlop, *Donovan*, 423.
36. Dunlop, *Donovan*, 423–5. Donovan and Eifler's combined weight was around 450 pounds. Dunlop conducted interviews with four of the officers involved in the Donovan visit: Eifler, Curl, Nicol Smith, and John Coughlin.
37. Vincent L. Curl Personnel File, Box 214, Entry 224, RG 226, NARA II.
38. Legion of Merit Award Citation, Oct. 12, 1944, Curl Personnel File.
39. Dunlop, *Donovan*, 425.
40. Donovan may have made up his mind to replace Eifler as early as October. Sacquety, "Organizational Evolution," 107n.
41. Yu, *OSS in China*, 134.
42. Sacquety, "Organizational Evolution," 114.
43. Moon and Eifler, *Deadliest Colonel*, 178.
44. See Moon and Eifler, *Deadliest Colonel*, esp. 120, 266–7.
45. Detachment 101 was credited with killing some 5,500 Japanese and rescuing 574 Allied personnel at a minimal cost in American lives. Sacquety,

"Organizational Evolution," 280–1; Peers and Brelis, *Behind the Burma Road*, 252–5.

46. Sacquety, "Organizational Evolution," 106. In 1944 and 1945, Eifler would run two missions impossible. Both were ill-conceived and fruitless. One was to stalk and possibly kill Werner Heisenberg, one of Germany's principal atomic scientists, and the other was to launch a guerrilla war in Japanese-occupied Korea from mini semisubmersibles. See *NAPKO Project*; Nicholas Reynolds, "Missions Impossible: The Stranger than Fiction Career of the OSS's Carl Eifler," *World War II Magazine*, August 2019, 30–39.
47. Stilwell had been the American China-Burma-India commander. After he was relieved in 1944, his command was split into China and Burma-India. Peers reported to Sultan, the Burma-India commander.
48. Hilsman, *American Guerrilla*, 289–96, is a discussion of Detachment 101's contribution to the war effort, concluding that guerrillas can never be more than a complement to conventional forces in a war between states.

14: The OSS, the NKVD, and the FBI

1. John R. Deane, *The Strange Alliance: The Story of Our Efforts at Wartime Cooperation with Russia* (New York: Viking, 1947), 50–51.
2. See for example Deane to Marshall, October 22, 1942, Section 1, Part 1, Box 371, RG 218.
3. Brown, *Last Hero*, 419–22, is a good discussion of these plans, with lengthy quotes from official memoranda. Beyond reporting atmospherics, those odd jobs would presumably become clear once the officer had assumed his position. A number of American companies were operating in the Soviet Union as part of the Lend-Lease Program.
4. Harriman to Roosevelt, March 18, 1944, Box 11, Map Room Papers, FDRL. In this cable, Harriman reviews the history of Donovan's visit. See also W. Averell Harriman and Elie Abel, *Special Envoy to Churchill and Stalin 1941–1946* (New York: Random House, 1975), 291–2.
5. Harriman to Roosevelt, March 18, 1944.
6. Earlier initiatives to establish a connection with Soviet intelligence had not borne fruit. Allen Weinstein and Alexander Vassiliev, *The Haunted Wood: Soviet Espionage in America—The Stalin Era* (New York: Random House, 1999), 239–40.
7. Jenny and Sherry Thompson, *The Kremlinologist: Llewellyn E. Thompson, America's Man in Cold War Moscow* (Baltimore, MD: Johns Hopkins University Press, 2018), 31.
8. Harriman and Abel, *Special Envoy*, 292. According to Deane, Roosevelt spoke to Stalin at the Tehran Conference about the need for Soviet assistance against Japan and was rewarded with "the first hint that Russian help would be forthcoming." Deane, *Strange Alliance*, 40–41.

9. [Donovan], "Memorandum of Conversation at the Commissariat for Internal Affairs," December 27, 1943, in "Memoranda for the President: OSS-NKVD Liaison," CIA Historical Review Program Release, 1993, accessed September 17, 2020, CIA, https://cia.gov/static/96846efedb50b770061aca98ae18d2c0/Memoranda-OSS-NVKD-Liaison.pdf. The Soviet version tracks with Donovan's 1944 and Deane's postwar versions: "Report by G. Ovakimyan of 28.12.43 on Conversation Held by Fitin and Ovakimyan with Donovan and General Deane 27.12.43," White Notebook No. 1, 85, Vassiliev Papers, LoC.
10. Deane, *Strange Alliance,* 51.
11. [Donovan], "Memorandum of Conversation."
12. Deane, *Strange Alliance,* 53.
13. Waller, *Wild Bill Donovan,* 224.
14. Charles E. Bohlen, *Witness to History 1929–1969* (New York: Norton, 1973), 155–6.
15. The exact date of Donovan's departure is not clear. Waller, *Wild Bill Donovan,* 224, places it on January 6, 1944. Deane and Harriman suggest a later departure.
16. "Red Agent Seized by FBI Men Here," *New York Times,* May 6, 1941.
17. These operations are the subject of Weinstein and Vassiliev, *Haunted Wood,* and John Earl Haynes, Harvey Klehr, and Alexander Vassilliev, *Spies: The Rise and Fall of the KGB in America* (New Haven: Yale University Press, 2010).
18. Hiss was a special assistant to the leadership at State, privy to the formation of US foreign policy. He was primarily handled by the GRU, but was well-known to the NKVD. White was second only to the secretary at Treasury for much of the war, drafting key policies. Both have been the subjects of biographies and lasting controversy even though the evidence of their espionage is now overwhelming.
19. There were at least twelve, and likely more. Haynes et al., *Spies,* 293, 328.
20. Bradley, *Very Principled Boy,* 37.
21. Not until after World War II did Soviet spies or Americans who spied for them face long jail terms—or even execution—if apprehended.
22. Bradley, *Very Principled Boy,* 68–73, describes the process. Lee can be said to have taken his social cover to the next level by having an affair with one of his contacts, Mary Price.
23. "Message from 'Koch' dated 17.1.44," White Notebook #1, 92–93. "Koch" was Lee's code name.
24. Donovan's purchase of a Soviet codebook from the Finns late in 1944 was not the act of a man who now believed in Soviet good will. See for example: Bradley F. Smith, *Sharing Secrets with Stalin: How the Allies Traded Intelligence, 1941–1945* (Lawrence, KS: University Press of Press, 1996), 232–3.
25. A dispute concerning OSS operations against foreign embassies in Washington is cited as evidence of "visceral hatred" between Donovan and Hoover. Waller, *Wild Bill Donovan,* 127. See also Donald Downes, *The Scarlet*

Thread: Adventures in Wartime Espionage (Cabin John, MD: Wildside Press, 2020), 93–94.

26. Hoover to Donovan, April 5, 1943, and Donovan to Hoover, April 9, 1943, Part 6, William J. Donovan File, accessed Mar. 28, 2022, https://vault.fbi.gov/William%20J%20Donovan%20/. To similar effect, see a courteous exchange of notes between Donovan and Hoover in November 1942 about "The War This Week." 30, Part 3, William J. Donovan File, accessed Mar. 28, 2022, https://vault.fbi.gov/William%20J%20Donovan%20/.
27. Hoover to Hopkins, February 10, 1944, printed in Whitehead, *FBI Story*, 228. Whitehead had access to FBI files when he wrote this book. Hoover wrote the foreword.
28. "Message from 'Koch' dated 17.1.44."
29. Harriman and Abel, *Special Envoy*, 294–5; "Message from 'Koch' dated 17.1.44."
30. Bradley, *Very Principled Boy*, 88. See also: "Transcript of a Conversation Held by P. M. Fitin, Chief of the First Dir. of the USSR NKGB, and Ovakimyan, Deputy Chief of the First Dir. of the USSR NKGB, with Deane on 7.04.44," White Notebook #1, 88. Some have read a threat into Hoover's memo—namely, that he himself might leak the information. See Waller, *Wild Bill Donovan*, 224–5.
31. "Franklin D. Roosevelt: Day by Day," March 7, 1944, accessed September 17, 2020, http://www.fdrlibrary.marist.edu/daybyday/daylog/march-7th-1944/.
32. Roosevelt to Harriman, March 15, 1944, Box 11, Map Room Papers, FDRL.
33. Harriman to Roosevelt, March 18, 1944; Harriman and Abel, *Special Envoy*, 294. They had not actually penetrated the NKVD in the same sense that the NKVD had penetrated the US government; they had simply begun to establish a liaison relationship.
34. Harriman and Abel, *Special Envoy*, 294.
35. This is clear from various entries in Vassiliev, White Notebook #1, 95–102; Deane, *Strange Alliance*, 57.
36. "'Koch' 11.12.44," Vassiliev White Notebook #1, 97.
37. "Message from 'Koch' dated 17.1.44."
38. For context, see: Smith, *Sharing Secrets with Stalin*. Smith argues that no one ally could produce the information needed to defeat Germany; reluctant allies needed to overcome their suspicions of each other for the duration of the war—and did so to a limited extent.
39. CINRAD stood for Communist Infiltration of Radiation Laboratory, and COMRAP for Comintern Apparatus, the former to defend American technology, the latter to defend against political subversion. Well-placed white-collar spies like Duncan Lee were safe from the FBI for the time being.
40. Tim Weiner, *Enemies: A History of the FBI* (New York: Random House, 2013), 97; Powers, *Secrecy and Power*, 254–5. Powers, who had early access

to FBI files, states the FY 1941 number as 1,596, and the corresponding 1944 number as 4,886.

41. See for example Weiner, *Enemies*, 83–86, 122; Batvinis, *Origins*, 94–95. Biddle was more successful than Jackson, but Hoover circumvented and outlasted both. Japanese Americans were not automatically included in the CDP; Hoover was against the administration's 1942 relocation program, which targeted ethnic groups regardless of citizenship. Powers, *Secrecy and Power*, 249–50.
42. Roosevelt to Jackson, May 21, 1940, Jackson Folder, Box 57, PSF, FDRL.
43. Batvinis, *Origins*, 127–34, 258, is a thorough discussion of this issue.
44. Powers, *Secrecy and Power*, 257.
45. Batvinis, *Origins*, 226–56, is an excellent description of the case.
46. See for example: Samuel A. Tower, "FBI's Hidden Struggles against Spies Continues: J. Edgar Hoover," *New York Times*, February 11, 1945, a useful contemporary summary of the FBI's wartime accomplishments according to Hoover.
47. FBI, "SIS Annual Report 1942–3," in FBI, *History of the SIS Division*, part 5, 7–8, accessed Sept. 1, 2020, https://vault.fbi.gov/special-intelligence-service.
48. On the actual (and limited) danger from Nazi spies in Latin America, see David P. Mowry, *German Clandestine Activities in South America in World War II* (Ft. Meade, MD: National Security Agency, 1989), esp. v, 7–14, 57–58.
49. Thomas D. Schoonover, *Hitler's Man in Havana: Heinz Lüning and Nazi Espionage in Latin America* (Lexington, KY: University of Kentucky Press, 2008). Unless otherwise noted, Schoonover is the source for information on Lüning.
50. Anon., "J. E. Hoover Praises Cuba on Spy Arrest," *New York Times*, Nov. 1, 1942.
51. FBI, *SIS History*, part 3, 378–9. Of two leads described in this source, one turned out to be an unwitting Chilean, the other an Argentine whom the Abwehr never actually used. Per Schoonover, the FBI initially feared that Lüning could be "a master spy," instrumental in U-boat successes in the Caribbean and the Gulf of Mexico. Schoonover, *Hitler's Man in Havana*, 11. Lüning did not in fact supply any actionable information to the Abwehr.
52. Schoonover, *Hitler's Man in Havana*, xvii. The Bureau's disproportionate response to the Lüning case was evidence that the US exaggerated the Nazi threat in Latin America. Mowry, *German Clandestine Activities in South America in World War II*, esp. v.
53. FBI, *SIS History*, part 3, 378.
54. FBI, "SIS Annual Report 1943–1944," 63.
55. FBI, *SIS History*, part 1, 63–65.
56. FBI, *SIS History*, part 1, 65–68.
57. FBI, "SIS Annual Report 1943–1944," 8–9.
58. FBI, *SIS History*, part 1, 68–69.
59. FBI, *SIS History*, part 1, 69. Berle did not, however, change his mind about the principal thrust of SIS's work. After the war he was effusive in his praise

of SIS's work, but primarily for its achievements in counterintelligence, "breaking one espionage ring after another." Berle to Hoover, September 17, 1946, quoted at length in FBI, *SIS History*, part 4, 659. The same source, from pages 645–67, contains testimonials from Foreign Service officers.

60. Powers, *Secrecy and Power*, 238, 247, 544n.
61. Powers, *Secrecy and Power*, 237, 252. Powers made an approximate count of the reports in the FDRL.
62. See Waller, *Wild Bill Donovan*, 317. See also Katz, "The OSS and the Development of the Research and Analysis Branch," in Chalou, ed., *Secrets War*, esp. 46.
63. See for example folders for "Justice—J. Edgar Hoover," Box 57, PSF, FDRL; "Office of Strategic Services: Donovan Reports," Boxes 147–154, PSF, FDRL.
64. In 1944 General Marshall chided the president for seldom taking time to read the Magic Summaries. Marshall to Roosevelt, Feb. 12, 1944, copy in SRH 111, NCML.
65. Even absent enemy action, long-distance wartime travel was risky.
66. This was James Grafton Rogers's term for one of Donovan's brainstorms.

15: Breaking Codes, Forging Links

1. Assistant Secretary of State Adolf Berle, the target of a BSC intrigue, happily joined in the campaign, which would end in a démarche to the British ambassador in March 1942 that curtailed many of BSC's activities. Smith, *Ultra-Magic Deals*, 37.
2. West, ed., *The Guy Liddell Diaries*, vol. 1, 116. The FBI would apply what Clegg learned to its operations in Latin America.
3. See Ewan Montagu, *Beyond Top Secret Ultra* (London: Peter Davies, 1977), 75–81, for the point of view of the British officer on the case, and Batvinis, *Hoover's Secret War*, 141–58, for that of an FBI case officer. Popov's German handlers asked him for information about Pearl Harbor a few months before the Japanese attack. Batvinis shows that the FBI did inform the navy of the German's request, absolving the Bureau of any implied blame. See also Aldrich and Cormac, *Black Door*, 100. However, one unevaluated bit of information was unlikely to have changed the outcome. Pearl Harbor was a systemic failure on many levels.
4. C. T. Edwards, "Exchange of Information with the United States Authorities," Sept. 1, 1940, CAB 81/98, TNA. Strong almost certainly exceeded his brief, making commitments that he was not authorized to make. Budiansky, *Battle of Wits*, 173.
5. Budiansky, *Battle of Wits*, 172–74, is a good overview.
6. Abraham Sinkov, NSA Oral History, May 1979, National Security Agency, accessed Oct. 31, 2020, https://media.defense.gov/2021/Jul/15/2002763513/-1/-1/0/NSA-OH-02-79-SINKOV.PDF, is a good description of the various phases of the trip. See also David Sherman, *The First*

Americans: The 1941 US Codebreaking Mission to Bletchley Park (Ft. Meade, MD: National Security Agency, Center for Cryptologic History, 2016).

7. Prescott H. Currier, NSA Oral History, Nov. 14, 1980, National Security Agency, accessed Oct. 31, 2020, https:// https://media.defense.gov/2021/Jul/15/2002763432/-1/-1/0/NSA-OH-02-72-CURRIER.PDF. The suspicion was that he might have other, unnamed assistants, perhaps including a native speaker of Japanese.
8. Prescott H. Currier, "My 'Purple' Trip to England," *Cryptologia* 20, no. 3 (July 1996) in Deavours et al., *Selections from Cryptologia*, 288.
9. Currier, NSA Oral History, 51.
10. Currier, NSA Oral History, 34, 53, 55. He implied that the fact that Sinkov and Rosen both happened to be Jewish while he and Weeks were Gentiles was another limiting factor.
11. There has been some dispute about the number of machines transported across the Atlantic. Smith, *Ultra-Magic Deals*, 55–56. The number of machines is almost immaterial; the key fact is that the Americans were sharing hardware at all.
12. He was the former foreign secretary who met Donovan in London in 1940.
13. Currier, NSA Oral History, 60.
14. Landis Gores, *Ultra: I Was There* (Morrisville, NC: Lulu, 2008), vol. 1, 7–8.
15. Jeffery, *MI6*, 319. The deed named him as the owner. In 1939, GC&CS had something like one hundred employees.
16. Prescott Currier and John Tiltman, "Presentation Given to Members of the Cryptanalysis Field," ca. 1974–5, 150, appended to Currier, NSA Oral History. Edward Travis, who would succeed Denniston as head of GC&CS, was also present. For a description of Denniston and his work, see Robin Denniston, *Thirty Secret Years: A. G. Denniston's Work in Signals Intelligence, 1914–1944* (Clifton-upon-Teme, UK: Polperro Heritage Press, 2007). The lifelong athlete Denniston played field hockey for Scotland at the Olympics before World War I.
17. Tiltman would even move to the United States after the war, work for the US government, and die in Hawaii.
18. The British were acutely aware of the difference between the two delegations. Sinkov and Rosen compared notes with Currier and Weeks, but only in a general way. Currier, NSA Oral History, 70.
19. Michael Kerrigan, *Enigma Code Breakers* (London: Amber Books, 2018), 139, 148.
20. John Tiltman, NSA Oral History, Dec. 17, 1978, National Security Agency, accessed Oct. 31, 2020, https://media.defense.gov/2021/Jul/15/2002763524/-1/-1/0/NSA-OH-07-78-TILTMAN.pdf.
21. F. H. Hinsley, "Introduction: The Influence of Ultra in the Second World War," in *Codebreakers: The Inside Story of Bletchley Park*, eds. F. H. Hinsley and Alan Stripp (Oxford: Oxford University Press, 1993), 2. Hinsley also

wrote and edited an exhaustive official history of British intelligence in World War II. Keegan, *Intelligence in War*, 160–1, discusses which German codes were broken, and when and how they were broken.

22. Sinkov, NSA Oral History, 4.
23. Denniston to "Director" [Menzies?], Aug. 5, 1941, HW 14/45, TNA. Budiansky, *Battle of Wits*, 178, discusses the presentation to the chiefs of staff and Churchill. See also Jeffery, *MI6*, 443.
24. Tiltman, NSA Oral History, 6.
25. Sinkov, NSA Oral History, 10.
26. Weeks to Denniston, Mar. 3, 1941, HW 14/45, TNA.
27. Budiansky, *Battle of Wits*, 359–62, uses the term "Rapid Analytic Machinery" and explains how they worked and why they were not exactly lineal ancestors of early computers.
28. Currier and Tiltman, "Presentation," ca. 1974–5.
29. Currier, NSA Oral History, 65–66.
30. Leo Rosen, NSA Oral History, Aug. 26, 1984, National Security Agency, accessed Oct. 31, 2020, https://media.defense.gov/2021/Jul/15/2002763514/-1/-1/0/NSA-OH-16-84-ROSEN.pdf.
31. GC&CS took their first steps thanks to the Polish codebreakers who had made the first inroads into German codes and shared their findings with their French and British counterparts before the Nazis invaded.
32. The same was true on modern ships of the Royal Navy in 1941. American sailors were taken aback to discover that the brand-new battleship *Prince of Wales* did not carry typewriters. When the US transferred the fifty overage destroyers to Britain, they apparently had to be "de-modernized" to conform to British standards. Farago, *Tenth Fleet*, 152.
33. Currier, NSA Oral History, 68.
34. Smith, *Ultra-Magic Deals*, 60–61.
35. Denniston to Director, Aug. 8, 1941, HW 14/45, TNA; Budiansky, *Battle of Wits*, 208. SIS might have initiated the letter, but Op-20-G would have supported the request.
36. Smith, *Ultra-Magic Deals*, 89; Denniston, *Thirty Secret Years*, 21.
37. Denniston, Notes, Aug. 14, 1941, HW 14/45, TNA. He later commented that "this feud between army and navy in USA" seemed incredible to him. Denniston, *Thirty Secret Years*, 18.
38. [Denniston?], "Notes on Conference Held August 15th," HW 14/45, TNA.
39. [Denniston?], "Minutes of Conference, August 16, 1941," HW 14/45, TNA.
40. Denniston, *Thirty Secret Years*, 21.
41. Denniston to A. J. L. Murray, Nov. 3, 1941, HW 14/45, TNA; Budiansky, *Battle of Wits*, 211.
42. Denniston, "Impression," [? Sept. 9, 1941], HW 14/45, TNA. He noted that his proposal for Anglo-American cooperation had been generally accepted by the army but not by the navy. In Denniston to Murray, Nov. 3, 1941, he commented favorably on Friedman's reputation.

43. Denniston, *Thirty Secret Years*, 19–20, 76–77, describes his father's medical problems in 1941 and how treatment almost bankrupted him, a sad commentary on the low pay that the head of Bletchley Park received. His net monthly pay was roughly equivalent to $140. A comparable American official would have earned roughly $500 a month in 1941.
44. Alfred Dillwyn Knox was a brilliant but difficult British codebreaker. Denniston, "Impression," [? Sept. 9, 1941]; Denniston to E. G. Hastings, Oct. 9, 1941, HW 14/15, TNA: "As to the famous Mrs. D. I have sent her nearly all she has asked for and asked her to prove her methods by success where we have failed." She would continue to exasperate and even humiliate the British in 1942. GC&CS officer John Tiltman remembered going to a meeting with her that turned out to be a confrontation, after which he concluded that "she was rather like a conventional witch." Tiltman, NSA Oral History, 9.
45. Denniston to Murray, Nov. 3, 1941.

16: Admiral Dönitz's Unintended Contribution to Allied Victory

1. Walter R. Borneman, *The Admirals* (New York: Back Bay, 2013), 92. The British perception of King is captured by Patrick Beesly, *Very Special Intelligence: The Story of the Admiralty's Operational Intelligence Centre 1939–1945* (New York: Ballantine, 1977), 111: the admiral was determined that the US Navy "should not again play second fiddle to the Royal Navy."
2. Churchill, *Their Finest Hour*, 599. This was quite an admission considering all the other setbacks Churchill faced during the war. Military historian John Keegan approvingly quotes a comment by naval historian Clay Blair that Churchill's fears were exaggerated, that the Germans could not have won the Battle of the Atlantic because the economic power of the Allies was too great; the *Kriegsmarine* would never be able to sink enough ships fast enough. This is not helpful for the historian interested in what the participants were thinking at the time. John Keegan, *Intelligence in War: The Value—and Limitations—of What the Military Can Learn About the Enemy* (New York: Knopf, 2003), 242.
3. Murray and Millett, *War to Be Won*, 244. Churchill, *Their Finest Hour*, 714, offers comparable statistics for 1940, including seventy-nine ships sunk in December for a total of 357,314 tons.
4. Churchill, *Grand Alliance*, 150.
5. Murray and Millett, *War to Be Won*, 246, conclude that "Ultra's contribution to the anti-submarine battle now became the most significant intelligence victory of the war, and the only episode in which intelligence alone had a decisive impact on military operations." Remembering Midway, Joe Rochefort might have disagreed that it was the *most* significant intelligence victory. Recent scholarship has modified Murray and Millett's judgment somewhat, arguing that while signals intelligence was an important factor,

it was one among many. John Ferris, *Behind the Enigma: The Authorized History of GCHQ, Britain's Secret Cyber-Intelligence Agency* (New York and London: Bloomsbury, 2020), esp. 241–2.

6. Washington to C.S.S., Nov. 27, 1941, HW 14/45, TNA.
7. C.S.S. to Washington, Dec. 1, 1941, HW 14/45, TNA. The British may have believed that they were being forthcoming, but once British and American codebreakers began to work more closely together, it became clear how much more there was to share.
8. Winston S. Churchill, *The Hinge of Fate* (Boston: Houghton Mifflin, 1950), 126; Murray and Millett, *War to Be Won*, 250.
9. A telling story that made the rounds among naval aviators in Florida was that the Coast Guard found a ticket stub from a Miami Beach theater in the pocket of a dead German sailor after his submarine sank. True or not, it reflected American perceptions of German prowess that were not far off the mark.
10. Churchill, *Hinge of Fate*, 117–20.
11. Borneman, *Admirals*, 177.
12. David Kohnen, *Commanders Winn and Knowles: Winning the U-Boat War with Intelligence, 1939–1943* (Krakow, POL: Enigma Press, 1999), 45–48.
13. Churchill, *Hinge of Fate*, 118.
14. Tiltman, NSA Oral History, 8–9. He was presumably referring to solved Enigma messages especially relating to the German Navy.
15. Kohnen, *Commanders Winn and Knowles*, 56.
16. Kohnen, *Commanders Winn and Knowles*, 51–52.
17. Kohnen, *Commanders Winn and Knowles*, 49.
18. Farago, *Tenth Fleet*, 76, 203.
19. Smith, *Ultra-Magic Deals*, 78.
20. FNU Barrett, "The Americans, the Navy Department and U-Boat Tracking" in "Operational Intelligence Centres, Formation and History," [1945?], ADM 223/286, TNA. This appears to be part of a Royal Navy war report written soon after the receipt of a corresponding US Navy War Report in May 1945. This document quotes earlier reports by Winn, who made additional comments and corrections in his own hand. As Jason S. Ridler points out in *Mavericks of War: The Unconventional, Unorthodox Innovators and Thinkers, Scholars, and Outsiders Who Mastered the Art of War* (Guilford, CT: Stackpole Books, 2018), the British often seemed best at unorthodox ways of war.
21. David Syrett, ed., *The Battle of the Atlantic and Signals Intelligence* (London: Routledge, 2018) is a collection of the Tracking Room's remarkable reports.
22. Beesly, *Very Special Intelligence*, 58–59.
23. Kohnen, *Commanders Winn and Knowles*, 76–77.
24. Barrett, "The Americans, the Navy Department and U-Boat Tracking."
25. Beesly, *Very Special Intelligence*, 114. Beesly was Winn's wartime deputy, and presumably heard this story from him.

26. Barrett, "The Americans, the Navy Department and U-Boat Tracking."
27. Beesly, *Very Special Intelligence*, 114.
28. Barrett, "The Americans, the Navy Department and U-Boat Tracking."
29. Beesly, *Very Special Intelligence*, 114.
30. Farago, *Tenth Fleet*, 205; Kohnen, *Commanders Winn and Knowles*, 78; Kenneth A. Knowles, Aug. 19, 1986, NSA Oral History, NCML.
31. Knowles, Oral History Interview, 2. Most covers portrayed sailors, but after Knowles left for Texas, the September 1, 1941, cover portrayed a female dancer from the Ziegfeld Follies in a bathing suit and yachting cap.
32. Knowles, Oral History Interview, 3.
33. Farago, *Tenth Fleet*, 205. Eachus and Ely were at Bletchley from July to October. They received "complete wiring diagrams and blueprints of the actual [British] bombes." Stephen Budiansky, "Bletchley Park and the Birth of the Very Special Relationship," in *Action This Day*, eds. Michael Smith and Ralph Erskine (London: Bantam, 2001), 226.
34. Knowles, Oral History Interview, 5–6.
35. Kohnen, *Commanders Winn and Knowles*, 84–116, contains a wealth of detailed information on its operations.
36. One two-by-eight-by-seven-foot bombe would cost roughly $48,000—at a time when a brand-new destroyer cost about $6,000,000. In 1943, the navy would order ninety-three bombes. Ralph Erskine, "Breaking German Naval Enigma on Both Sides of the Atlantic," in Smith and Erskine, eds., *Action This Day*, esp. 191–2. The German Navy had upgraded to a four-rotor system in early 1942.
37. This agreement has been called both the Travis-Wenger agreement (after Edward Travis, Denniston's replacement as head of GC&CS, and Joe Wenger of Op-20-G) and the Travis-Holden agreement (after Captain Carl F. Holden, the then director of Naval Communications). Erskine, "Breaking German Naval Enigma," in Smith and Erskine, eds., *Action This Day*, esp. 191. Under the agreement, the Japanese Navy would remain the primary responsibility of the US Navy. See also Benson, *History of U.S. Communications Intelligence*, esp. 61, and Budiansky, *Battle of Wits*, 237–9.
38. The US would come to have more—and faster—bombes than the British, able to process the sophisticated four-rotor U-boat traffic by the end of 1943. Kohnen, *Commanders Winn and Knowles*, 72, 104.
39. Knowles, Oral History Interview, 7. See also Kenneth A. Knowles, "The American View" in "Ultra and the Battle of the Atlantic," Symposium at USNA, October 28, 1977, National Security Agency, accessed Nov. 20, 2020, https://www.nsa.gov/portals/75/documents/news-features/declassified-documents/cryptologic-spectrum/Ultra.pdf.
40. Farago's *Tenth Fleet* is an excellent description of the subject, based in part on the author's firsthand experience. Knowles incorporated his own firsthand insights when he reviewed Farago's book: Kenneth A. Knowles, "*The Tenth Fleet* by Ladislas Farago," book review, *Studies in Intelligence* 7,

no. 2 (Spring 1963): A19–A23. The British broke into the four-rotor U-boat code in December 1942; the American bombes came online in the spring and summer of 1943. Paul Kennedy, *Engineers of Victory: The Problem Solvers Who Turned the Tide in the Second World War* (New York: Random House, 2013), 5–73, discusses the Atlantic War and the Allies' array of weapons.

41. Knowles, "The American View," 14; Knowles, NSA Oral History. Exactly which sinkings prompted Winn's comment is not clear from either Knowles's interview or Beesly, *Very Special Intelligence*, 195–1. David Syrett, *The Defeat of the German U-Boats: The Battle of the Atlantic* (Columbia, SC: University of South Carolina Press, 1994), 145–80, 271–2, is a detailed description of the battles in the summer of 1943.
42. Knowles, "The American View," 14. To the same effect, see Beesly, *Very Special Intelligence*, 195–6.
43. Over time, the British were increasingly inclined to the American point of view, accepting the need to eliminate the threat, not just avoid it, even if it entailed some risk to Ultra. See for example Beesly, *Very Special Intelligence*, 196.
44. Knowles, "*Tenth Fleet*" book review.
45. This correlated with British success at rerouting convoys out of harm's way.
46. Keegan, *Intelligence in War*, 236, offers a good discussion of this ongoing process and how the German position was not unreasonable. To read Enigma messages, the Allies had to have the machine itself (or an analog), the rotors, the settings, and the bigrams—the codes that designated grid squares in the ocean—all of which the British eventually acquired.
47. In early 1942, the B-Dienst broke into Naval Cipher No. 3, a joint system used by the British, Americans, and Canadians for routing convoys. Smith, *Ultra-Magic Deals*, 118. Warner, *Rise and Fall of Intelligence*, 92–93, 103–4, makes the point that, like their allies, the Germans enjoyed occasional success at the tactical and operational levels, but could not compete with American and British intelligence at the strategic level.
48. Budiansky, *Battle of Wits*, 248–50, 290–4. The British had been careful to not reference information that could have come only from a compromised German message in general message traffic.
49. During World War II, both the US Navy and Army used the multirotor devices originally designed by Friedman and Rowlett that the enemy never broke. Especially while in enemy waters, US Navy submariners sent far fewer messages than their German counterparts, who were required to report almost daily. More air cover also might have helped the Germans. The Allies' advantage came from combined arms, especially coordination among air and sea forces. See Theodore R. Roscoe, *United States Submarine Operations in World War II* (Annapolis, MD: Naval Institute Press, 1949) and Mucklow, *SIGABA/ECM II Cipher Machine*.
50. Knowles, "*Tenth Fleet*" book review.

51. Though parts of the building were still under construction, some spaces were ready for tenants beginning in mid-1942.
52. Telford Taylor, NSA Oral History, Jan. 22, 1985, National Security Agency, accessed Dec. 4, 2020, https://media.defense.gov/2021/Jul/15/2002763518/-1/-1/0/NSA-OH-01-85-TAYLOR.pdf.
53. A pioneer in artificial intelligence, Turing played a key role in applying his knowledge to codebreaking. At Bletchley Park, he was for a while the head of Hut 8, which was responsible for German naval codes.
54. Benson, *History of U.S. Communications Intelligence*, 97.
55. Smith, *Ultra-Magic Deals*, 136–41. The army and the navy both had a stake in its development at Bell Laboratories.
56. Tiltman, NSA Oral History, Dec. 17, 1978, 16.
57. Benson, *History of U.S. Communications Intelligence*, 99–100. The Strong-Tiltman meetings occurred in late December, roughly the same time as Clarke approached Taylor. Budiansky, *Battle of Wits*, 297, has Tiltman generally advocating for collaboration.
58. Quoted in Smith, *Ultra-Magic Deals*, 138.
59. A well-documented version of the story is Benson, *History of U.S. Communications Intelligence*, 97–103.
60. Benson, *History of U.S. Communications Intelligence*, 103. In December 1942 the army had contracted with AT&T to build bombes. The navy was contracting with National Cash Register for the same purpose. Friedman turned down a navy offer to join in the NCR contract. Budiansky, *Battle of Wits*, 297.
61. Tiltman, NSA Oral History, Dec. 17, 1978, 11.
62. Smith, *Ultra-Magic Deals*, 145.
63. Benson, *History of U.S. Communications Intelligence*, 107.
64. Taylor to Clarke, Apr. 5, 1943, quoted at length in Budiansky, *Battle of Wits*, 298. See also Benson, *History of U.S. Communications Intelligence*, 103–8.
65. Colin MacKinnon, ed., "Bletchley Park Diary, William F. Friedman," 2013, accessed Dec. 10, 2020, Colin MacKinnon (website), https://colinmackinnon.com/attachments/The_Bletchley_Park_Diary_of_William_F._Friedman_E.pdf. The original is at NCML. I have also relied on MacKinnon, "William Friedman's Bletchley Park Diary."
66. For McCormack's official explanation, see Alfred McCormack, "War Experience of Alfred McCormack," 9–10, SRH 185. See also Louis T. Stone Jr., "Memorandum Describing American Liaison," Oct. 12, 1945, SRH 153, NCML.
67. MacKinnon, ed., "Bletchley Park Diary, William F. Friedman," 4; Taylor, NSA Oral History.
68. Denniston was the GC&CS managing director through 1941. In early 1942, he became duty director (C) for civilian traffic and moved to the London office, while Edmund Travis took over at Bletchley as deputy director (S) for military traffic. In 1944, Travis would become director of GC&CS.

69. As implied above, the army's dispute with GC&CS was not over diplomatic traffic, but over German military traffic and the technology used to break it.
70. Alastair Denniston, "Informal Memorandum by Cmdr. Denniston Outlining His Original Concept of the American Liaison," May 1943, reproduced in Denniston, *Thirty Secret Years*, 158–62.
71. Taylor, NSA Oral History.
72. London to Washington, May 21, 1943, reproduced in Denniston, *Thirty Secret Years*, 123.
73. London to Washington, May 21, 1943; London to Washington, May 22, 1943, reproduced in Denniston, *Thirty Secret Years*, 137–41.
74. London to Washington, June 6, 1943, reproduced in Denniston, *Thirty Secret Years*, 144–5.
75. D. R. Nicoll, "Sir Edward Wilfrid Harry Travis," *Oxford Dictionary of National Biography*, accessed Dec. 18, 2020, https://doi.org/10.1093/ref:odnb/61098; F. H. Hinsley, "Alexander Guthrie (Alastair) Denniston," *Oxford Dictionary of National Biography*, accessed Dec. 18, 2020, https://doi.org/10.1093/ref:odnb/32783.
76. MacKinnon, ed., "Bletchley Park Diary, William F. Friedman," 13.
77. For Friedman's detailed and highly technical report, see: William F. Friedman, "Report on E Operations of the GC&CS at Bletchley Park," Aug. 12, 1943, accessed Dec. 10, 2020, Colin MacKinnon (website), https://colinmackinnon.com/attachments/Report_bw_200_dpi_all_pages.pdf. US Navy codebreakers had, of course, already established a foothold at Bletchley and were cooperating successfully with the British. Friedman's unique background enabled him to see and understand more than they had.
78. MacKinnon, ed., "Bletchley Park Diary, William F. Friedman," 31 passim.
79. MacKinnon, ed., "Bletchley Park Diary, William F. Friedman," 73. As director of naval communications, Redman oversaw the work of Op-20-G, Wenger's home base.
80. The original title of the document was "Agreement between the British Government Code and Cipher School and U.S. War Department." It has also been known as the BRUSA agreement, an abbreviation the British originally used for a British-USA communications circuit. See MacKinnon, "Bletchley Park Diary, William F. Friedman," 6n, and Benson, *History of U.S. Communications Intelligence*, 108–9. The agreement was signed by Strong and Travis on May 17, but not ratified by Marshall's office until June. The inclusion of Japanese and German traffic made the agreement two-sided, with each party contributing its flagship product. The agreement covered intelligence and counterintelligence traffic but not diplomatic or naval traffic.
81. MacKinnon, ed., "Bletchley Park Diary, William F. Friedman," 79.
82. MacKinnon, ed., "Bletchley Park Diary, William F. Friedman," 123.

17: Intelligence and the Main Event

1. MacKinnon, ed., "Bletchley Park Diary," 155–7.
2. Taylor's approach was to trust but verify. To ensure important material was getting through, he violated the agreement once or twice, sending it through both British and American channels. Taylor, NSA Oral History.
3. Asa Briggs, *Secret Days: Codebreaking in Bletchley Park* (S. Yorkshire, UK: Frontline Books, 2011), 23.
4. Alfred Friendly, "Confessions of a Code Breaker," *Washington Post,* Oct. 27, 1974. See also Gores, *Ultra: I Was There,* esp. vol. 1, 3–90.
5. Under various agreements negotiated in 1942 and 1943, OSS obtained access to counterintelligence decrypts. OSS would not, however, receive access to other Axis traffic. Benson, *History of U.S. Communications Intelligence,* 109; Smith, *Ultra-Magic Deals,* 131–72.
6. Section V's duties included: developing all-source information on hostile intelligence services, running double agents overseas, and controlling the analysis and use of intercepts. Naftali, *X-2,* 82; Muggeridge, *Chronicles,* esp. 401. "Counterintelligence" is the more general term for this function; "counterespionage" is the specific term for combatting enemy spies.
7. Jimmy Burns, *Papa Spy: Love, Faith, and Betrayal in Wartime Spain* (London: Bloomsbury, 2010), 158–9.
8. MI6 to MI5 teleprinter message, Jan. 26, 1941, Ángel Alcázar de Velasco File, KV 2–3535, TNA.
9. Extracts from Foreign Office Files, January 1941, in Alcázar File, KV 2–3535.
10. Burns, *Papa Spy,* 166–7; Philby, *My Silent War,* 60–61.
11. J. C. Masterman, *The Double-Cross System: The Incredible True Story of How Nazi Spies Were Turned into Double Agents* (New Haven: Yale University Press, 1972), 57–58.
12. Holt, *Deceivers,* 163. Masterman in *Double-Cross System,* 59, contented himself with a discreet reference to "a study of secret sources" that had unmasked the spy.
13. Masterman expanded this list to seven goals. Masterman, *Double-Cross System,* 58. See also Ben Macintyre, *Double Cross: The True Story of the D-Day Spies* (New York: Broadway Paperbacks, 2013).
14. See for example Naftali, *X-2,* 46–47, 67. Donovan barely tolerated those who wanted to conduct background investigations of employees. Some members of MI6 also held the opinion that counterintelligence was less important in wartime. See for example Philby, *My Silent War,* 59.
15. Naftali, *X-2,* 13, 66–67.
16. Naftali, *X-2,* 80–81.
17. Donovan to Strong, Sept. 7, 1942, quoted in Naftali, *X-2,* 101.
18. Naftali, *X-2,* 139–40; Robin W. Winks, *Cloak & Gown: Scholars in the Secret War, 1939–1961* (New York: William Morrow, 1987), 261.

19. Naftali, *X-2*, 141–2.
20. Winks, *Cloak & Gown*, 251, 262.
21. Naftali, *X-2*, 95–96, 174, 195–6.
22. Winks, *Cloak & Gown*, 263.
23. Philby, *My Silent War*, 85.
24. Winks, *Cloak & Gown*, 286–7; Muggeridge, *Chronicles*, 398–9.
25. Cowgill and Pearson eventually formed a strong personal friendship that included their families. Winks, *Cloak & Gown*, 288–9; Naftali, *X-2*, 212.
26. Muggeridge, *Chronicles*, 401. Naftali, *X-2*, 213, and Winks, *Cloak & Gown*, 267–8, describe a similar process for the Americans.
27. For example, a 1943 decrypt revealed a German plan to deceive Eisenhower about Axis troop movements in Tunisia. Naftali, *X-2*, 219. See also Brown, ed., *Secret War Report*, 89.
28. Winks, *Cloak & Gown*, 274.
29. Cutler, *Counterspy*, 13.
30. Brown, ed., *Secret War Report*, 89; Naftali, *X-2*, 300. Donovan signed the order at the urging of his longtime assistant, James R. Murphy.
31. Cutler, *Counterspy*, 13.
32. See chapter 11 prior and Nelson D. Lankford, *The Last American Aristocrat: The Biography of Ambassador David K. E. Bruce* (Boston: Little, Brown, 1996), 130. Initially known as "Special Activities/Bruce," it had by now morphed into the Secret Intelligence Branch.
33. Nelson D. Lankford, ed., *OSS against the Reich: The World War II Diaries of Colonel David K. E. Bruce* (Kent, OH: Kent State University Press, 1991), 17.
34. Lankford, ed., *OSS against the Reich*, 23–24.
35. For the months before D-day, see US Office of Strategic Services, "Report on OSS Activities for the Month of June 1944," OSS Monthly Activity Reports—May [*sic*] 1944, ERR, accessed Jan. 19, 2021, CIA, https://www.cia.gov/readingroom/docs/CIA-RDP13X00001R000100140008-1.pdf.
36. Brown, ed., *Secret War Report*, 335. Lankford, ed., *OSS against the Reich*, puts the total at 2,900 in July 1944. MacPherson, *American Intelligence in War-Time London* is a detailed scholarly study of the base. The unpublished, more specific war report is US Office of Strategic Services, "History of London Station," Roll 10, Microfilm 1623, RG 226, NARA II. OSS London's original mission was maintaining relations with foreign intelligence agencies and channeling useful reports back to Washington.
37. See chapter 1 prior.
38. See chapter 1 prior. One of the best German cases was run by the Czechoslovak secret service. František Moravec, *Master of Spies: The Memoirs of General František Moravec* (New York: Doubleday, 1975), 165–75. See also Jeffery, *MI6*, 475–506, and Olson, *Last Hope Island*, esp. chapter 10.
39. Section V's Philby was just one member of a large spy ring. Similarly, Duncan Lee was but one of many Soviet spies in the US government. Richard Sorge was a communist posing as a loyal German correspondent in Tokyo.

40. Moravec, *Master of Spies*, 194–211. See also Callum MacDonald, *The Assassination of Reinhard Heydrich* (Edinburgh: Birlinn, 2007) and Douglas Dodds-Parker, *Setting Europe Ablaze* (Windlesham, UK: Springwood Books, 1983), 96. Neither SIS nor SOE favored targeted killings after the Heydrich operation. Jeffery, *MI6*, 538–40.
41. While committing unconscionable errors in tradecraft, like dressing two agents in matching outfits and ignoring danger signals built into communication plans, SOE fell victim to a well-run German double agent operation. See for example: M. R. D. Foot, *SOE: The Special Operations Executive 1940–46* (London: London Bridge, 1984), 130–4, and Olson, *Last Hope Island*, esp. chapters 11, 15, and 16. The Dutch debacle and its effects ripple through the classic memoir by SOE codemaker Leo Marks, *Between Silk and Cyanide: A Codemaker's War, 1941–1945* (London: Free Press, 2000). On page 148, he recounts how SOE oversold its capabilities to OSS.
42. Dodds-Parker, *Setting Europe Ablaze*, 104. Though the date of the visit is not clear, it may have been in June 1942.
43. London to Gibraltar, Jul. 19, 1942, HS 8-13, TNA.
44. Compare for example Dodds-Parker, *Setting Europe Ablaze*, 86–87, with Lovell, *Of Spies and Stratagems*, 23, 29–66. The two services developed almost identical gadgets that ran the gamut from the murderous to the ridiculous.
45. CD [Hambro] Minute on DCDO to CD, Jan 12, 1943, HS 8–37, TNA.
46. Draft of letter from Gubbins to Cadogan, Sept. 1943, HS 8–7, TNA. See also chapter 6 prior for the 1942 agreement that divided the world geographically, basically granting OSS rights to parts of the world that the British were less interested in.
47. Occupied by Germany in 1941, Yugoslavia was the scene of a three-way guerrilla war among Josep Tito's communists, Dragoljub Mihailović's noncommunist irregulars, and the Nazis. The British and the Americans struggled over whether they should support one or both of the anti-Nazi groups. Smith, *OSS*, 129–62, is an admirable overview. See also Franklin Lindsay, *Beacons in the Night: With the OSS and Tito's Partisans in Wartime Yugoslavia* (Stanford, CA: Stanford University Press, 1993).
48. The diplomat Maclean was a temporary brigadier, favored by Churchill and generally considered to be the senior British officer in Yugoslavia. Waller, *Wild Bill Donovan*, 186–7, describes the personal conflict. Donovan's attack on Maclean included the threat that, if pressed, the Americans would treat the British like the enemy. Even after seventy-five years, the hostility is almost palpable in files such as Cairo to London, Nov. 20, 1943, HS 8–7, TNA, which begins "General Donovan has demanded . . ." Another document in the same file attests to SOE's sense that Donovan could not be trusted to honor Anglo-American agreements. CD to Cadogan draft, Sept. 1943, HS 8–7, TNA.

49. Brown, ed., *Secret War Report,* 334–62, is an overview of "the principal OSS effort of the European war." R&A supplied "necessary basic information" about France and Germany. OSS, "European Theater Report Digest, November, 1943" in OSS Monthly Activity Reports, European Theater, 1943, ERR, accessed Jan. 19, 2021, CIA, https://www.cia.gov/readingroom/docs/CIA-RDP13X00001R000100140002-7.pdf. R&A also assessed the various European resistance movements with varying degrees of success. See also MacPherson, *American Intelligence in War-Time London,* chapters 4 and 5; Christof Mauch, *The Shadow War against Hitler: The Covert Operations of America's Wartime Secret Intelligence Service* (New York: Columbia University Press, 2003), esp. 63–106. MacPherson sees R&A trying and failing to be relevant; Mauch sees R&A making a difference in its analyses of strategic bombing. Mauch, *Shadow War,* 137 et seq., describes MO black propaganda but seems reluctant to draw conclusions about its value. Lankford, ed., *OSS against the Reich,* 73, is a 1944 discussion of MO. Cutler, *Counterspy,* 16 discusses how X-2 prepared for D-day, noting how the branch extracted information from detailed—and decrypted—German messages.
50. In January 1943, SOE and SO London agreed on "operational arrangements." Wyman W. Irwin, "A Special Force: Origin and Development of the Jedburgh Project in Support of Operation Overlord" (MA diss., U.S. Army Command and General Staff College, Leavenworth, KS, 1991); MacPherson, *American Intelligence in War-Time London,* 70–80. SO London opted to organize along the same lines as SOE.
51. Quoted in Abrutat, *Vanguard,* 83–84. For Spartan, see Irwin, "A Special Force," 42, 54–55.
52. In the spring of 1944, SO/London was still very much the junior partner in the SOE/SO relationship due to the fact that it had deployed a comparatively small number of agents to the continent—less than twenty at that point. The Jedburghs would change that situation. They were originally intended to supplement larger SOE/SO operational groups with an almost identical mission. See Troy J. Sacquety, "OSS, Office of Strategic Services, A Primer on the Special Operations Branches and Detachments of the Office of Strategic Services" in Sacquety, ed. *OSS, Office of Strategic Services, Primer & Manuals* (Ft. Bragg, NC: U.S. Army Special Operations Command, 2013).
53. Irwin, "A Special Force," 67–68. Bruce wrote that the choice was work with SOE on Jedburgh operations or forgo cooperation with SOE. At this stage of the war. OSS would have found it difficult to go it alone.
54. Hall, *You're Stepping on My Cloak and Dagger,* 62. Other firsthand accounts include: Ib Melchior, *Case by Case: A U.S. Army Counterintelligence Agent in World War II* (Novato, CA: Presidio, 1993) and William J. Morgan, *The O.S.S. and I* (New York: Curtis Books, 1957).
55. Hall, *You're Stepping on My Cloak and Dagger,* 32. Hall was the cousin of Virginia Hall, a well-known OSS operative.

56. Joseph F. Haskell to Supreme Allied Commander [Eisenhower], Jan. 11, 1944, HS 6, TNA. This file contains foundational material on the OSS-SOE relationship.
57. The Overlord operations plan called first for Allied air forces to bomb bridges and rail yards, then for the Resistance to sabotage the railroads, phone lines, and power grid before ambushing German troops moving toward the coast. Thereafter, the Resistance would shift to "a more general campaign of guerrilla warfare." Will Irwin, *The Jedburghs: The Secret History of the Allied Special Forces, France 1944* (New York: Public Affairs, 2005), 68.
58. Max Hastings, *Das Reich: The March of the 2nd SS Panzer Division through France, June 1944* (Minneapolis, MN: Zenith Press, 2013), 131–2.
59. MacPherson, *American Intelligence in War-Time London*, 59–60.
60. See for examples OSS, "European Theater Report Digest, October 1943" and "European Report Digest, December 1943" in OSS Monthly Activity Reports, European Theater, 1943, ERR, accessed Jan. 19, 2021, CIA, https://www.cia.gov/readingroom/docs/CIA-RDP13X00001R000100140002-7.pdf. The agreement was reached on October 30 at a meeting attended by Bruce and a number of senior officers.
61. Quoted in OSS, "European Theater of Operations Report January 1944," OSS Monthly Activity Reports—January 1944, ERR, accessed Jan. 19. 2021, CIA, https://www.cia.gov/readingroom/docs/CIA-RDP13X00001R000100140005-4.pdf.
62. Frances P. Miller, *Man from the Valley, Memoirs of a 20th Century Virginian* (Chapel Hill, NC: University of North Carolina Press, 1971), 112–3. David Abrutat, *Vanguard*, 82–83, is an up-to-date overview of Sussex.
63. OSS, "Report on OSS Activities for the Month of May 1944," OSS Monthly Activity Reports—May 1944, ERR, accessed Jan. 19, 2021, CIA, https://www.cia.gov/readingroom/docs/CIA-RDP13X00001R000100140008-1.pdf.
64. OSS, "Report on OSS Activities for the Month of June 1944" in OSS Monthly Activity Reports May [*sic*], 1944, ERR, accessed Jan. 19, 2021, CIA, https://www.cia.gov/readingroom/docs/CIA-RDP13X00001R000100140008-1.pdf; Jeffery, *MI6*, 538. In the words of one SHAEF officer, spotting Panzer Lehr was enough to justify the effort that had gone into the program. Miller, *Man from the Valley*, 113.
65. Abrutat, *Vanguard*, is an overview of all intelligence and reconnaissance before D-day. See also F. H. Hinsley, *British Intelligence in the Second World War*, abridged ed. (Cambridge: Cambridge University Press, 1993), 436–7. Lankford, ed., *OSS against the Reich*, 72, cites Bruce on information from the French Resistance. See also Douglas J. Porch, *The French Secret Services, From the Dreyfus Affair to the Gulf War* (New York: Farrar, Straus & Giroux, 1995), 245–9.
66. American and British codebreakers both contributed to identifying most of the German formations in France. See David Kenyon, *Bletchley Park and*

D-Day (New Haven: Yale University Press, 2019), esp. 148. "History of the Special Branch," SRH 035, 59, proudly reported "the publication, about ten days before . . . D-Day of the complete German order of battle" in Normandy.

67. Quoted in Boyd, *Hitler's Japanese Confidant*, 105–6.
68. Ferris, *Behind the Enigma*, 259–60. See also Holt, *Deceivers*, 565–7.
69. Eisenhower made a special point of extending his personal thanks to the army codebreakers who, in Rowlett's words, "had produced information on . . . German fortifications . . . vital to the success of the invasion. . . ." Boyd, *Hitler's Japanese Confidant*, 179.
70. Miller, *Man from the Valley*, 112. Miller was writing from memory: "Though I took no notes, I will never forget his words." His memory tracks with another meeting that Donovan called after D-day, as well as Bruce's diary.
71. Miller, *Man from the Valley*, 115; Lankford, ed., *OSS against the Reich*, 106.
72. Waller, *Wild Bill Donovan*, 239. This was consistent with the elaborate precautions taken to protect Overlord, including travel and communications bans. Hinsley, *British Intelligence*, 442.
73. This was the apparent result of an appeal by Donovan to Forrestal: WJD to Forrestal, May 9, 1944, requesting him to write to the navy commander in chief in Europe and "ask if he could take care of David Bruce and myself." Box 2, Cave Brown Papers. Forrestal's cable is quoted in Waller, *Wild Bill Donovan*, 239. Donovan's friend and protector Secretary Knox had died in April 1944 of a heart attack; Undersecretary Forrestal replaced him.
74. Lankford, ed., *OSS against the Reich*, 47. This is Bruce's contemporary diary, an excellent primary source for this period.
75. Lankford, ed., *OSS against the Reich*, 63.
76. Lankford, ed., *OSS against the Reich*, 56.
77. The deception was known by the overall codename Bodyguard, and had two main parts, Fortitude North, the notion that the British and Americans planned to invade Norway, and Fortitude South, aimed at convincing the Germans that Normandy was a feint. To support Bodyguard, military intelligence officers created the fictitious First US Army Group, supposedly under the command of then lieutenant general George S. Patton, and MI5 double agents fed misleading reports to the Abwehr. The deception was bolstered by years of intricate work, mostly by British officers who invented and sustained mythical units that made the Anglo-American forces seem about a third stronger than they actually were. Bodyguard has been hailed as "the most successful strategic deception of all time." Holt, *Deceivers*, 590. Ferris, *Behind the Enigma*, 258, 261, generally concurs, noting the close cooperation between the deceivers and Bletchley Park. Other works qualify those judgments somewhat. Mary Kathryn Barbier, *D-Day Deception: Operation Fortitude and the Normandy Invasion* (Mechanicsburg, PA: Stackpole Books, 2007) analyzes the questions of success and effect. (Were the Germans deceived? If so, did it make a difference?) Peter Caddick-Adams, *Sand and Steel: The D-Day Invasion and the Liberation of France*

(Oxford: Oxford University Press, 2019), shows how difficult it is to isolate the effect of intelligence on this battle. See also Kenyon, *Bletchley Park and D-Day,* esp. 157–179 and Nigel West, *Codeword Overlord: Axis Espionage and the D-Day Landings* (Stroud, UK: History Press, 2019). By inflating the strength of the Anglo-American armies, the least the Allied deceivers did was to enable the German belief that Eisenhower had enough forces for a second landing.

78. Lankford, ed., *OSS against the Reich,* 65.
79. This oft-told story of Donovan's D+1 adventure appears in a letter that Bruce wrote in 1958 and in an after-dinner speech by Bruce in 1971. Though it may have been embellished, it is not inconsistent with his diary. The letter is described in Lankford, ed., *OSS against the Reich,* 220; the speech is quoted at length in Smith, *OSS,* 184–5.
80. Lankford, ed., *OSS against the Reich,* 65–66.
81. Lankford, ed., *OSS against the Reich,* 220. Recent works have suggested that Bradley was not the good-natured soldier's general portrayed in early histories. See for example Murray and Millett, *War to be Won,* 418.
82. According to their monthly situation report, "the Allied forces on the beachhead were supported by [OSS] agents . . . who provided intelligence on [the] enemy order of battle and contributed to the great eruption of sabotage and open fighting which impeded the German . . . counter blow." OSS, "Report on OSS Activities for the Month of June 1944."
83. See Lankford, ed., *OSS against the Reich,* 70–71, for Bruce's detailed diary entry.
84. Quoted in Lankford, *Last American Aristocrat,* 159. That officer was Col. Benjamin A. Dickson, the First Army G-2 officer.
85. Stimson Diary, entry for June 15, 1944; WJD to FDR, June 14, 1944, describes his D-day adventure, Box 2, Cave Brown Papers. See also Waller, *Wild Bill Donovan,* 248.
86. US War Department, Strategic Services Unit, *War Report of the OSS* (New York: Walker, 1976), vol. 2, ix. Bruce was aware of what he called "the strict orders against D[onovan] being anywhere near Normandy." Quoted in Lankford, ed., *OSS against the Reich,* 220. Donovan's attitude was "Do as I say, not as I do"; in September 1944, he was outraged by the conduct of OSS officers who were captured by the Germans while on a joyride near the German-Luxembourg border, and later moved to discipline one of the survivors, claiming that he was "absent without leave and had no orders whatever permitting him to make [that] trip." Waller, *Wild Bill Donovan,* 293–5; Peter Finn, *A Guest of the Reich: The Story of American Heiress Gertrude Legendre's Dramatic Captivity and Escape from Nazi Germany* (New York: Pantheon Books, 2019), esp. 182. One of the OSS officers was an X-2 officer, indoctrinated into the Ultra secret. An example of Churchill's attitude comes from the case of Air Cdre. Ronald Ivelaw-Chapman, RAF, who flew a mission over France in May 1944 even though he was

fully briefed on the plans for D-day. When he was shot down, Churchill immediately ordered that no effort be spared to rescue him, and that if he could not be rescued, he must be killed. David Stafford, *Churchill and Secret Service* (New York: Overlook Press, 1997), 290; M. R. D. Foot and J. M. Langley, *MI9: Escape and Evasion. 1939–1945* (Boston and Toronto: Little Brown, 1980), 211.

18: A Dream Come True

1. William J. Casey, *The Secret War against Hitler* (Washington, DC: Regnery, 1988), 186, differentiates between "tactical" intelligence, what an army unit obtained locally from short-range patrols, prisoners of war, or aerial reconnaissance, and "strategic" intelligence from deep penetrations and agent operations.
2. Distinguished historian Max Hastings, in *Secret War*, 488–90, concludes that "the most ruthless and cynical operations run by all intelligence services [in World War II] were those involving short-range spies—locally recruited civilians dispatched to report what they could see behind the enemy's front line." His evidence is largely anecdotal and focuses on German and Soviet practices. His American examples are unproductive operations in Belgium in the desperate days after the Battle of the Bulge and in Germany in the spring of 1945. As Bruce was about to learn, the army had a more positive view of OSS operations in France in 1944.
3. Lankford, ed., *OSS against the Reich*, 89.
4. OSS London, "ETO Theater Report," Jul. 1, 1944, quoted in MacPherson, *American Intelligence in War-Time London*, 85–86. The report concluded that "OSS is therefore in a position where it must give full support to its field units, even though they may have been considered as subsidiary and incidental to the long-range activities of the organization." The OSS teams at the division level were a mix of SO and SI officers.
5. Reports of the numbers of such leads vary. A conservative estimate is that X-2 started with one hundred names. MacPherson, *American Intelligence in War-Time London*, 195. X-2 was also beginning to double some of these agents back against the Germans, a mission which it learned to perform well. Reynolds, "The 'Scholastic' Marine Who Won a Secret War."
6. OSS, "Summary of OSS Activities during August 1944," OSS Activities Aug. 1944, ERR, accessed Feb. 3, 2021, CIA, https://www.cia.gov/readingroom/document/cia-rdp13x00001r000100140010-8.
7. Lankford, ed., *OSS against the Reich*, 89, 97.
8. Lankford, ed., *OSS against the Reich*, 122–3.
9. Smith, *OSS*, 177–8; Brown, *Last Hero*, 324–6. The agents in Penny Farthing were not unlike line-crossers and were for the most part volunteers. For a remarkable inside look, see Hélène Deschamps with Karyn Monget, *Spyglass: An Autobiography* (New York: Henry Holt, 1995), 238–81.

10. OSS, "Summary of OSS Activities during August 1944"; Brown, *Last Hero*, 586–7. According to Brown, the other services that contributed intelligence were the French and British—presumably resistance sources reporting to military intelligence.
11. Quoted in Porch, *French Secret Services*, 558. As noted above, the US military relied mostly on OSS to handle spies who did not speak English.
12. Waller, *Wild Bill Donovan*, 256–8, 264–7; Smith, *OSS*, 103. Donovan claimed to be personally fond of the Pope, who is generally not regarded as a hail-fellow-well-met. On May 10, the Pope granted an audience to the senior SS man in Italy, Karl Wolff, who claimed that he was committed to ending the war. Not unlike the Pope, Weizsäcker turned out to be another fence-sitter; he did not live up to his OSS codename.
13. Quoted in Lankford, *Last American Aristocrat*, 155. Much the same story is told from Hemingway's point of view in Reynolds, *Writer, Sailor, Soldier, Spy*, 165–71.
14. Lankford, ed., *OSS against the Reich*, 166. Settling scores would be a continuing part of France's liberation from the Nazis.
15. Lankford, ed., *OSS against the Reich*, 163–4. Initially inclined to cut off and surround the French capital, Eisenhower hesitated to attack the city itself.
16. Lankford, ed., *OSS against the Reich*, 164–5. There is no record of more conventional clandestine reporting from within Paris, like a Sussex team or a resistance cell with a transmitter.
17. Lankford, ed., *OSS against the Reich*, 170.
18. Reynolds, "The 'Scholastic' Marine Who Won a Secret War," 26, quoting Frank Holcomb.
19. Ronald C. Rosbottom, *When Paris Went Dark: The City of Light under German Occupation, 1940–1944* (New York: Back Bay Books, 2014), 310–47, describes the liberation and the controversies surrounding it, including whether Choltitz deserves credit for saving the city from destruction.
20. Lankford, ed., *OSS against the Reich*, 181.
21. Lankford, ed., *OSS against the Reich*, 185, 189. Casey in *Secret War*, 174, had much the same impression.
22. OSS, "Summary of OSS Activities during August 1944."
23. T. J. Betts to J. Haskell, Jul. 27, 1944, in OSS-European Theater of War/Operation Overlord, ERR, accessed Feb. 10, 2021, CIA, https://www.cia.gov/readingroom/docs/CIA-RDP13X00001R000100220002-8.pdf.
24. See G. L. King to 2671st Special Reconnaissance Battalion, Jun. 2, 1945 in OSS-European Theater of War/Operation Overlord.
25. Eisenhower to Gubbins and Bruce, May 31, 1945, reproduced in Brown, ed., *Secret War Report*, 461–2.
26. Hastings, *Das Reich*, 220. Hastings's book is a case study of the extent to which the Resistance, supported by the SOE and the OSS, impeded German movement. He concludes that other factors held equal or greater sway.
27. Casey, *Secret War*, 172.

28. OSS, "Summary of OSS Activities during September 1944," OSS Activities Sept. 1944, ERR, accessed Feb. 18, 2021, CIA, https://www.cia.gov/reading room/document/cia-rdp13x00001r000100140011-7. This report cites the address as 79 Champs-Élysées.
29. Quoted in Central Intelligence Agency, Center for the Study of Intelligence, *OSS Exhibition Catalogue* (Washington, DC: US Government Printing Office, 2015), 15.
30. Robert H. Alcorn, *No Bugles for Spies* (New York: Popular Library, 1964), 95.
31. While she was working for the Department of State before the war, Hall's supervisor noted approvingly that she had little interest in fashion or cosmetics, but disapprovingly that she did not care much for office work or typing either. Cara Moore Lebonick, "The Secrets of the Office of Strategic Services Personnel Records: Spotlight on Virginia Hall," Apr. 16, 2020, accessed Feb. 17, 2021, *The Text Message* (blog), National Archives, https://text-message.blogs.archives.gov/2020/04/16/the-secrets-of-the-office-of-strategic-services-personnel-records/. According to Donovan, "Only a small percentage of the [OSS] women ever went overseas, and a still smaller percentage was assigned to actual operations behind enemy lines." Quoted in Elizabeth McIntosh, *Sisterhood of Spies: The Women of the OSS* (Annapolis, MD: Naval Institute Press, 2009), 11. McIntosh herself was on active duty in China in morale operations. Ann Todd, *Operation Black Mail: One Woman's Covert War Against the Imperial Japanese Army* (Annapolis, MD: Naval Institute Press, 2017). A number of women were hired as clerks and then distinguished themselves in the foreign field, such as the X-2 operative Betty Lussier. Betty Lussier, *Intrepid Woman: Betty Lussier's Secret War, 1942–1945* (Annapolis, MD: Naval Institute Press, 2010).
32. Virginia Hall, "Activity Report," Sept. 30, 1944, accessed Feb. 17, 2021, National Archives Catalog, https://catalog.archives.gov/id/595661. Hall notes that she personally did not take part in any of the teams' many sabotage operations.
33. OSS, "Summary of OSS Activities during August 1944."
34. Casey, *Secret War*, 170.
35. OSS, "Summary of OSS Activities during September 1944."
36. See for example: Lankford, ed., *OSS against the Reich*, 190. On October 29, Bruce was working on a list of possible replacements for ETO staff, starting with himself.
37. Casey, *Secret War*, 177.
38. Quoted in Joseph E. Persico, *Casey: The Lives and Secrets of William J. Casey: From the OSS to the CIA* (New York: Penguin, 1991), 41.
39. MacPherson, *American Intelligence in War-Time London*, is a detailed analysis of how OSS London functioned.
40. Persico, *Casey*, 58. Casey, *Secret War*, 22–23, is more upbeat about his first day at work in London, and places his arrival in late October 1943, as does

his OSS personnel file. "William J. Casey," OSS Personnel Files, RG 226, NARA II, College Park, MD.

41. Quoted in Persico, *Casey*, 58.
42. Quoted in Persico, *Casey*, 57, 61.
43. Donovan, "Future OSS Operations in Central Europe," Sept. 2, 1944, quoted in MacPherson, *American Intelligence*, 161, and Waller, *Wild Bill Donovan*, 269.
44. Waller, *Wild Bill Donovan*, 268–9. The future Supreme Court justice's job was to develop ways to use the European labor movement against Germany. An early indication of the shift in priorities appeared in Bruce's diary on August 30: "We are now to make the penetration of Germany our prime objective." Lankford, ed., *OSS against the Reich*, 182. MacPherson, *American Intelligence in War-Time London*, 166, has Casey drafting a memo on the subject on September 11.
45. Casey to Donovan, Oct. 12, 1944, attaching "OSS Program against Germany," reproduced in Casey, *Secret War*, 251–9.
46. Chief, SI, ETO [Casey] to CO, OSS, ETO [Forgan], "Final Report on SI Operations into Germany," Jul. 24, 1945, reproduced in Casey, *Secret War*, 279–96.
47. Brown, ed., *Secret War Report*, 515–6, reviews the situation and notes that the brunt of the German offensive hit the First Army, whose G-2, Colonel Dickson, had angrily refused to allow an OSS Detachment on his turf. A careful discussion of the information and analysis at First Army in December 1944 is in David W. Hogan Jr., *A Command Post at War: First Army Headquarters in Europe, 1943–1945* (Washington, DC: US Army, Center of Military History, 2000), 205–11.
48. See for example Hinsley, *British Intelligence*, 563–6. Hinsley discusses the evidence and judges that "the Sigint [signals intelligence] was not conclusive." For a different point of view, see Murray and Millett, *War to Be Won*, 464, citing the indicators that should have put SHAEF on alert.
49. Casey, *Secret War*, 184.
50. Persico, *Casey*, 69. See also Casey, *Secret War*, 184–5. Waller, *Wild Bill Donovan*, 271, is an overview from Donovan's perspective. According to Bruce's diary, he was in France during the Battle of the Bulge. Lankford, ed., *OSS against the Reich*, 201.
51. Quoted in Persico, *Casey*, 66. Lord would go on to obtain a law degree from Yale but would spend most of his life writing bestselling history books, including *A Night to Remember* (1955), about the *Titanic*.
52. Casey, *Secret War*, 185.
53. Persico, *Casey*, 70; Casey, *Secret War*, 185. See also Foot, *SOE*, 206–7. The Soviets had run a productive network known as the Red Orchestra inside Germany earlier in the war. By 1944, they apparently had few, if any, well-placed spies inside the Reich. See for example Christopher Andrew and Oleg Gordievsky, *KGB: The Inside Story* (New York: Harper Perennial, 1991), 276–7.

54. Casey, *Secret War*, 188, 194–5.
55. Casey, *Secret War*, 186–93; Brown, ed., *Secret War Report*, 541–5.
56. Persico, *Casey*, 75. Brown, ed., *Secret War Report*, 544–5, describes the luxurious accommodations.
57. In charge of SI for ETO, Casey oversaw detachments that were now in France but had originated in the Mediterranean Theater of Operations (MEDTO), like Henry Hyde's. Casey gave Hyde a good deal of latitude in running his own operations. From its base in Bari, Italy, OSS MEDTO ran operations from Italy into Austria. See Tofte to Casey, "S.I. Operations against Germany & Austria," Dec. 26, 1944, reproduced in Casey, *Secret War*, 259–62.
58. Casey, *Secret War*, 198. The final total for all OSS penetrations into Germany in 1945 was on the order of one hundred missions. [Casey], "Final Report."
59. [Casey], "Final Report."
60. "Remarks of William J. Casey . . . before the OSS/Donovan Symposium," Sept. 19, 1986, ERR, accessed Feb. 15, 2021, CIA, https://www.cia.gov/readingroom/docs/CIA-RDP88G01116R000500550009-8.pdf.
61. [Casey], "Final Report."
62. Douglas Waller, *Disciples: The World War II Missions of the CIA Directors Who Fought for Wild Bill Donovan* (New York: Simon and Schuster, 2015), 379, is an account based on Casey's reports and letters home in 1945. Casey, *Secret War*, 211–3 is the same story, written years later.
63. [Casey], "Final Report."
64. Persico, *Casey*, 68, 86.

19: Allen Dulles's Nearly Private War

1. Lankford, ed., *OSS against the Reich*, 195.
2. On Dulles's background see James Srodes, *Allen Dulles: Master of Spies* (Washington, DC: Regnery, 1999) and Peter Grose, *Gentleman Spy: The Life of Allen Dulles* (Boston: Houghton Mifflin, 1994). See also Waller, *Disciples* and Stephen Kinzer, "When a C.I.A. Director Had Scores of Affairs," *New York Times*, Nov. 10, 2012. Allen's sister Eleanor famously commented that he had had at least one hundred extramarital affairs in his lifetime.
3. Srodes, *Dulles*, 230.
4. Allen W. Dulles, *The Secret Surrender* (Guilford, CT: Lyons Press, 2006), 14–15.
5. Mayer had been a COI officer before the public information function had been hived off.
6. Bancroft, *Autobiography*, 129. Chapter 17 of Bancroft's memoir is the source of the information in this paragraph. While she writing many years after the fact, she was able to obtain access to some of her reports as well as letters and journals. She includes the text of at least one letter from Dulles in the book. Bancroft's extensive 1979 and 1980 oral history is generally

consistent with her autobiography. [Mary Bancroft], "The Reminiscences of Mary Bancroft," Oral History Research Office, Columbia University, 2001.

7. Srodes, *Dulles*, 229, quoting another tenant.
8. Dulles appears to have had an inkling that his relationship with Bancroft was inappropriate, even by the standards of the day. He commented that he planned to keep her salary low for the sake of appearances in case of an investigation after the war. Bancroft, *Autobiography*, 137.
9. Bancroft, *Autobiography*, 148.
10. Prussia had its own police forces, including both regular and political police, which were absorbed into the Third Reich's security apparatus.
11. Bancroft, *Autobiography*, 164; Allen W. Dulles, *Germany's Underground: The Anti-Nazi Resistance* (New York: Da Capo Press, 2000), 129. A carefully sourced biographical sketch is in Prof. Peter Hoffmann's introduction to the 1998 edition of Hans B. Gisevius, *To the Bitter End: An Insider's Account of the Plot to Kill Hitler, 1933–1944* (New York: Da Capo Press, 1998), vii–xi.
12. Jeffery, *MI6*, 380–2. The agent was Halina Szymańska, who cultivated Gisevius as an asset, obtaining information that was "very sound." The excerpts Jeffery quotes seem too general to have been particularly useful, especially when compared to Ultra or Magic. For a different but unsourced view, see Nigel West, *MI6*, 116–7, which has Admiral Canaris sending Gisevius to Bern to transmit "top-grade political intelligence" to London through Szymańska. Canaris was director of the Abwehr.
13. Gisevius, *To the Bitter End*, 481, meaning a source to exploit.
14. Dulles stated that Gisevius neither requested nor received any payment or preferential treatment from the US; he was not an OSS agent. Dulles, *Germany's Underground*, 128.
15. There was a loose gentleman's agreement between OSS and MI6 not to poach each other's agents but to share information of mutual interest.
16. Dulles's tradecraft was at best adequate. He liked to make fun of Donovan for not keeping track of secret documents, but Dulles's own habit of holding secret meetings at home with multiple sources was an invitation to disaster, one that, thankfully, was apparently never accepted. Waller, *Wild Bill Donovan*, 118–9. Dulles used a variety of means to communicate with Washington that, by his own admission, included less than secure telephone links. The Swiss provided a scrambler, which meant that they, and probably the Germans, could intercept his calls. Dulles, *Secret Surrender*, 16–17, is his general description of communications challenges. Dulles, *Germany's Underground*, 130–1, elaborates on the compromise reported by Gisevius, and notes that no one was betrayed by a message originating in Bern.
17. Bancroft, *Autobiography*, 164. The manuscript would eventually be published as Gisevius, *To the Bitter End*, in German and English as well as other European languages.
18. Bern to Washington (2–133), Jan. 27, 1944, in Neal H. Petersen, ed., *From Hitler's Doorstep: The Wartime Intelligence Reports of Allen Dulles,*

1942–1945 (University Park, PA: Pennsylvania State University Press, 1996), 205–6.

19. See for example Nicholas Stargardt, *The German War: A Nation under Arms, 1939–1945* (New York: Basic Books, 2017), a compelling exploration of German behavior during the war. Peter Hoffmann, *German Resistance to Hitler* (Cambridge, MA: Harvard University Press, 1988), is a classic overview of the Resistance.
20. Bern to Washington (2–134), Jan. 27, 1944, in Petersen, ed., *Hitler's Doorstep,* 206–7.
21. See for example Mauch, *Shadow War,* 50, citing messages from Washington to Bern.
22. The policy was first announced in early 1943. Originally published in 1974, Anne Armstrong, *Unconditional Surrender: The Impact of Casablanca Policy on World War II* (Chicago: Barakaldo eBooks, 2020), is a thorough if somewhat dated discussion. Chapter 1 discusses Roosevelt's views.
23. Bern to Washington, Apr. 7, 1944, in Petersen, ed., *Hitler's Doorstep,* 264–5. A summary of Dulles's early reporting is Magruder to Berle, May 17, 1944, in US Department of State, *Foreign Relations of the United States: Diplomatic Papers,* 1944, General, vol. 1, eds. E. Ralph Perkins and S. Everett Gleason (Washington, DC: US Government Printing Office, 1966), doc. 280 (hereinafter cited as *FRUS: Diplomatic Papers*).
24. Bern to Washington (4–16), Jul. 13, 1945, and Bern to Washington, Jul. 15, 1944, both in Petersen, ed., *Hitler's Doorstep,* 331–5. The policy of unconditional surrender arguably worked to Hitler's advantage, motivating some Germans to fight on. But it was only one factor among many. Hitler's hold on Germany would remain strong until the day of his death. See Stargardt, *German War,* 482–544.
25. Dulles, *Germany's Underground,* 131–2; Gisevius, *To the Bitter End,* 483.
26. Leahy, *I Was There,* 249.
27. "The Breakers connection was established in order to secure intelligence[,] and for more than a year this contact has proven itself to be a real value to us." Bern to London, Jul. 26, 1944, in Petersen, ed., *Hitler's Doorstep,* 349–50.
28. See for example Magruder to Berle, May 17, 1944, and Donovan to SecState, Jul. 18, 1944, in *FRUS: Diplomatic Papers,* 1944, General, vol. 1, docs. 280 and 301.
29. See for example Bern to Washington, Jul. 22, 1944, and Jul. 24, 1944, in Petersen, ed., *Hitler's Doorstep,* 341–2, 346–8.
30. See for example Petersen, ed., *Hitler's Doorstep,* 17.
31. Waller, *Wild Bill Donovan,* 317. OSS reports became dead letters, especially when the president was traveling and only a limited number of messages could be sent forward. Boyd, *Hitler's Japanese Confidant,* 103.
32. Shepardson to Bern, Apr. 28, 1943 quoted in Srodes, *Dulles,* 268–9. Many in G-2 would have scoffed at anything that came from OSS; Shepardson added that OSS still had complete confidence in Dulles.

33. Shepardson to Dulles, Jan. 25, 1944, quoted in Grose, *Gentleman Spy,* 189.
34. See for example West, *MI6,* 224–5. In the Venlo incident the Germans dangled supposed anti-Nazi conspirators.
35. The best account of this case is one of the earliest: Anthony Quibble, "Alias George Wood," *Studies in Intelligence* 10, no. 1 (Winter 1966). See also Lucas Delattre, *Betraying Hitler: The Story of Fritz Kolbe, the Most Important Spy of the Second World War* (London: Atlantic Books, 2005) and Greg Bradsher, "A Time to Act: The Beginning of the Fritz Kolbe Story, 1900–1943," *Prologue* 34, no. 1 (Spring 2002).
36. Quibble, "Alias George Wood," 70.
37. Bern to London, Aug. 21, 1943, in Petersen, ed., *Hitler's Doorstep,* 106–7.
38. Dansey is described in Lawrence D. Stokes, "Secret Intelligence and Anti-Nazi Resistance: The Mysterious Exile of Gottfried Reinhold Treviranus," *International History Review* 28, no. 1 (2006): 59–61. See also Jeffery, *MI6,* 58.
39. This allegation originated with Gisevius before he started working closely with Dulles. Jeffery, *MI6,* 511.
40. Philby, *My Silent War,* 92–96.
41. This was both a means for checking earlier work—the original text confirmed that the Ultra methods were working—and for breaking into messages that had not been decrypted.
42. Quibble, "Alias George Wood," 75.
43. Norman Pearson to David Bruce, Nov. 23, 1943, quoted in Delattre, *Betraying Hitler,* 128–30.
44. Quibble, "Alias George Wood," 76.
45. McCormack's attitude toward other American agencies could be blistering. After listing OSS among them, he criticized "the penchant of certain agencies to bring forth a stream of classified reports . . . that are not much more than irresponsible guesswork." Alfred McCormack, "Origins, Functions, and Problems of the Special Branch, MIS," Apr. 15, 1943, 25, SRH 116, NCML.
46. The British would eventually unmask the ambassador's butler, who was stealing documents from his safe at home and selling them to the Germans. Jeffery, *MI6,* 503–4. The Germans did not entirely trust the source, whom they codenamed "Cicero." They paid him well, giving him the enormous sum of 200,000 pounds sterling—but in counterfeit notes. He was arrested when he tried to use them after the war.
47. Bern to Washington, Dec. 29, 1943, in Petersen, ed., *Hitler's Doorstep,* 185–6.
48. Delattre, *Betraying Hitler,* 166–7. OSS chose to highlight the atmospherics on life in Germany for Roosevelt—Kolbe's impressions—rather than conclusions based on the documents he passed.
49. It is unclear whether this was a formal review or a cable-by-cable exercise, or both. Brown, *Last Hero,* 280, reports that McCormack worked as part of

a panel that included General Strong of G-2 and James Murphy, the head of OSS/X-2.

50. Quoted in Delattre, *Betraying Hitler*, 170.
51. Quibble, "Alias George Wood," 82, 86. McCormack successfully took control of the dissemination of Wood's material with exceptions for dissemination to the White House and the Secretary of State.
52. Bern to Washington, Feb. 16, 1945, in Petersen, ed., *Hitler's Doorstep*, 449.
53. Bancroft, *Autobiography*, 134. Dulles would remain in touch with both Gisevius and Kolbe after the war, helping them to the extent possible.
54. Donovan to Roosevelt, Dec. 1, 1944, quoted in Delattre, *Betraying Hitler*, 195–6.
55. Roosevelt to Donovan, n.d., quoted in Delattre, *Betraying Hitler*, 196.
56. Dulles, *Secret Surrender*, 58.
57. Dulles, *Secret Surrender*, 62.
58. This was the Vessel case, supposedly vetted by the OSS Reporting Board before Donovan's office forwarded reports to Roosevelt in January and February 1945. Brown, *Last Hero*, 683–703; Waller, *Wild Bill Donovan*, 298–302.
59. Dulles, *Secret Surrender*, 64.
60. Dulles, *Secret Surrender*, 73.
61. Bern to Washington, Mar. 5, 1945, in Petersen, ed., *Hitler's Doorstep*, 462–4.
62. Dulles, *Secret Surrender*, 79.
63. Quoted in Allen W. Dulles and Gero von Gaevernitz, "The First German Surrender, The End of the Italian Campaign," May 22, 1945, 8 in Operation Sunrise, Box 8, Entry A 1190, RG 226, NARA II. See also Dulles, *Secret Surrender*, esp. 56, essentially a later, longer version of the 1945 report.
64. Dulles, *Secret Surrender*, 76.
65. Bern to Washington, London, Paris, and Caserta, Mar. 9, 1945, in Petersen, ed., *Hitler's Doorstep*, 467–9. Dulles wrote later that Wolff's morals mattered less than his ability to bring about a surrender. Dulles and Gaevernitz, "The First German Surrender," 8.
66. Bern to Washington, et al., Mar. 9, 1945.
67. Some of the evidence is circumstantial, from the way in which the incoming messages were edited and forwarded, over Donovan's signature, to the White House. See for example the OSS file "Germany Sunrise 16,107," esp. 24–37, accessed Mar. 18, 2021, https://www.cia.gov/readingroom/docs/CIA-RDP13X00001R000100460004-0.pdf. See also Casey, *Secret War*, 200–1.
68. Winston S. Churchill, *Triumph and Tragedy* (Boston: Houghton Mifflin, 1953), 442.
69. Leahy, *I Was There*, 334.
70. Bern to London, Apr. 3, 1945, in Petersen, ed., *Hitler's Doorstep*, 488–9.
71. The radio operator was Vaclav Hradecky, a Czech in his midtwenties who spoke perfect German. He had been a prisoner of the SS at Dachau, where

he was starved and beaten. He carried an OSS radio, transmission schedule, and one-time pad with him to Italy. Dulles, *Secret Surrender*, 120–1; Bern to Caserta, Apr. 13, 1945, in Petersen, ed., *Hitler's Doorstep*, 499.

72. Harriman and Abel, *Special Envoy*, 439–40.
73. Troy, *Donovan and the CIA*, 265. See also Waller, *Wild Bill Donovan*, 318–9.
74. Dulles, *Secret Surrender*, 122–3.
75. Casey, *Secret War*, 200–1.
76. Srodes, *Dulles*, 187, describes them as friendly acquaintances but not friends, and concludes that Dulles irritated Donovan.
77. Dulles, *Secret Surrender*, 123.
78. Bern to Caserta, Apr. 18, 1945, in Petersen, ed., *Hitler's Doorstep*, 501–2.
79. Dulles, *Secret Surrender*, 129.
80. Putzell [OSS executive officer] to Conway [Truman's secretary], Apr. 19, 1945, in "Germany Sunrise 16,107."
81. JCS to OSS, Apr. 20, 1945 in "Germany Sunrise 16,107"; Bern to Washington, Apr. 21, 1945, in Petersen, ed., *Hitler's Doorstep*, 508–9.
82. Bern to Washington, Apr. 21, 1945, in Petersen, ed., *Hitler's Doorstep*, 508–9.
83. Dulles, *Secret Surrender*, 139–40.
84. Dulles, *Secret Surrender*, 142.
85. Bern to Caserta and Washington, Apr. 24, 1945, in Petersen, ed., *Hitler's Doorstep*, 511–2.
86. Smith and Agarossi, *Operation Sunrise*, 144. See also Dulles, *Secret Surrender*, 163.
87. Bern to Washington, Apr. 27, 1945, in Petersen, ed., *Hitler's Doorstep*, 515–6.
88. Paris to Washington, May 6, 1945, in Petersen, ed., *Hitler's Doorstep*, 520–1.
89. Dulles, *Secret Surrender*, 209, 217–8.
90. Lemnitzer to Donovan, May 15, 1945, in "Germany Sunrise, 16,107."
91. Paris to Washington, May 6, 1945, in Petersen, ed., *Hitler's Doorstep*, 520–1.
92. Dulles, *Secret Surrender*, 218. The OSS War Report elaborated on this argument. Brown, ed., *Secret War Report*, 255.
93. See for examples: Murray and Millett, *A War to Be Won*, and Smith and Agarossi, *Operation Sunrise*, esp. 3–7.
94. Though Dulles made no commitments to Wolff, the SS general did benefit from Sunrise. He was not one of the accused but a witness at Nuremberg. In 1949 he was given a light sentence after denazification hearings in the British zone of occupation. In those proceedings, he benefited from affidavits by Dulles and others who described his role in Sunrise. In 1964, he was tried by a West German court for his complicity in the Holocaust and sent to prison. Smith and Agarossi, *Secret Surrender*, 188–91. A more recent article makes the dubious case that Dulles was part of a conspiracy to shield Wolff. Kerstin von Lingen, "Conspiracy of Silence: How the 'Old Boys' of American Intelligence Shielded SS General Karl Wolff from Prosecution," *Holocaust and Genocide Studies* 22, no. 1 (Spring 2008).
95. Dulles, *Secret Surrender*, 218–9.

20: When Doing "Swell Work" Wasn't Enough

1. Benson, *History of U.S. Communications Intelligence*, 90–91. See also Burns, *Quest for Cryptologic Centralization*, esp. 14–28.
2. Benson, *History of U.S. Communications Intelligence*, 134.
3. Strong would die on January 11, 1946, after an operation. "Maj. Gen. Strong, G-2 Ex-Head, Dies," *New York Times*, Jan. 12, 1946.
4. "Gen. Bissell Heads Army Intelligence," *New York Times*, Feb. 5, 1944.
5. McCormack, "War Experience," SRH 185, 33–36. Much the same description of the Bissell briefings would surface in a newspaper column, perhaps stemming from an intentional leak. Joseph and Stewart Alsop, "An American Secret Service," *Washington Post*, Jan. 13, 1946. Stewart Alsop served in OSS during the war.
6. Benson, *History of U.S. Communications Intelligence*, 143–8. The relationship between the codebreaking and intelligence staffs remained a contentious issue throughout the war. On July 1, 1943, after a number of name changes, what had been Friedman's SIS became known as the Signal Security Agency (SSA) but remained under the Signal Corps. In late 1944, the SSA came under the operational control of the G-2 but remained under the administrative control of the Corps. On September 15, 1945, this cumbersome arrangement ended as the director of Military Intelligence assumed full control of the SSA, which was renamed the Army Security Agency (ASA). This meant that that codebreakers and intelligence officers were finally in the same chain of command. G-2, "The Achievements of the Signal Security Agency in World War II," Feb. 20, 1946, 7–9, SRH 349, NCML.
7. McCormack, "War Experience," SRH 185, 47–53. As director of intelligence, McCormack continued to produce his Magic Summaries, but his influence was waning. By June 1945, Bissell had had enough of outsiders like McCormack and his lawyers; he decreed that henceforth he would only take in regular officers. Bissell, "Officer Replacement Policy," Jun. 13, 1945 in SRH 141, NCML.
8. Quoted in Benson, *History of U.S. Communication Intelligence*, 146.
9. Kahn, *Codebreakers*, 604, 607. Dewey claimed that at least twelve senators knew the whole story about the codes and Pearl Harbor. Carter Clarke, "Statement for Record of Participation of Brig. Gen. Carter W. Clarke, GSC, in the Transmittal of Letters from Gen. George C. Marshall to Gov. Thomas E. Dewey," [1944?], SRH 043, NCML. See also William S. White, "Marshall Says Dewey Kept Code Secret out of Politics," *New York Times*, Dec. 8, 1945. Marshall mentioned additional concerns about "Washington 'whispering'" after the Yamamoto shootdown.
10. Kahn, *Codebreakers*, 604.
11. The main source for this story is Clarke's "Statement for Record." It is buttressed by David Kahn, "Carter W. Clarke Interview," Dec. 6, 1963, David Kahn Archive, NCML Clarke did not hesitate to faithfully execute his

orders even though he would almost certainly have chosen Dewey over the president he called "the old con man." Clarke to Kahn, Feb. 3, 1987, David Kahn Archive.

12. Carter, "Statement for Record."
13. Marshall to Dewey, Sept. 27, 1944, reprinted in Kahn, *Codebreakers*, 605–7.
14. See Kahn, *Reader of Gentlemen's Mail*, 161–2. Yardley's first book, *The American Black Chamber*, revealed the secrets of his work against Japanese codes in the 1920s and sold so well that he wanted to write a sequel.
15. Quoted in Kahn, *Codebreakers*, 608.
16. This complicated story is told clearly in Waller, *Wild Bill Donovan*, 157–9.
17. See for example: Stimson Diary, entry for Nov. 10, 1943. In April 1945, Stimson would side with army leadership against Donovan's plans. David F. Rudgers, *Creating the Secret State: The Origins of the Central Intelligence Agency, 1943–1947* (Lawrence, KS: University Press of Kansas, 2000), 194.
18. Quoted in Rudgers, *Secret State*, 19–20; Troy, *Donovan and the CIA*, 218–9. Apparently requested by Walter Bedell Smith, then chief of staff to Eisenhower, Donovan's paper was a think piece intended to help Smith envision the postwar military.
19. Troy, *Donovan and the CIA*, 220.
20. Quoted in Troy, *Donovan and the CIA*, 222.
21. Carter to Roosevelt, Oct. 26, 1944, Donovan Folder, Box 153, PSF, FDRL.
22. Roosevelt to Donovan, Oct. 31, 1944, Donovan Folder, Box 153, PSF, FDRL. See also Troy, *Donovan and the CIA*, 226.
23. Donovan to Roosevelt, Nov. 18 1944, Donovan Folder, Box 153, PSF, FDRL.
24. Phillips Payson O'Brien, *The Second Most Powerful Man in the World: The Life of Admiral William D. Leahy, Roosevelt's Chief of Staff* (New York: Dutton, 2019).
25. Roosevelt made this clear in more than one way. In December 1944, when he created the five-star rank for America's leading admirals and generals, Leahy was first on the list, followed by Marshall, King, MacArthur, and Nimitz.
26. Leahy, *I Was There*, 189.
27. O'Brien, *Second Most Powerful Man*, 378–9, overstates the case, but only slightly.
28. O'Brien, *Second Most Powerful Man*, 377; see also chapters 12 and 14 prior.
29. FDR to WDL, Nov. 22, 1944, Donovan Folder, Box 153, PSF, FDRL.
30. Troy, *Donovan and the CIA*, 225, 253.
31. Quoted in Burns, *Quest for Cryptologic Centralization*, 18–19.
32. Benson, *History of U.S. Communications Intelligence*, 136–8. The ANCIB was empowered to coordinate navy and army plans and operations, formulate joint agreements, and negotiate with other intelligence organizations.
33. Walter Trohan, "New Deal Plans Super Spy System, Sleuths Would Snoop on U.S. and the World," *Chicago Daily Tribune*, Feb. 9, 1945. The second half of this article reprinted Donovan's Nov. 18, 1944, memorandum. The story

also appeared in the affiliated *Washington Times-Herald*, whose version Donovan read at home that morning. Rudgers, *Secret State*, 25–27; Waller, *Wild Bill Donovan*, 304–5.

34. Walter Trohan, "Super-Spy Ideas Denounced as New Deal OGPU," *Chicago Daily Tribune*, Feb. 10, 1945; Walter Trohan, "Army Submits Own Plans for Super-Spy Unit, Opposes Control by Donovan Agency," *Chicago Daily Tribune*, Feb. 11, 1945. The affiliated *Washington Times-Herald* played the story down somewhat, but kept it above the fold on the front page. Walter Trohan, "Army, Navy Want Control of 'Spy' Setup; Generals, Admirals Declare War on OSS," *Washington Times-Herald*, Feb. 11, 1945.
35. Walter Trohan, *Political Animals: Thirty-Eight Years of Washington-Watching by the Chicago Tribune's Veteran Observer* (New York: Doubleday, 1975), 3–5, describes McCormick. This was not the *Tribune*'s first story to expose sensitive information. In the mistaken belief that censors had cleared it, the *Tribune* had run stories after Midway touting US foreknowledge of Japanese plans—and then blindly defended its action as an assertion of First Amendment rights. See Elliott Carlson, *Stanley Johnston's Blunder: The Reporter Who Spilled the Secret behind the U.S. Navy's Victory at Midway* (Annapolis, MD: Naval Institute Press, 2017).
36. Donovan to JCS, Feb. 15, 1945, quoted in Troy, *Donovan and the CIA*, 258–9.
37. Rudgers, *Secret State*, 29. Trohan's claim is reported in Troy, *Donovan and the CIA*, vi. Troy met and corresponded with Trohan about the leak. Trohan and Early had known each other for years—at least since 1934—and were on reasonably good terms. Trohan, *Political Animals*, 201. When Early died at an early age, Trohan literally wrote his obituary. Walter Trohan, "A Great American Passes," *Carbuilder*, October 1951. In the February 9 story ("New Deal Plans Super Spy System, Sleuths Would Snoop on U.S. and the World"), Trohan claimed to have obtained Donovan's memorandum on February 8. This was when Early was at Yalta. If he leaked the documents to Trohan, he would have done so (or arranged to do so) before he left. See also Linda Lotridge Levin, *The Making of FDR: The Story of Stephen T. Early, America's First Modern Press Secretary* (Amherst, NY: Prometheus Books, 2008), chapter 23.
38. "Franklin D. Roosevelt: Day by Day," entry for Mar. 15, 1945, accessed Apr. 22, 2021, http://www.fdrlibrary.marist.edu/daybyday/daylog/march-15th-1945/. Waller, *Wild Bill Donovan*, 312–3. Roosevelt commented to the press that they had not discussed Donovan's plans. Press Summary, Page 74, Part 3, Donovan File, accessed Apr. 30, 2021, https://vault.fbi.gov/William%20J%20Donovan%20/William%20J%20Donovan%20Part%203%20of%207/view.
39. Lubin to Roosevelt, Apr. 4, 1945, and Roosevelt to Donovan, Apr. 5, 1945, Donovan Folder, Box 153, PSF, FDRL; Waller, *Wild Bill Donovan*, 313–4. Waller notes that this was a bold bureaucratic move, another end run by Donovan.

40. The results of the poll emerged from a meeting of cabinet officers on April 12 and constituted a lukewarm endorsement for a postwar intelligence agency, a topic they wanted to defer until after the war. Rudgers, *Secret State,* 31. A survey by the House Appropriations Committee in late spring, apparently a reaction to the Trohan articles, brought confirmation from MacArthur and Nimitz that OSS was not active in their theaters, a mixed endorsement from the Joint Chiefs, and positive reviews from CBI, ETO, and MEDTO, the commands where OSS had been most active. Troy, *Donovan and the CIA,* 283.
41. Richard E. Schroeder, *The Foundation of the CIA: Harry Truman, the Missouri Gang, and the Origins of the Cold War* (Columbia, MO: University of Missouri Press, 2017), 84.
42. Rudgers, *Secret State,* 35. On April 30, Donovan forwarded another copy of his November 18 memorandum to the White House.
43. "World Bank," *Washington Post,* Jun. 20, 1946, records the tale that, when Smith offered to resign, Roosevelt told him he was so indispensable that he would assign a marine guard to make sure that he did not leave government service.
44. Harry S. Truman, *1945: Year of Decisions* (New York: Konecky & Konecky, 1955), 98–99.
45. Truman Memorandum, May 12, 1945, quoted in Schroeder, *Foundation of the CIA,* 80. This was a refrain during the Truman presidency; no part of American intelligence should function like the Gestapo, especially not the FBI. See for example Weiner, *Enemies,* 134.
46. Truman wanted contact with Hoover to come through the attorney general or Brig. Gen. Harry H. Vaughan, a longtime associate who served as military aide to the president. This was a demotion for Hoover. Though not a Roosevelt intimate, he had had direct access to FDR.
47. [FBI], "Bureau Memo," May 2, 1945, 80, part 3, Donovan File, accessed Apr. 30, 2021, https://vault.fbi.gov/William%20J%20Donovan%20/William%20J%20Donovan%20Part%203%20of%207/view. Biddle briefed Hoover's deputy, Edward Tamm, who apparently created the memo.
48. Quoted in Troy, *Donovan and the CIA,* 270. Donovan's authorized biographer, Richard Dunlop, records a version of this meeting that is flattering to Donovan but not to Truman, who reportedly thanked Donovan for his wartime service before lecturing him. Dunlop, *Donovan,* 468. Dunlop quotes a full paragraph of the lecture without sourcing.
49. Troy, *Donovan and the CIA,* 270.
50. See for example: Donovan to Truman, Aug. 25, 1945, in C. T. Thorne Jr. and D. S. Patterson, eds., *Foreign Relations of the United States, 1945–1950: Emergence of the Intelligence Establishment* (Washington, DC: Department of State / US Government Printing Office, 1996), doc. 3.
51. A copy of the report and related documents is in OSS/Donovan Folder, Rose A. Conway Files, Harry S. Truman Library (HSTL), available at:

https://www.scribd.com/doc/284321062/Park-Report-Memorandum-for-the-Record-Colonel-Park-s-Comments-on-OSS-Declassified-Top-Secret-Report-12-March-1945 (accessed Apr. 26, 2021). Clayton Bissell, "Memorandum for the Record," Mar. 12, 1945, records that Park presented the report to him and that Bissell thought it inappropriate to comment. Park signed an undated memorandum claiming that Roosevelt had commissioned the report. Handwritten dates on the report itself are 12/44 and 3/30/45, suggesting that it might have been compiled in December 1944 and delivered to the White House on or about March 30, 1945. FBI files reference a version dated Apr. 13, 1945. File Summary, 77, Part 3, Donovan File, accessed Apr. 30, 2021, https://vault.fbi.gov/William%20J%20Donovan%20/William%20J%20Donovan%20Part%203%20of%207/view. Waller, *Wild Bill Donovan,* 335–6, concludes that Donovan rival John Grombach played a major role in compiling the report.

52. Troy, *Donovan and the CIA,* 282, points out the overlap between Trohan's stories and the Park Report. Trohan overstated the case somewhat. MacArthur resisted intelligence operations of any kind that were not under his control. Nimitz generally cooperated with national assets, and was willing to consider the use of OSS. The only part of Asia where OSS was actively committed remained CBI.
53. "Army Used 'Reds,' House Group Told," *New York Times,* Jul. 19, 1945.
54. It was not for lack of trying. With Donovan's approval, Colonel Eifler had a far-fetched plan to run operations against Japan with Korean commandos. *NAPKO Project of OSS*; Moon and Eifler, *Deadliest Colonel,* 216–32. OSS Bern was also monitoring possible Japanese peace feelers in Europe from May to August 1946. Petersen, ed., *Hitler's Doorstep,* 523–4.
55. Richard B. Frank, *Downfall: The End of the Imperial Japanese Empire* (New York: Penguin, 2001), esp. 104–7, 211–3, 238. Frank devotes much of the book to the role of diplomatic—Magic—and military—Ultra—traffic. Originally used by the British for German decrypts, the US adopted the Ultra designation and used it in the Pacific, especially late in the war.
56. See Frank, *Downfall,* esp. 257, 343, 348; Max Hastings, *Retribution: The Battle for Japan, 1944–45* (New York: Vintage Books, 2009), 444–81. The Japanese government could have prevented the bombing of Nagasaki by promptly surrendering after the bombing of Hiroshima. But it was not yet prepared to do so. See also Truman, *1945,* 420-1.
57. Op-20-3-G50, "Japan's Surrender Maneuvers," Aug. 29, 1945, SRH 090, NCML.
58. Waller, *Wild Bill Donovan,* 332.
59. Truman, *1945,* 437.
60. Truman, *1945,* 484, 486–7, enumerating some of the agencies affected.
61. Rudgers, *Secret State,* 35, 41–2.
62. Bess Furman and Tillman Durdin, "U.S. Cloak and Dagger Exploits and Secret Blows in China Bared," *New York Times,* Sept. 14, 1945.

63. Meyer Berger, "Wainwright Tells of Long Captivity," *New York Times*, Sept. 14, 1945; "OSS 'Underground Railway' Plan Saved U.S. Fliers in Axis Areas," *New York Times*, Sept. 16, 1945. At great risk to themselves, small OSS detachments entered prisoner-of-war camps after V-E and V-J Days and began the process of liberating Allied prisoners of war. See example Hilsman, *American Guerrilla*, 229–36.
64. Forrest Davis, "The Secret History of a Surrender," *Saturday Evening Post*, September 22, 1945.
65. Donovan to Truman, Sept. 13, 1945, reprinted in Michael Warner, ed., *The CIA under Harry Truman* (Washington, DC: Center for the Study of Intelligence, 1994), 3. Donovan's memo of Sept. 4, 1945, to senior White House advisor Judge Samuel Rosenman denounced Budget Director Smith's plans as "absurd" and lectured Rosenman: "it's time for us to grow up . . . and realize that the new responsibilities we have . . . require an adequate intelligence system." Donovan to Rosenman, Sept. 4, 1945, in Thorne and Patterson, ed., *Emergence of the Intelligence Establishment*, doc. 6.
66. Smith Diary, entry for Sept. 13, 1945, quoted in Troy, *Donovan and the CIA*, 296.
67. Other agencies were liquidated in September as well, including the War Refugee Board and the Office of War Information, many of whose employees had started in OSS. The Joint Chiefs wanted more time to study the order—as well as demanding prompt action! Troy, *Donovan and the CIA*, 300–1.
68. Leahy to Stimson and Forrestal, Sept. 19, 1945, in Warner, ed., *CIA under Harry Truman*, 5–9. The document was to be routed through the Navy and War secretaries to the president.
69. Leahy to Stimson and Forrestal, Sept. 19, 1945, in Warner, ed., *CIA under Harry Truman*, 5–9.
70. A copy of the executive order is in Warner, ed., *CIA under Harry Truman*, 11–13.
71. Troy, *Donovan and the CIA*, 302, quoting from Smith's diary.
72. Truman to Byrnes, Sept. 20, 1945, in Thorne and Patterson, eds., *Emergence of the Intelligence Establishment*, doc. 20.
73. Stone to Smith, Sept. 20, 1945, in Thorne and Patterson, eds., *Emergence of the Intelligence Establishment*, doc. 16. Waller, *Wild Bill Donovan*, 338, has Donovan leaving town so that he would not be at his desk when the memo arrived. Dunlop, *Donovan*, 473, has an alternate version, reporting that Stone delivered the message to Donovan, who accepted it with "stoic grace."

21: An End and a Beginning

1. Ramsey to Martin, Sept. 24, 1945, in Thorne and Patterson, ed., *Emergence of the Intelligence Establishment*, doc. 18. According to Ramsey, it was an

oversight that the executive order allowed only ten days to liquidate OSS, rather than the three or four weeks recommended earlier. OSS was not the only agency liquidated in September 1945.

2. Like McCormack, McCarthy was a wartime phenomenon. He joined the army when France collapsed in 1940 and had a brilliant five-year career in Washington, rising to the rank of colonel. Like Marshall, he was a graduate of the Virginia Military Institute. He was at State from August to October 1945. After leaving Washington, he went on to an equally brilliant career as a filmmaker, eventually producing the film *Patton*.
3. Washington to London, Sept. 12, 1945, in Thorne and Patterson, eds., *Emergence of the Intelligence Establishment*, doc. 73.
4. MacPherson, *American Intelligence in War-Time London*, 134–5, even disputes the value of its study of strategic bombing, which is often cited as one of R&A's strengths. Warner, *America's First Intelligence Agency*, 12.
5. Katz, *Foreign Intelligence*, 29. Most former members of R&A would not stay at State but soon returned to academic pursuits, where they helped stimulate the growth of interdisciplinary research, area studies, and Sovietology, as well as the sense that academics should engage outside the classroom. Katz, "The OSS and the Development of the Research and Analysis Branch" in Chalou, ed., *Secrets War*, 46–47.
6. See chapter 13 prior. Magruder was on record as favoring an OSS that focused more on intelligence and less on paramilitary operations.
7. McCloy to Magruder, Sept. 26, 1945, in Thorne and Patterson, eds., *Emergence of the Intelligence Establishment*, doc. 95.
8. Rudgers, *Secret State*, 45–46. Some 1,362 former members of the Research and Analysis Branch would shift to State, while 9,028 other members of OSS shifted to War. It appears that many employees did not physically move.
9. "General William J. Donovan Selected OSS Documents, 1941–1945," n.d., Microfilm Roll List and Index, RG 226, NARA II; Patricia Eames, ed., "OSS Project: General Donovan's Files," Volunteer Ventures, *Record*, May 1998, National Archives, accessed Jun. 5, 2021, https://www.archives.gov/publications/record/1998/05/oss-project.html. Eames notes that Donovan's and Putzell's hands and fingerprints literally appear in the frames. In 1945 the line between government and personal papers was not always clear. That Donovan did not delegate the work, and worked at night, suggests that he was aware that this initiative was irregular.
10. "Donovan Gives Farewell Talk to OSS Crew," *Washington Post*, Sept. 30, 1945.
11. Dunlop, *Donovan*, 473–4; Waller, *Disciples*, 394–5. On the same day, Donovan sent a letter to the OSS distribution list, enclosing a copy of his farewell address and Truman's letter. He also offered "former members of OSS" a chance to purchase a commemorative lapel pin. Les Hughes, "OSS Veteran's Certificate and Lapel Pin," Insigne, 2003, accessed May 7, 2021, http://www.insigne.org/OSS-pin.htm.

12. As the Joint Staff's Ludwell L. Montague stated plainly, "Without William Donovan's initiative, in 1941 and again in 1944, there would have been no Central Intelligence Agency." Quoted in Troy, *Donovan and the CIA*, 251.
13. Hoover to Clark, Sept. 6, 1945, in Thorne and Patterson, eds., *Emergence of the Intelligence Establishment*, doc. 8.
14. Hoover to Clark, Sept. 21, 1945, cover letter of Clark to Truman, draft memorandum, n.d., in Thorne and Patterson, eds., *Emergence of the Intelligence Establishment*, doc. 17.
15. Hoover to Clark, Sept. 27, 1945, in Thorne and Patterson, eds., *Emergence of the Intelligence Establishment*, doc. 19.
16. Miles to Stone, Sept. 19, 1945, and Appleby to Clark, Oct. 31, 1945, in Thorne and Patterson, eds., *Emergence of the Intelligence Establishment*, docs. 11 and 37.
17. Chiles to Hoover, Oct. 2, 1945, in Thorne and Patterson, eds., *Emergence of the Intelligence Establishment*, doc. 22.
18. See for example Ladd to Hoover, Nov. 5, 1945, in Thorne and Patterson, eds., *Emergence of the Intelligence Establishment*, doc. 43.
19. Hoover to Vaughan, Nov. 8, 1945, HSTL, National Archives, accessed Apr. 30, 2021, https://trumanlibrary.gov/node/317534. Lee's treachery emerged first from the testimony of a 1945 defector. The FBI subsequently worked with the codebreaking establishment to exploit Soviet messages that Arlington Hall systematically collected during World War II and started to analyze in 1943. Robert L. Benson and Michael Warner, eds., *Venona: Soviet Espionage and the American Response 1939–1957,* (Washington, DC: NSA and CIA, 1996), esp. xiii–xvii.
20. Truman to Byrnes, Sept. 20, 1945, in Thorne and Patterson, eds., *Emergence of the Intelligence Establishment*, doc. 15. The commonsense reading of Truman's letter is that he simply wanted State to take the lead in developing a plan.
21. The clearest expression of McCormack's ideas is in Rudgers, *Secret State*, esp. 83–87. See also the summary in Souers to Clifford, Dec. 27, 1945, in Warner, ed., *CIA under Harry Truman*, 17–19.
22. Quoted in Troy, *Donovan and the CIA*, 335.
23. Rudgers, *Secret State*, 86; T. B. Inglis, "Memorandum for Information," Nov. 2, 1945, in Thorne and Patterson, eds., *Emergence of the Intelligence Establishment*, doc. 41. Admiral Inglis sat next to McCormack at a lunch.
24. H. Thebaud, "Memorandum for General Marshall and Admiral Leahy," Aug. 22, 1945, National Security Agency, accessed May 10, 2021, https://media.defense.gov/2021/Jul/14/2002762212/-1/-1/0/marshall_king_memo.pdf. They included an exception for the FBI, but only for "criminal activities in the U.S." This prohibition was intended to apply to cryptanalytic work and was not apparently a bar to receiving Magic-Ultra reports. In December, State would join the ANCIB and be represented by McCormack.
25. Smith, *Ultra-Magic Deals*, 212–3. Smith offers a detailed and readable discussion of these developments.

26. Claims that Magic and Ultra shortened the war by any given period of months or years are problematic. For discussions of this topic, see Budiansky, *Battle of Wits*, 332–3; Hinsley, "The Influence of Ultra in the Second World War," in Hinsley and Stripp, eds., *Codebreakers*, 11–13. Hastings, *Secret War*, esp. 544–9, is a useful summing-up of the value of various kinds of intelligence during the war.
27. "Joint Meeting of ANCIB and ANCICC," Oct. 15, 1945, accessed May 10, 2021, National Security Agency, https://media.defense.gov/2021/Jul/14/2002762210/-1/-1/0/ARMY_NAVY_19451015_MTG.PDF.
28. "Joint Meeting of ANCIB and ANCICC," Nov. 1, 1945, National Security Agency, accessed May 10, 2021, https://media.defense.gov/2021/Jul/14/2002762203/-1/-1/0/ARMY_NAVY_19451101_MTG.pdf. This agreement would be ratified in 1946 and is usually known as "UKUSA."
29. See for example Smith, *Ultra-Magic Deals*, 210. The author speculates that development of this "long-term intelligence shield" might have made Truman more willing to let OSS go. There is no evidence that this thought did or did not occur to Truman. Unquestionably, King and Marshall placed a much higher premium on Magic and Ultra than on anything OSS ever produced, and registered no objection to the wartime agency's demise.
30. O'Brien, *Second Most Powerful Man*, 357–8. Specific threats, like the Soviet Union, were apparently not discussed at this time.
31. The Committee would eventually produce 39 volumes of proceedings containing hundreds of exhibits and testimony from 151 witnesses. See Nelson, *Pearl Harbor*, 437–54, and Prange, *At Dawn We Slept*, esp. 582–738.
32. Quoted in Troy, *Donovan and the CIA*, 316.
33. Quoted in Rudgers, *Secret State*, 63–64.
34. The process is described in detail in Rudgers, *Secret State*, 47–92, and in even more detail in Troy, *Donovan and the CIA*, 305–49.
35. See for example Ladd to Hoover, Nov. 5, 1945, in Thorne and Patterson, eds., *Emergence of the Intelligence Establishment*, doc. 43.
36. "Minutes of Meeting," Nov. 14, 1945, in Thorne and Patterson, eds., *Emergence of the Intelligence Establishment*, doc. 45.
37. Srodes, *Dulles*, 370. In 1944 Donovan had denied Dulles's request to coordinate operations in the European theater for much the same reason: he believed that Dulles had been unable to keep a firm grip on OSS Bern as it grew in 1944 and 1945. Srodes, *Dulles*, 316. Lovell, *Of Spies and Stratagems*, 209, is a loosely sourced reference to a postwar meeting between Donovan and Dulles, at which Donovan reportedly told Dulles that he was good at collecting intelligence but should not run an intelligence agency.
38. Magruder to Lovett, n.d., in Thorne and Patterson, eds., *Emergence of the Intelligence Establishment*, doc. 34. Concerns about possible communist infiltration at State appear to have influenced Admiral Leahy's thinking against that department. O'Brien, *Second Most Powerful Man*, 381. The timing, late November 1945, coincided with the FBI's continuing

investigation into Soviet espionage, targeting at least one very prominent officer at State, Alger Hiss.

39. Magruder to Irwin, Jan. 15, 1946, in Warner, ed., *CIA under Harry Truman*, 21–23.
40. Smith to Truman, Nov. 28, 1945, in Thorne and Patterson, eds., *Emergence of the Intelligence Establishment*, doc. 58.
41. Miles to Hoelscher, Jan. 3, 1946, Thorne and Patterson, eds., *Emergence of the Intelligence Establishment*, doc. 65.
42. This process is perfectly described in Schroeder, *Foundation of the CIA*, esp. 85–91.
43. Troy, *Donovan and the CIA*, 344–6, 464–5, describes the process and includes a copy of Truman's directive.
44. Troy, *Donovan and the CIA*, 346.
45. Smith to Rosenman, Jan. 10, 1946, in Thorne and Patterson, eds., *Emergence of the Intelligence Establishment*, doc. 70.

Epilogue

1. See for example Troy Sacquety, "History in the 'Raw,'" *Veritas* 5, no. 3 (2009), and Troy Sacquety, "The OSS Influence on Special Forces," *Veritas* 14, no. 2 (2018). Sacquety does not dispute the tie between OSS and SF, but questions its extent. He also points out that OSS was not the only precursor to SF.
2. See Moon and Eifler, *Deadliest Colonel*, 238–44, 247–68, for Eifler's own testimony about his postwar years.
3. Persico, *Casey*, 202.
4. "Resignation of Col. McCormack Laid to Reorganization Feud," *Washington Post*, Apr. 25, 1946. This was a continuation of the feud that started in the fall of 1945 and was not unlike his clashes with army regulars during the war. "Newly Named Intelligence Chief Gets DSM," *Washington Post*, Nov. 24, 1945, describes his award. In 1947, then secretary of state Marshall would create a standalone Bureau of Intelligence and Research, and place OSS veteran William A. Eddy at its helm.
5. [McCormack], "Biographical Sketch," 1954, in SRH 185.
6. "Newly Named Intelligence Chief Gets DSM," *Washington Post*, Nov. 24, 1945.
7. Troy, *Donovan and the CIA*, 345, has the last in a series of three photos of the occasion.
8. Bancroft, *Autobiography of a Spy*, 141, 242–3.
9. Waller, *Wild Bill Donovan*, 349–50, offers a good overview of the transition.
10. SSU would complete a narrative about OSS, which was eventually published as US Office of Strategic Services, *War Report of the OSS* (New York: Walker, 1976). Published the same year, Brown, ed., *Secret War Report*, is a shorter version of the same document with commentary by Brown.

11. Srodes, *Dulles*, 371–2.
12. Persico, *Casey*, 576.
13. Hoover to Ruth Donovan, Feb. 9, 1959, in Part 7, Donovan File.
14. "J. Edgar Hoover," accessed Mar. 27, 2022, https://www.legion.org/distinguishedservicemedal/1946/j-edgar-hoover. The Legion noted the contrast with World War I, when German agents committed dramatic acts of sabotage like that at Black Tom, New Jersey, in 1916. Hoover would no doubt have welcomed official recognition as well.
15. Quoted in Fagone, *Woman Who Smashed Codes*, 329–30, 333. MacKinnon, "William Friedman's Bletchley Park Diary," 4n–5n, refers to the letter from Friedman's psychiatrist that MacKinnon found among Elizebeth Friedman's papers citing insomnia, depression, and a struggle with alcohol as his maladies.
16. Fagone, *Woman Who Smashed Codes*, 322–3.
17. Friedman to Clarke, "Personal Correspondence with Brigadier General Carter W. Clarke," Apr. 23, 1952, National Security Agency, accessed May 23, 2021, https://nsa.gov/ Portals/75/documents/news-features/declassified-documents/friedmandocuments/correspondence/FOLDER_364/41734719077395.pdf.
18. Clarke to Friedman, Dec. 4, 1950, "Personal Correspondence." Clarke believed that Donovan had been devious and disorganized. [Forrest Pogue, ed.], "Major General Carter W. Clarke," Interview, Jul. 6, 1959, G. C. Marshall Foundation, Lexington, VA.
19. Clarke Interview, Dec. 6, 1963, David Kahn Papers.
20. Clarke to Kahn, Feb. 3, 1987, David Kahn Papers. Clarke was referring to the state of affairs before the summaries came online.
21. Layton et al., *"And I Was There."* In addition to describing the Battle of Midway from the Pearl Harbor perspective, Layton's book contained an unproven allegation: that Roosevelt and Churchill had a secret agreement for America to come to Britain's aid if Japan attacked her possessions.
22. An admirable summary of the evidence, the context, and the literature is: Tom Johnson, "What Every Cryptologist Should Know about Pearl Harbor," *Cryptologic Quarterly* 6, no. 2 (September 1987). Safford's contentions are examined—and discounted to zero—in Robert J. Hanyok, "The Pearl Harbor Warning That Never Was," *Naval History* 23, no. 2 (April 2009).
23. Carlson, *Joe Rochefort's War*, 434–7.
24. Carlson, *Joe Rochefort's War*, 432.
25. See for example Edwin McDowell, "Officer Who Broke Japanese War Codes Gets Belated Honor," *New York Times*, Nov. 17, 1985.
26. Elizabeth Kastor, "Medal Ends 44-Year Campaign," *Washington Post*, May 31, 1986.
27. Carlson, *Joe Rochefort's War*, 455–6.
28. "Alfred McCormack," editorial, *Washington Post*, July 13, 1956.

SELECT BIBLIOGRAPHY OF BOOKS AND ARTICLES

Abrutat, David. *Vanguard: The True Stories of the Reconnaissance and Intelligence Missions behind D-Day.* Annapolis, MD: Naval Institute, 2019.

Aldrich, Richard J., and Rory Cormac. *The Black Door: Spies, Secret Intelligence, and British Prime Ministers*. London: William Collins, 2016.

Aldrich, Richard J., Rory Cormac, and Michael S. Goodman, eds. *Spying on the World: The Declassified Documents of the Joint Intelligence Committee, 1936–2013.* Edinburgh: Edinburgh University Press, 2014.

Andrew, Christopher. *For the President's Eyes Only: Secret Intelligence and the American Presidency from Washington to Bush.* New York: HarperCollins, 1995.

Anonymous. *Napko Project of OSS.* Seoul, Korea: South Korean Patriots and Veterans Administration Agency, 2001.

Atkinson, Rick. *An Army at Dawn: The War in North Africa, 1942–1943.* New York: Henry Holt, 2002.

Bancroft, Mary. *Autobiography of a Spy.* New York: William Morrow, 1983.

Barbier, Mary Kathryn. *D-Day Deception: Operation Fortitude and the Normandy Invasion.* Mechanicsburg, PA: Stackpole Books, 2007.

Batvinis, Raymond. *Hoover's Secret War against Axis Spies: FBI Counterespionage during World War II.* Lawrence, KS: University Press of Kansas, 2014.

——. *The Origins of FBI Counterintelligence.* Lawrence, KS: University Press of Kansas, 2007.

Beaulac, Willard L. *Career Ambassador.* New York: McMillan, 1951.

——. *Franco: Silent Ally in World War II.* Carbondale, IL: Southern Illinois University Press, 1986.

Beesly, Patrick. *Very Special Admiral: The Life of Admiral J. H. Godfrey, CB.* London: Hamish Hamilton, 1980.

——. *Very Special Intelligence: The Story of the Admiralty's Operational Intelligence Centre 1939–1945.* New York: Ballantine Books, 1981.

Bennett, Gill. *Churchill's Man of Mystery: Desmond Morton and the World of Intelligence*. London: Routledge, 2007.

Benson, Robert L. *A History of U.S. Communications Intelligence during World War II: Policy and Administration*. Ft. Meade, MD: National Security Agency, 1997.

Bidwell, Bruce W. *History of the Military Intelligence Division, Department of the Army General Staff: 1775–1941*. Frederick, MD: University Publications of America, 1986.

Bird, Kai. *The Chairman: John J. McCloy & the Making of the American Establishment*. New York: Simon and Schuster, 1992.

Boghardt, Thomas. *The Zimmermann Telegram: Intelligence, Diplomacy, and America's Entry into World War I*. Annapolis, MD: Naval Institute Press, 2012.

Boyd, Carl. *Hitler's Japanese Confidant: General Ōshima Hiroshi and MAGIC Intelligence, 1941–1945*. Lawrenceville, KS: University Press of Kansas, 1993.

Bradley, Mark. *A Very Principled Boy: The Life of Duncan Lee, Red Spy and Cold Warrior*. New York: Basic Books, 2014.

Bradsher, Greg. "A Time to Act: The Beginning of the Fritz Kolbe Story, 1900–1943." *Prologue* 34, no. 1 (Spring 2002).

Breaks, Katherine. "'Ladies of the OSS: The Apron Strings of Intelligence in World War II." *American Intelligence Journal* 13, no. 3 (Summer 1992): 91–6.

Breitman, Richard, and Norman J. W. Goda. *Hitler's Shadow: Nazi War Criminals, U.S. Intelligence, and the Cold War*. Washington, DC: National Archives and Records Administration, 2010.

Brown, Anthony Cave. *Bodyguard of Lies*. New York: Harper & Row, 1975.

——. *The Last Hero: Wild Bill Donovan*. New York: Times Books, 1982.

——. *"C": The Secret Life of Sir Stewart Graham Menzies, Spymaster to Winston Churchill*. New York: Macmillan, 1987.

Brown, Anthony Cave, ed. *The Secret War Report of the OSS*. New York: Berkley, 1976.

Bryce, Ivar. *You Only Live Once: Memories of Ian Fleming*. London: Weidenfeld & Nicolson, 1975.

Budiansky, Stephen. *Battle of Wits: The Complete Story of Codebreaking in World War II*. New York: Simon and Schuster, 2002.

Bureau of the Budget. *The United States at War: Development and Administration of the War Program by the Federal Government*. Washington, DC: US Government Printing Office, 1946.

Burns, Thomas L. *The Quest for Cryptologic Centralization and the Establishment of NSA: 1940–1952*. Ft. Meade, MD: National Security Agency, Center for Cryptologic History, 2005.

Carlson, Elliot. *Joe Rochefort's War: The Odyssey of the Codebreaker Who Outwitted Yamamoto at Midway*. Annapolis, MD: Naval Institute Press, 2011.

———. *Stanley Johnston's Blunder: The Reporter Who Spilled the Secret behind the U.S. Navy's Victory at Midway.* Annapolis, MD: Naval Institute Press, 2017.

Casey, William J. *The Secret War against Hitler.* Washington, DC: Regnery, 1988.

Chalou, George C., ed. *The Secrets War: The Office of Strategic Services in World War II.* Washington, DC: National Archives and Records Administration, 1992.

Charles, Douglas M. "'Before the Colonel Arrived': Hoover, Donovan, Roosevelt, and the Origins of American Central Intelligence, 1940–41." *Intelligence and National Security* 20, no. 2 (2005).

Churchill, Winston S. *The Gathering Storm.* New York: Houghton Mifflin, 1948.

———. *Their Finest Hour.* Boston: Houghton Mifflin, 1949.

Clark, Mark W. *Calculated Risk.* New York: Enigma Books, 2007.

Clark, Ronald W. *The Man Who Broke Purple: The Life of Colonel William F. Friedman, Who Deciphered the Japanese Code in World War II.* Boston: Little, Brown, 1977.

Cline, Ray. *Secrets, Spies, and Scholars.* Washington, DC: Acropolis Books, 1976.

Colville, John. *The Fringes of Power: Downing Street Diaries, 1939–1955.* London: Hodder & Stoughton, 1985.

Coon, Carleton S. *A North Africa Story: The Anthropologist as OSS Agent, 1941–1943.* Ipswich, MA: Gambit, 1980.

Cull, Nicholas J. *Selling War: The British Propaganda Campaign against American "Neutrality" in World War II.* New York and Oxford: Oxford University Press, 1995.

Currier, Prescott H. "My 'Purple' Trip to England in 1941." *Cryptologia* 20, no. 3 (July 1996).

Cutler, Richard W. *Counterspy: Memoirs of a Counterintelligence Officer in World War II and the Cold War.* Washington, DC: Potomac Books, 2004.

Danchev, Alex, ed. *Establishing the Anglo-American Alliance: The Second World War Diaries of Brigadier Vivian Dykes.* London: Brassey's, 1990.

Davis, Forrest. "The Secret History of a Surrender." *Saturday Evening Post,* September 22, 1945.

Dawidoff, Nicholas. *The Catcher Was a Spy: The Mysterious Life of Moe Berg.* New York: Vintage Books, 1995.

Deane, John R. *The Strange Alliance: The Story of Our Efforts at Wartime Cooperation with Russia.* New York: Viking, 1947.

Delattre, Lucas. *Betraying Hitler: The Story of Fritz Kolbe, the Most Important Spy of the Second World War.* London: Atlantic Books, 2005.

Denniston, Robin. *Thirty Secret Years: A. G. Denniston's Work in Signals Intelligence, 1914–1944.* Clifton-upon-Teme, UK: Polpero Heritage Press, 2007.

Department of Defense. *The "Magic" Background of Pearl Harbor.* Washington, DC: US Government Printing Office, 1977.

Dodds-Parker, Douglas. *Setting Europe Ablaze.* Windlesham, UK: Springwood Books, 1983.

Donovan, William J., and Edgar A. Mowrer. *Fifth Column Lessons for America*. Washington, DC: American Council on Public Affairs, 1941.

Dorwart, Jeffery M. *Conflict of Duty: The U.S. Navy's Intelligence Dilemma, 1919–1945*. Annapolis, MD: Naval Institute Press, 1983.

——. "The Roosevelt-Astor Espionage Ring." *New York History* 62, no. 3 (July 1981): 307–22.

Drea, Edward J. *MacArthur's Ultra: Codebreaking and the War against Japan, 1942–1945* (Lawrence, KS: University Press of Kansas, 1992).

Dulles, Allen W. *Germany's Underground: The Anti-Nazi Resistance*. New York: Da Capo Press, 2000.

——. *The Secret Surrender*. Guilford, CT: Lyons Press, 2006.

Dunlop, Richard. *Donovan: America's Master Spy*. Chicago: Rand McNally, 1982.

Erbelding, Rebecca. *Rescue Board: The Untold Story of America's Efforts to Save the Jews of Europe*. New York: Doubleday, 2018.

Fagone, Jason. *The Woman Who Smashed Codes: A True Story of Love, Spies, and the Unlikely Heroine Who Outwitted America's Enemies*. New York: Dey Street, 2017.

Farago, Ladislas. *The Tenth Fleet*. New York: Paperback Library, 1964.

Ferris, John. *Behind the Enigma: The Authorized History of GCHQ, Britain's Secret Cyber-Intelligence Agency*. New York and London: Bloomsbury, 2020.

Finn, Peter. *A Guest of the Reich: The Story of American Heiress Gertrude Legendre's Dramatic Captivity and Escape from Nazi Germany*. New York: Pantheon Books, 2019.

Foot, M. R. D. *SOE: The Special Operations Executive 1940–46*. London: London Bridge, 1984.

Ford, Corey. *Donovan of OSS*. Boston: Little, Brown, 1970.

Fowler, W. B. *British-American Relations 1917–1918: The Role of Sir William Wiseman*. Princeton, NJ: Princeton University Press, 1969.

Frank, Richard B. *Downfall: The End of the Imperial Japanese Empire*. New York: Penguin, 1999.

Friedman, William [anonymously]. *The Friedman Legacy: A Tribute to William and Elizebeth Friedman*. Ft. Meade, MD: National Security Agency, Center for Cryptologic History, 2006.

Gilbert, James L., and John P. Finnegan, eds. *U.S. Army Signals Intelligence in World War II: A Documentary History*. Washington, DC: US Army, Center of Military History, 1993.

Gisevius, Hans B. *To the Bitter End: An Insider's Account of the Plot to Kill Hitler, 1933–1944*. New York: Da Capo Press, 1998.

Godfrey, John H. *The Naval Memoirs of Admiral J. H. Godfrey*. Vol. 5, part 1. London: Privately published, ca. 1964.

Gores, Landis. *Ultra: I Was There*. Morrisville, NC: Lulu, 2008.

Greenhut, Jeffrey. *A Brief History of Naval Cryptanalysis*. Washington, DC: National Security Agency, nd.

Grose, Peter. *Gentleman Spy: The Life of Allen Dulles*. Boston: Houghton Mifflin, 1994.

Hall, Roger. *You're Stepping on My Cloak and Dagger*. Annapolis, MD: Naval Institute Press, 2004.

Hamm, Diane L., ed. *Military Intelligence: Its Heroes and Legends*. Arlington, VA: US Army Intelligence and Security Command, Deputy Chief of Staff, Operations, History Office, 1987.

Hastings, Max. *Das Reich: The March of the 2nd SS Panzer Division through France, June 1944*. Minneapolis, MN: Zenith Press, 2013.

——. *Finest Years: Churchill as Warlord 1940–45*. London: HarperPress, 2009.

——. *Retribution: The Battle for Japan, 1944–45*. New York: Vintage, 2009.

——. *The Secret War: Spies, Ciphers, and Guerrillas 1939–1945*. New York: HarperCollins, 2016.

Hayashi, Brian M. *Asian American Spies: How Asian Americans Helped Win the Allied Victory*. New York and Oxford: Oxford University Press, 2021.

Haynes, John Earl, Harvey Klehr, and Alexander Vassilliev. *Spies: The Rise and Fall of the KGB in America*. New Haven: Yale University Press, 2010.

Heuvel, William J. vanden. *Hope and History: A Memoir of Tumultuous Times*. Ithaca, NY: Cornell University Press, 2019.

Hilsman, Roger. *American Guerrilla: My War behind Enemy Lines*. Washington, DC: Brassey's, 1990.

Hinsley, F. H. *British Intelligence in the Second World War*. Abridged ed. Cambridge: Cambridge University Press, 1993.

Hinsley, F. H., and Alan Stripp, eds. *Codebreakers: The Inside Story of Bletchley Park*. Oxford: Oxford University Press, 1993.

Hogan, David W. Jr. *A Command Post at War: First Army Headquarters in Europe, 1943–1945*. Washington, D.C.: US Army, Center of Military History, 2000.

Holmes, W. J. *Double-Edged Secrets*. New York: Berkeley Books, 1981.

Holt, Thaddeus. *The Deceivers: Allied Military Deception in the Second World War*. New York: Skyhorse, 2007.

Hough, Frank O., Verle E. Ludwig, and Henry I. Shaw Jr. *Pearl Harbor to Guadalcanal*. History of U.S. Marine Corps Operations in World War II, vol. 1. Washington, DC: US Government Printing Office, 1958.

Howarth, Patrick. *Intelligence Chief Extraordinary: The Life of the Ninth Duke of Portland*. London: Bodley Head, 1986.

Hyde, H. Montgomery. *Room 3603: The Story of the British Intelligence Center in New York during World War II*. New York: Farrar, Straus, 1962.

Iatrides, John O., ed. *Ambassador MacVeagh Reports: Greece, 1933–1947*. Princeton, NJ: Princeton University Press, 1980.

Irwin, Will. *The Jedburghs: The Secret History of the Allied Special Forces, France 1944*. New York: Public Affairs, 2005.

Jeffery, Keith. *MI6: The History of the Secret Intelligence Service, 1909–1949*. London: Bloomsbury, 2010.

Jeffreys-Jones, Rhodri. *The Nazi Spy Ring in America: Hitler's Agents, the FBI, and the Case That Stirred the Nation.* Washington, DC: Georgetown University Press, 2020.

Johnson, Kevin Wade. *The Neglected Giant: Agnes Meyer Driscoll.* Ft. Meade, MD: National Security Agency, Center for Cryptologic History, 2015.

Kahn, David. *The Codebreakers.* New York: Macmillan, 1967.

——. "Pearl Harbor and the Inadequacy of Cryptanalysis." *Cryptologia* 15, no. 4 (October 1991).

——. *The Reader of Gentlemen's Mail: Herbert O. Yardley and the Birth of American Codebreaking.* New Haven: Yale University Press, 2004.

——. "Roosevelt, Magic and Ultra." *Cryptologia* 16, no. 4 (October 1992).

Katz, Barry M. *Foreign Intelligence: Research and Analysis in the Office of Strategic Services 1942–1945.* Cambridge, MA: Harvard University Press, 1989.

Keegan, John. *Intelligence in War.* New York: Knopf, 2003.

Kennedy, Paul. *Engineers of Victory: The Problem Solvers Who Turned the Tide in the Second World War.* New York: Random House, 2013.

Kenyon, David. *Bletchley Park and D-Day.* New Haven: Yale University Press, 2019.

Kinzer, Stephen. *The Brothers: John Foster Dulles, Allen Dulles, and Their Secret World War.* New York: Time Books / Henry Holt, 2013.

Kohnen, David. *Commanders Winn and Knowles: Winning the U-Boat War with Intelligence, 1939–1943.* Krakow, POL: Enigma Press, 1999.

——. "Tombstone of Victory: Tracking the U-505 from German Commerce Raider to American War Memorial 1944–1954." *Journal of America's Military Past* 32, no. 3 (Winter 2007).

Kruh, Louis. "Stimson, the Black Chamber, and the 'Gentlemen's Mail' Quote." *Cryptologia* 12, no. 2 (April 1988).

Langer, William L. *In and Out of the Ivory Tower.* New York: Neale Watson Academic Publications, 1977.

Lankford, Nelson D. *The Last American Aristocrat: The Biography of Ambassador David K. E. Bruce.* Boston: Little, Brown, 1996.

Lankford, Nelson D., ed. *OSS against the Reich: The World War II Diaries of Colonel David K. E. Bruce.* Kent, OH: Kent State University Press, 1991.

Layton, Edwin T., Roger T. Pineau, and John Costello. *"And I Was There": Pearl Harbor and Midway—Breaking the Secrets.* New York: Quill / William Morrow, 1985.

Leahy, William D. *I Was There.* New York: Whittlesey House, 1950.

Lee, Bruce. *Marching Orders: The Untold Story of World War II.* New York: Crown, 1995.

Leutze, James, ed. *The London Observer: The Journal of General Raymond E. Lee 1940–1941.* London: Hutchinson, 1971.

Lippman, Thomas W. *Arabian Knight: Colonel Bill Eddy USMC and the Rise of American Power in the Middle East.* Vista, CA: Selwa Press, 2008.

Lovell, Stanley P. *Of Spies and Stratagems: Incredible Secrets of World War II Revealed by a Master Spy*. New York: Pocket Books, 1964.

Lycett, Andrew. *Ian Fleming: The Man behind James Bond*. Atlanta: Turner Publishing, 1995.

Macintyre, Ben. *Double Cross: The True Story of the D-Day Spies*. New York: Broadway Paperbacks, 2013.

MacKinnon, Colin. "William Friedman's Bletchley Park Diary: A New Source for the History of Anglo-American Intelligence Cooperation." *Intelligence and National Security* 20:4 (2005).

MacPherson, Nelson. *American Intelligence in War-Time London: The Story of the OSS*. London: Frank Cass, 2003.

Masterman, J. C. *The Double-Cross System: The Incredible True Story of How Nazi Spies Were Turned into Double Agents*. New Haven: Yale University Press, 1972.

Mattingly, Robert E. *Herringbone Cloak: GI Dagger Marines of the OSS*. Washington, DC: USMC History and Museums Division, 1989.

Mauch, Christof. *The Shadow War against Hitler: The Covert Operations of America's Wartime Secret Intelligence Service*. New York: Columbia University Press, 2003.

McIntosh, Elizabeth Peet. *Sisterhood of Spies: The Women of the OSS*. Annapolis, MD: Naval Institute Press, 2009.

McLachlan, Donald. *Room 39: A Study in Naval Intelligence*. New York: Atheneum, 1968.

Montagu, Ewen. *Beyond Top Secret Ultra*. London: Peter Davies, 1977.

Moon, Thomas, and Carl F. Eifler. *The Deadliest Colonel*. New York: Vantage Press, 1975.

Moon, Tom. *This Grim and Savage Game: O.S.S. and the Beginning of U.S. Covert Operations in World War II*. Los Angeles: Burning Gate Press, 1991.

Morgan, William J. *The O.S.S. and I*. New York: Norton, 1957.

Mowrer, Edgar Ansel. *Triumph and Turmoil: A Personal History of Our Time*. London: Allen and Unwin, 1970.

Mowry, David P. *German Clandestine Activities in South America in World War II*. Ft. Meade, MD: National Security Agency, 1989.

Mundy, Liza. *Code Girls: The Untold Story of the American Women Code Breakers of World War II*. New York: Hachette, 2018.

Murphy, Robert. *Diplomat among Warriors*. New York: Pyramid Books, 1965.

Murray, Williamson, and Allan R. Millett, *A War to Be Won: Fighting the Second World War, 1937–1945*. Cambridge, MA, and London: Belknap Press of Harvard University Press, 2000.

Murrow, Edward R. *This Is London*. New York: Simon and Schuster, 1941.

Naftali, Timothy. "X-2 and the Apprenticeship of American Counterespionage, 1942–1944." PhD diss., Harvard University, 1993.

Nasaw, David. *The Patriarch: The Remarkable Life and Turbulent Times of Joseph P. Kennedy*. New York: Penguin, 2012.

O'Brien, Phillips Payson. *The Second Most Powerful Man in the World: The Life of Admiral William D. Leahy, Roosevelt's Chief of Staff.* New York: Dutton, 2019.

O'Donnell, Patrick K. *They Dared Return: The True Story of Jewish Spies behind the Lines in Nazi Germany.* New York: Da Capo Press, 2009.

Olson, Lynne. *Last Hope Island: Britain, Occupied Europe, and the Brotherhood That Helped Turn the Tide of War.* New York: Random House, 2017.

Packard, Wyman H. *A Century of U. S. Naval Intelligence.* Washington, DC: Department of the Navy, 1996.

Parker, Frederick D. *Pearl Harbor Revisited: U.S. Navy Communications Intelligence 1924–1941.* Ft. Meade, MD: National Security Agency, Center for Cryptologic History, 2013.

——. *A Priceless Advantage: U.S. Navy Communications Intelligence and the Battles of Coral Sea, Midway, and the Aleutians.* Ft. Meade, MD: National Security Agency, Center for Cryptologic History, 2017.

Parshall, Jonathan, and Anthony Tully. *Shattered Sword: The Untold Story of the Battle of Midway.* Sterling, VA: Potomac Books, 2007.

Persico, Joseph E. *Casey: The Lives and Secrets of William J. Casey: From the OSS to the CIA.* New York: Penguin, 1991.

——. *Piercing the Reich: The Penetration of Nazi Germany by American Secret Agents during World War II.* New York: Ballantine Books, 1979.

Petersen, Neal H., ed. *From Hitler's Doorstep: The Wartime Intelligence Reports of Allen Dulles, 1942–1945.* University Park, PA: Pennsylvania State University Press, 1996.

Philby, Kim. *My Silent War.* New York: Ballantine, 1983.

Powers, Richard Gid. *Secrecy and Power: The Life of J. Edgar Hoover.* New York: Free Press, 1987.

Prados, John. *Combined Fleet Decoded: The Secret History of American Intelligence and the Japanese Navy in World War II.* Annapolis, MD: Naval Institute Press, 1995.

Prange, Gordon W. *At Dawn We Slept: The Untold Story of Pearl Harbor.* New York: McGraw-Hill, 1981.

Prange, Gordon W., Donald M. Goldstein, and Katherine V. Dillon. *Miracle at Midway.* New York: Penguin, 1982.

Prettiman, C. A. "The Many Lives of William Alfred Eddy." *Princeton University Library Chronicle* 53, no. 2 (Winter 1992).

Reischauer, Edwin O. *My Life between Japan and America.* New York: Harper & Row, 1986.

Reynolds, Nicholas. "The 'Scholastic' Marine Who Won a Secret War: Frank Holcomb, the OSS, and American Double-Cross Operations in Europe." *Marine Corps History* 6, no. 1 (Summer 2020).

——. *Writer, Sailor, Soldier, Spy: Ernest Hemingway's Secret Adventures, 1935–1961.* New York: William Morrow, 2017.

Ridler, Jason S. *Mavericks of War: The Unconventional, Unorthodox Innovators and Thinkers, Scholars, and Outsiders Who Mastered the Art of War.* Guilford, CT: Stackpole Books, 2018.

Riebling, Mark. *Wedge: The Secret War between the FBI and the CIA* (New York: Knopf, 1994).

Roberts, Andrew. *Churchill: Walking with Destiny.* New York: Viking, 2018.

Rogg, Jeffrey. "The Spy and the State: The History and Theory of American Civil-Intelligence Relations." PhD diss., Ohio State University, 2020.

Rowlett, Frank B. *The Story of Magic: Memoirs of an American Cryptologic Pioneer.* Foreword and epilogue by David Kahn. Laguna Hills, CA: Aegean Park Press, 1998.

Rudgers, David F. *Creating the Secret State: The Origins of the Central Intelligence Agency, 1943–1947.* Lawrence, KS: University Press of Kansas, 2000.

Sacquety, Troy. *The OSS in Burma: Jungle War against the Japanese.* Lawrence, KS: University Press of Kansas, 2014.

——. "The OSS Influence on Special Forces." *Veritas* 14, no. 2 (2018).

Sacquety, Troy J., ed. *OSS, Office of Strategic Services: Primer & Manuals.* Ft. Bragg, NC: US Army Special Operations Command, 2013.

Schlesinger, Arthur M. Jr. *A Life in the 20th Century: Innocent Beginnings, 1917–1950.* Boston and New York: Houghton Mifflin, 2000.

Schroeder, Richard E. *The Foundation of the CIA: Harry Truman, The Missouri Gang, and the Origins of the Cold War.* Columbia, MO: University of Missouri Press, 2017.

Schwarz, Jordan A. *Liberal: Adolf A. Berle and the Vision of an American Era.* New York and London: Free Press, 1987.

Sheldon, Rose Mary M. "William F. Friedman: A Very Private Cryptographer and His Collection." *Cryptologic Quarterly* 34, no. 1 (2015).

Sherman, David. *The First Americans: The 1941 US Codebreaking Mission to Bletchley Park.* Ft. Meade, MD: National Security Agency, Center for Cryptologic History, 2016

Sherwood, Robert E. *Roosevelt and Hopkins: An Intimate History.* New York: Harper, 1948.

Smith, Amanda, ed. *Hostage to Fortune: The Letters of Joseph P. Kennedy.* New York: Viking, 2001.

Smith, Bradley F. *Sharing Secrets with Stalin: How the Allies Traded Intelligence, 1941–1945.* Lawrenceville, KS: University Press of Kansas, 1996.

——. *The Ultra-Magic Deals and the Most Secret Special Relationship, 1940–1946.* Shrewsbury, UK: Airlife, 1993.

Smith, Bradley F., and Elena Agarossi. *Operation Sunrise: The Secret Surrender.* New York: Basic Books, 1979.

Smith, Michael, and Ralph Erskine, eds. *Action This Day.* (London: Bantam, 2001).

Smith, Richard Harris. *OSS: The Secret History of America's First Central Intelligence Agency*. Berkeley, CA: University of California Press, 1972.

Spector, Ronald H. *Eagle against the Sun: The American War with Japan*. New York: Free Press, 1985.

Spence, Richard. "Englishmen in New York: The SIS American Station, 1915–21." *Intelligence and National Security* 19, no. 3 (2004).

Sperber, A. M. *Murrow: His Life and Times*. New York: Fordham University Press, 1998.

Srodes, James. *Allen Dulles: Master of Spies*. Washington, DC: Regnery, 1999.

Stafford, David. *Churchill and Secret Service*. New York: Overlook Press, 1997.

Stephenson, William S., ed. *British Security Coordination: The Secret History of British Intelligence in the Americas, 1940–1945*. New York: Fromm International, 1999.

Stevenson, William. *A Man Called Intrepid*. New York: Harcourt, Brace, Jovanovich, 1976.

Stimson, Henry L., and McGeorge Bundy. *On Active Service in Peace and War*. New York: Octagon Books, 1971.

Stout, Mark. "The Pond: Running Agents for State, War, and the CIA: The Hazards of Private Spy Operations." *Studies in Intelligence* 48, no. 3 (2004).

Sullivan, Brian R. "'A Highly Commendable Action': William J. Donovan's Intelligence Mission for Mussolini and Roosevelt, December 1935–February 1936." *Intelligence and National Security* 6, no. 2 (1991).

Syrett, David. *The Defeat of the German U-Boats: The Battle of the Atlantic*. Columbia, S.C.: University of South Carolina Press, 1994.

Syrett, David, ed., *The Battle of the Atlantic and Signals Intelligence*. London: Routledge, 2018.

Terasaki, Gwen. *Bridge to the Sun: A Memoir of Love and War*. Chapel Hill, NC: University of North Carolina Press, 1957.

Thorne, C. T. Jr., and D. S. Patterson, eds., *Foreign Relations of the United States, 1945–1950: Emergence of the Intelligence Establishment*. Washington, DC: Department of State / US Government Printing Office, 1996.

Todd, Ann. *OSS Operation Black Mail: One Woman's Covert War against the Imperial Japanese Army*. Annapolis, MD: Naval Institute Press, 2017.

Trohan, Walter. *Political Animals: Thirty-Eight Years of Washington-Watching by the Chicago Tribune's Veteran Observer*. New York: Doubleday, 1975.

Troy, Thomas F. *Donovan and the CIA: A History of the Establishment of the Central Intelligence Agency*. Frederick, MD: University Publications of America, 1981.

——. *Wild Bill and Intrepid: Donovan, Stephenson, and the Origin of CIA*. New Haven: Yale University Press, 1996.

Troy, Thomas F., ed. *Wartime Washington: The Secret OSS Journal of James Grafton Rogers 1942–1943*. Frederick, MD: University Publications of America, 1987.

Truman, Harry S. *1945: Year of Decisions*. New York: Konecky & Konecky, 1955.

Tully, Grace. *F.D.R.: My Boss.* New York: Scribner's, 1949.

Ulbrich, David J. *Preparing for Victory: Thomas Holcomb and the Making of the Modern Marine Corps, 1936–1943.* Annapolis, MD: Naval Institute Press, 2011.

United States Congress. *Pearl Harbor Attack: Hearings Before the Joint Committee on the Investigation of the Pearl Harbor Attack.* Washington, DC: US Government Printing Office, 1946.

Usdin, Steven T. *Bureau of Spies: The Secret Connections between Espionage and Journalism in Washington.* New York: Prometheus Books, 2018.

Van Der Rhoer, Edward. *Deadly Magic: A Personal Account of Communications Intelligence in World War II in the Pacific.* New York: Scribner's, 1978.

Walker, David A. "OSS and Operation Torch." *Journal of Contemporary History* 22, no. 4 (October 1987).

Waller, Douglas. *Disciples: The World War II Missions of the CIA Directors Who Fought for Wild Bill Donovan.* New York: Simon and Schuster, 2015.

——. *Wild Bill Donovan: The Spymaster Who Created the OSS and Modern American Espionage.* New York: Free Press, 2011.

Warner, Michael. *The Office of Strategic Services: America's First Intelligence Agency.* Washington, DC: Central Intelligence Agency, 2000.

——. *The Rise and Fall of Intelligence: An International Security History.* Washington, DC: Georgetown University Press, 2014.

Warner, Michael, ed., *The CIA under Harry Truman.* Washington, DC: Center for the Study of Intelligence, 1994.

Weiner, Tim. *Enemies:, A History of the FBI.* New York: Random House, 2013.

West, Nigel. *MI6: British Secret Intelligence Service Operations 1909–45.* New York: Random House, 1983.

West, Nigel, ed. *The Guy Liddell Diaries,* vol. 1, *1939–1942: MI5's Director of Counter-Espionage in World War II.* London: Frank Cass, 2005.

Wheeler-Bennett, Sir John. *Special Relationships: America in Peace and War.* London: Macmillan, 1975.

Wheeler-Bennett, Sir John, ed. *Action This Day: Working with Churchill.* New York: St. Martin's, 1969.

Winks, Robin W. *Cloak & Gown: Scholars in the Secret War, 1939–1961.* New York: William Morrow, 1987.

Yardley, Herbert O. *The American Black Chamber.* Annapolis, MD: Naval Institute Press, 1931.

Yu, Maochun. *OSS in China: Prelude to Cold War.* Annapolis, MD: Naval Institute Press, 2011.

Zacharias, Ellis M. *Secret Missions: The Story of an Intelligence Officer.* Annapolis, MD: Naval Institute Press, 1946.

INDEX